A £20·85

Manufa[illegible]y

Manufacturing Strategy

Process and content

Edited by

Christopher A. Voss

Professor of Operations Management
London Business School

CHAPMAN & HALL
University and Professional Division
London · Glasgow · Weinheim · New York · Tokyo · Melbourne · Madras

Published by Chapman & Hall, 2-6 Boundary Row, London SE1 8HN

Chapman & Hall, 2-6 Boundary Row, London SE1 8HN, UK

Blackie Academic & Professional, Wester Cleddens Road, Bishopbriggs, Glasgow G64 2NZ, UK

Chapman & Hall GmbH, Pappelallee 3, 69469 Weinheim, Germany

Chapman & Hall USA, One Penn Plaza, 41st Floor, New York, NY10119, USA

Chapman & Hall Japan, ITP - Japan, Kyowa Building, 3F, 2-2-1 Hirakawacho, Chiyoda-ku, Tokyo 102, Japan

Chapman & Hall Australia, Thomas Nelson Australia, 102 Dodds Street, South Melbourne, Victoria 3205, Australia

Chapman & Hall India, R. Seshadri, 32 Second Main Road, CIT East, Madras 600 035, India

First edition 1992
First published in paperback 1995

Typeset in 10/12 pt Meridien by Best-set Typesetter Ltd., Hong Kong
Printed in Great Britain by T.J Press (Padstow) Ltd, Cornwall

ISBN 0 412 43660 4 (HB) 0 412 63480 5 (PB)

A Catalogue record for this book is available from the British Library

∞ Printed on acid-free paper, manufactured in accordance with ANSI/NISO Z39.48-1992 and ANSI/NISO Z39.48-1984 (Permanence of Paper)

Contents

Contributors

Everett E. Adam, Jr, University of Missouri.

Carl Anderson, University of North Carolina.

Kimberly A. Bates, University of Minnesota, Curtis L. Carson School of Management.

Joseph D. Blackburn, Associate Dean and Director of Operations Roundtable, Owen Graduate School of Management, Vanderbilt University.

Stuart Chambers, Lecturer in Production/Operations Management, Warwick Business School, University of Warwick.

Attila Chikán, Budapest University of Economics.

Robert S. Collins, International Institute for Management Development.

Krisztina Demeter, Budapest University of Economics.

William J. Doll, Professor of MIS and Strategic Management, The University of Toledo.

Frank L. DuBois, The American University.

Terry Elrod, Vanderbilt University.

Craig A. Giffi, Deloitte and Touche, Cleveland, Ohio.

M.J. Gregory, University of Cambridge.

T.J. Hill, Professor of Management at the University of Bath and the author of *Manufacturing Strategy: Text and Cases*, published by Irwin (1980).

Daniel T. Jones, European Research Director, International Motor Vehicle Programme, and Professor of Motor Industry Management, Cardiff Business School, University of Wales College of Cardiff.

Francis Jouffroy, Partner, Eurosept, Industrial Management and Logistics, Paris.

Quarterman Lee, Vice President, The Leawood Group.

William B. Lindsley, Vanderbilt University.

Ann Marucheck, University of North Carolina.

Arnoud De Meyer, Associate Professor of Technology Management, INSEAD.

Susan D.A. Misterek, University of Minnesota, Curtis L. Carson School of Management.

Jinchiro Nakane, Waseda University.

Michael D. Oliff, International Institute for Management Development.

Ronald Pannesi, University of North Carolina.

K.W. Platts, University of Cambridge.

Aleda V. Roth, Fuqua School of Business, Duke University, North Carolina.

Roger G. Schroeder, University of Minnesota, Curtis L. Carson School of Management.

Gregory M. Seal, Director, Deloitte and Touche, Cleveland, Ohio.

Wickham Skinner, Harvard Business School.

Paul M. Swamidass, University of Missouri.

Jean-Claude Tarondeau, Professor, University of Paris X and ESSEC.

Thomas E. Vollman, International Institute for Management Development (visiting).

Mark A. Vonderembse, Professor of Information Systems and Operations Management, The University of Toledo.

C.A. Voss, Professor of Operations Management, London Business School and Chairman of UK Operations Management Association.

F.C. 'Ted' Weston, Jr, Computer Information Systems Department, College of Business, Colorado State University.

Anthony J. Zahorik, Vanderbilt University.

Introduction

The need for manufacturing decisions to be made in a strategic context has been recognized for many years. The first developments in the area that is now known as manufacturing strategy took place in the 1950s at the Harvard Business School, where it was found that in a particular industry there were a wide range of policies being followed. This led to the study of the manufacturing strategy from an industrial viewpoint, an essential part of which was the understanding of technology options within an industry. Given the technology of an industry it was found that companies could position themselves successfully with a number of different combinations of market and process decisions. This led to the famous series of industry-based manufacturing books based on the courses taught at the Harvard Business School. It is only the modesty of the average production and operations management professor that prevents our field from claiming that the industrial approaches popularized by Porter were in fact developed initially in the manufacturing strategy area.

The major event that led to the initial popularization and dissemination were the two seminal articles by Wickham Skinner. In the first, 'Manufacturing the missing link in corporate strategy' (Skinner, 1969) a comprehensive framework for looking at manufacturing strategies in the context of their industry and technology was developed. The excellence of the framework is evidenced by the fact that subsequent frameworks have tended to simplify rather than develop it. The second seminal article by Skinner was 'The focused factory' (Skinner, 1974). The concept of focus is key to manufacturing strategy and much work has taken place in trying to develop the concept further.

Since Skinner's initial publications there have been a number of notable advances to the area. The first were led by Hayes and Wheelwright (1984). They developed the area on a number of fronts. First, they developed the product process matrix. This is a simple framework that allows companies to examine the area of process choice from both a static and a dynamic point of view. Widely used in industry is their four-stage framework for the state of development of manufacturing strategy. Their two books also began the continuing work of fleshing out the individual decision areas in manufacturing strategy such as process choice, capacity strategy, choice of technology, the engineering manufacturing interface and development of infrastructure. The late Bill Abernathy was prominent in developing awareness of the importance of competing through manufacturing in an international context. Another prominent figure in the development of manufacturing strategy is Hill. He has led the way on operationalizing the implementation of manufacturing strategy (Hill, 1989). His frameworks and concepts such as his five-stage model and the concept of order-winning and order-qualifying criteria have proved very influential. A third stream has been learning from the manufacturing strategies of other countries, particularly Japan. Richard Schonberger (1982) has led this area and numerous others are now contributing. The industry-based approach has not been neglected. The work at MIT on the future of the automobile has led to some of the most interesting manufacturing strategy research (Altschuler, *et al.*, 1986).

A manufacturing strategy is made up of many decisions and each of these needs to be considered in the strategic context. A number of people have contributed to this area, including: Ettlie (1990), Jaikumar (1986) and Voss (1986) in manufacturing technology; Ferdows (1989) in international manufacturing; A. Roth, Nakane Miller, K. Ferdows and other in surveys of manufacturing practice in the various publications of the Boston Manufacturing Roundtable; Clark and Fujimoto in new product development. A full review of the literature of the area can be found in Anderson *et al.* (1989).

Research in manufacturing strategy has always been difficult to do. It often needs a cross-functional team and apart from surveys rarely lends itself to the highly quantitative research that is preferred in many universities. However, in the last few years there has been a surge in research due to the growing interest in the area, demand from industry and a realization of its importance to the performance of national economies as well as that of the firm. From the realization of the new-found importance of the area and the widespread work being done, the UK Operations Management Association invited the leading researchers and practitioners from around the world to a conference at the University of Warwick in June 1990. Eighty papers were presented and it was decided to publish a book presenting the latest thinking in the area. Many of the articles in the book are drawn from the conference, others are drawn from the best of current published research. The inputs reflect the thinking in a wide range of countries and the book has a strong focus on process and application. It is designed to be relevant to practitioners as well as academics through presenting much of the new thinking in an applied way.

The book is divided into five parts. The first part gives an overview of manufacturing strategy written by two of the main authorities in the area. The second part focuses on the process of manufacturing strategy formulation and implementation. This is a critical area linking theory and practice. The third part looks at industry and the international context. Manufacturing strategies must be developed in the context of global competition. In addition, we can learn from the development of manufacturing strategies in different countries worldwide including the new Eastern Europe. Finally, a manufacturing strategy is firmly rooted in its industry and technology. The motor industry is one that illustrates this well.

There is an increasing number of tools and frameworks that are proving useful in the analysis and formulation of manufacturing strategy. Part 4 examines a number of these including focus, flexibility, time-based competition and the theory of constraints. Finally, the book reviews manufacturing strategy in the context of needs for research.

REFERENCES

Altschuler A. *et al.* (1984) *The Future of the Automobile*, Allen & Unwin, Winchester, Mass.

Anderson, J., Cleveland, G. and Schroder, R.G. (1989) Operations strategy; a literature review. *Journal of Operations Management*, **8**(2).

Clark, K. and Fujimoto, T. Overlapping problem solving in new product development, in *Managing International Manufacturing* (ed. K. Ferdows), North Holland, Amsterdam.

Ettlie, J. *et al.* (1990) *Manufacturing Strategy, the Research Agenda for the Next Decade*, Kluwer, Norwell, Mass.

Ferdows, K. (ed.) (1989) *Managing International Manufacturing*, North Holland, Amsterdam.

Hayes, R. and Schmenner, R. (1978) How you should organise manufacturing. *Harvard Business Review*, Jan.–Feb.

Hayes, R. and Wheelwright, S. (1984) *Recovering our Competitive Edge, Competing through Manufacturing*, Wiley, New York.

Hill, T.J. (1989) *Manufacturing Strategy*, Irwin, Homewood.

Jaikumar, R. (1986) Post industrial manufacturing. *Harvard Business Review*, Nov.–Dec., 69–76.

Schonberger, R. (1982) *Japanese Manufacturing Techniques*, The Free Press, New York.

Skinner, W. (1969) Manufacturing – the missing link in corporate strategy. *Harvard Business Review*, May–June, 136–45.

Skinner, W. (1974) The focused factory, *Harvard Business Review*, May–June, 113–21.

Voss, C.A. (1986) Implementing manufacturing technology, a manufacturing strategy approach. *International Journal of Operations and Production Management*, **4**(2), 17–26.

PART ONE

Strategic Approaches to Manufacturing

As was stated in the Introduction, two of the major contributors to the pursuit of strategic approaches to manufacturing have been Wickham Skinner and Terry Hill. In this opening part of the book each has put forward his views. The subsequent parts of this book will examine the process of implementation, international approaches and the content of manufacturing strategy. It is appropriate that Terry Hill starts the book by presenting a framework for manufacturing strategy. Terry's five-part framework is widely used both by practitioners and by academics. It is referred to many times in this book and provides a baseline for structuring other contributions.

Wickham Skinner was the driving force behind the spread of the manufacturing strategy area. His original article that defined the area was entitled 'Manufacturing, the missing link in corporate strategy'. He has taken a new look at the subject in his chapter 'missing the links in manufacturing strategy'. Rather than describe his own particular approaches he has chosen to reflect on where we are now in the study and practice of the area and the issues facing us in the future.

CHAPTER 1

Incorporating manufacturing perspectives in corporate strategy

T.J. Hill

The impact of world competition on the US economy has been a stark reality. Although manufacturing output has increased from the early 1970s, the continuing decline in manufacturing trade has resulted in record US trade deficits.

1.1 WHY HAS THIS HAPPENED?

The reasons are many and varied. Some are unsubstantiated opinions, others supported by fact. Some will be more relevant to some sectors and companies than others. However, the growing US trade deficit in manufacturing, the relatively low US exports per capita and the inferior annual growth in GDP per employee/output per hour is the worrying outcome. Particular reasons include:

1. There has been a failure, conscious or otherwise, of industry and society at large to recognize the size of the competitive challenge, the impact it was having, and would have, on our very way of life, and to recognize the need to change.
2. The failure to appreciate the impact of increasing manufacturing capacity in the last 30 years. World manufacturing capacity up to the mid 1960s was, by and large, less than demand, and in this period, companies could sell all they made. The redressing of this capacity/demand imbalance heralded the growing prominence of marketing.

Source: Hill, T.J. (1990)
Published by Superconnections

One outcome was that the manufacturing perspective within the corporate strategy debate diminished with many companies becoming almost entirely marketing orientated. This position has, in turn, been compounded by an organizational failure to distinguish between marketing (the function) and market (the business). As a consequence, market opportunities are often explored from a single perspective leaving the aftermath for manufacturing to handle and resolve. The result is that the essential underpinning within manufacturing to support the essential needs of the market place has characteristically been inadequate. This is not so in our more successful competitors.

3. Top management's lack of experience in manufacturing has had further ramifications for a business. Considering that manufacturing accounts for some 70–80% of assets, expenditure and people, it is imperative that senior executives fully appreciate the arguments and counter arguments in manufacturing so as to ensure that the accompanying wide range of perspectives are taken into account when making important decisions. Once large investments have been made, then rarely does a company invest a second time to redress the mistake.
4. The emphasis within the production manager's role has, in turn, been directed towards short-term issues and tasks. The overriding pressures to meet day-to-day targets have reinforced the tendency of manufacturing executives to concern themselves with this feature to the exclusion of the important long term. No longer are the key issues solely confined to operational control and fine tuning the system. The need is for broad, business-oriented manufacturing managers but companies have developed too few of them. The use of specialists as the way to control our businesses has increasingly led to a reduction in the breadth of line managers' responsibilities which has narrowed their experience base. Furthermore, many manufacturing managers have been outgrown by specialist argument and found themselves unable to cope with the variety of demands placed upon them. The response by many has been to revert increasingly to their strengths and this has reinforced their short-term at the expense of their long-term role.

1.2 STRATEGIC DOMINANCE – PERSPECTIVES OVER TIME

In broad terms, the changing world demand/capacity balance has brought with it a change in the fortunes of different functions. Up to the mid-1960s, many industries had enjoyed a capacity/demand relationship, which favored manufacturing. This, together with post-war growth helped create the dominance of the manufacturing function in many corporations. As the demand/capacity balance began to even out, and selling into existing and new markets became more difficult, the power base of corporate influence began to swing away from manufacturing, and heralded the rise of marketing. By the mid-1970s, the impact of recessions and energy crises had, in turn, opened the door to the influence of accounting and finance. These varying fortunes, however, rarely seemed to be based upon what was best for the total good of the business, but more on which functional perspectives appeared to provide the key that would open the door to corporate success or salvation. The consequence was that those functions out in the cold were themselves left without due corporate influence.

However, in times of increasing world competition, an over-capacity in many sectors of manufacturing industry, an increasing scarcity of key resources and decreasing product life cycles, there is an overriding logic for businesses to incorporate all the key functional perspectives when determining policy decisions. Why then, does this not happen? In many organizations, manufacturing adopts or is required to take a reactive stance in corporate deliberations. Yet how can the perspective of the function that controls such a large slice of the assets, expenditure and people, thus underpinning the very wealth of the company, be omitted?

1.3 REASONS FOR MANUFACTURING'S REACTIVE ROLE IN CORPORATE STRATEGY

The fact that manufacturing executives have an exacting and critical corporate role to play is undisputed. Why then do they adopt their current role, and why does this situation exist and not appear to improve?

1.3.1 The production manager's view of himself

One of the major contributions to this situation is that production managers also see themselves holding a reactive corporate brief. They too define their role as being one which requires them to react as well as possible to all that is asked of the production system. As a consequence, they do not see that their role is to explain the different sets of manufacturing implications created by alternative policy decisions and changes in direction. They fail, by default, to contribute to the corporate level and hence to help the company arrive at decisions which embrace all the important business perspectives.

1.3.2 The company's view of the production manager's role

Production management's perceived role of reacting to the demands placed upon it is reinforced by the corporate view of manufacturing's contribution, and hence the qualities incumbents should possess. One major company, recognizing the important corporate contribution to be made by its manufacturing executives, undertook a series of tailor-made courses in manufacturing strategy. The first of these comprised 16 manufacturing vice presidents who, during the course, reflected as a group that their aggregate company service exceeded 300 years, yet the collective training they had received to help them prepare for their manufacturing executives' roles was less than 30 days.

1.3.3 Production managers are too late in the corporate debate

Production managers are often not involved in corporate policy decisions, until these decisions have started to take shape. This position, having been partially created by the last two points, will, once established, tend to become standard practice. The result is that production executives have less opportunity to contribute to decisions or strategy alternatives and, as a consequence, always appear to be complaining about the unrealistic demands made of them and the problems that invariably ensue.

1.3.4 The 'can't say "no"' syndrome

'Can't say no' is still the hallmark of the production culture. This helps no one. Manufacturing executives tend to respond to corporate needs and difficulties without evaluating the consequences or alternatives that need to be considered, and explaining these to others. Rarely is it a 'yes' or 'no' decision. The need to be able to say 'no' from a total business perspective and with sound, corporate-related arguments is an important part of any executive's role, and manufacturing is no exception.

1.3.5 Lack of language

On the whole, production managers do not have a history of explaining their function clearly and effectively to others within an organization. This is particularly the case in terms of the manufacturing strategy issues that need to be considered and the production consequences that will arise from the corporate decisions under discussion.

The reasons for this difference cannot, however, be wholly placed at the production manager's door. The knowledge-base concepts and language have not been adequately developed. Consequently, shared perspectives within manufacturing, let alone between functions, are not held, which adds fuel to the lack of interfunctional understanding.

1.3.6 Functional goals and measures

In many organizations, the managers of different functions are measured in terms of their departmental efficiency (an operational perspective) and not in terms of overall effectiveness (a business perspective). Furthermore, their career prospects are governed by their performance within the functional value system. As a consequence, this leads to managers making trade-offs which are sub-optimal for the business as a whole.

A typical accountant or salesman, when receiving a customer's order will look at one figure as being significant – the total value of the order placed. For them, sales value is the important measure of business relevance.

In the same situation, manufacturing managers will look at the order makeup. For them, the business relevance of the order is not its value, but its volume and product mix. It is this aspect which determines the ease with which the order can be made in terms of the process configurations already laid down. It is this that will reflect manufacturing's ability to meet the cost base of the products made and hence the profit margin that will result.

Where marketing is measured by the level of sales revenue achieved, each $1 of sales in their value system will carry the same weight as any other $1 of sales. Hence, all sales are deemed equal. In reality, however, this is not the case. Orders are not of the same value to a business. One significant outcome, therefore, is that a great business divide separates the two realities of the market place and manufacturing. The simplistic measure of total sales value disguises the product characteristics so essential to manufacturing's ability to sustain cost and profit margin structures in a business, and other order winners that create long-term growth potential.

1.3.7 Top management's view of strategy

The authors of business plans and corporate marketing reviews look outward from the business. Top executives associate themselves with these activities, seeing them as legitimate corporate strategy issues. They concentrate their attention on the external environment in which the business operates.

Manufacturing plans are built in line with the stated business needs and are based upon the internal dimensions of the process involved, and top executives are less likely to associate themselves with these activities. Typically, they request a manufacturing strategy statement from the production executive, without becoming involved in its structure and development. They assume that it is not an inherent part of their role, which increases the difficulties in establishing a corporate strategy through dialogue and understanding.

1.4 LINKING MANUFACTURING WITH CORPORATE MARKETING DECISIONS

What then is required is an orientation based neither on products/markets nor on manufacturing, but a strategy which embraces the market/production interface, so that the degree of fit between the proposed marketing strategy and manufacturing's ability to support it is known at the business level, and objectively resolved within corporate perspectives.

For this to take place, relevant internal information explaining the company's manufacturing capability needs to be available within a business, as well as the traditional marketing information which is primarily concerned with the customer and the market opportunities associated with the company's products. Furthermore, to be effective, the ownership of its use must be vested in top management. As with other functions, manufacturing strategy is not owned by manufacturing. It requires corporate ownership. Senior executives need to understand all the strategic inputs in the corporate debate, for without this understanding the resolution between conflicting or non-matching functional perspectives cannot be fully investigated. What results is that individual functions are left handling the trade offs involved.

There is no shortcut to moving forward. However, there are five basic steps to be taken. These provide an analytical, objective structure in which the corporate debate and consequent actions can be taken.

Table 1.1 outlines the steps involved and also illustrates the iterative nature of this debate.

The approach embodied in this methodology has been tested in upwards of 40 plants. In all instances, it has enabled companies to create an effective dialogue between marketing and manufacturing, to ask questions of marketing with manufacturing answers, thus identifying through debate and shared agreement the way in which different segments work and the manufacturing task which is involved. The key is for companies to identify in detail how products win orders in the marketplace which leads companies to recognize that the traditional way of describing markets is insufficient in itself. Typically, business descriptions of their activities are normally based upon products, customers or geographical area. Although this summarizes a company's activities, it gives insufficient insight in terms of how segments differ

Table 1.1 Framework for reflecting manufacturing strategy issues in corporate decisions (steps involved)

Corporate objectives	*Marketing strategy*	*How do products win orders in the market place?*	*Manufacturing strategy*	
			Process choice	*Infrastructure*
Growth Survival Profit Return on investment Other financial measures	Product markets and segments Range Mix Volumes Standardization versus customization Level of innovation Leader versus follower alternatives	Price Quality Delivery speed reliability Demand increases Colour range Product range Design leadership Technical support being supplied	Choice of alternative processes Trade-offs embodied in the process choice Process positioning Capacity size timing location Role of inventory in the process configuration	Function support Manufacturing planning and control systems Quality assurance and control Manufacturing systems engineering Clerical procedures Payment systems Work structuring Organizational structure

Note: Although the steps to be followed are given as finite points in a stated procedure, in reality the process will involve statement and restatement, for several of these aspects will impinge on each other.

from one to another in terms of the essential ways in which orders are won.

The work has also helped to further distinguish difference by identifying two sets of criteria.

1. *Qualifying criteria* – These enable a company to get into and stay in a marketplace. But these do not win orders. They qualify a company for a potential or existing customer's list of acceptable suppliers.
2. *Order winners* – Once the qualifying criteria have been achieved, manufacturing has to turn its attention to the way in which orders are won. Its aim is to identify the order winners relevant to different segments and then to provide these better than its competitors. Order winners, therefore, need to be not only identified but also weighted in terms of relative importance. In this way, identifying the critical characteristics of different markets becomes the signpost for manufacturing investment. And, as this is large, it ensures that a company can more consciously choose between the alternative investments involved.

One essential characteristic of developing sound and sustainable corporate strategies is interfunctional debate with differences being resolved, not within functions, but at the next level up, that of the business. In the past, many companies have developed their strategic statements without attempting to incorporate this key feature. The result is functional conflict and less effective corporate performance. In turn, the outcome is decline at both the corporate and national level.

The approach described here provides an effective vehicle for this essential but missing feature of corporate strategy debate to take place. Discussing the business is not only the appropriate perspective on which to build corporate strategy but is also an effective way of avoiding functional argument born of discussing functional rather than business perspectives.

1.5 CONCLUSIONS

In the past, manufacturing's role has typically been perceived as attempting to support whatever requests are made of it. The argument that 'the customer is always right' has held sway now for many years. But this cannot be so. The only perspective that can always be right is the business perspective. And, for this to be evaluated, the implication for manufacturing to meet varying market needs has to be both established and understood by all those responsible for strategic decisions. Unless, therefore, manufacturing is able to provide relevant inputs into the corporate strategy debate, this critical shortcoming will continue, and home and world markets continue to be lost.

REFERENCE

Hill, T.J. (1989) *Manufacturing strategy: Text and Cases*, Irwin, Homewood, IL.

CHAPTER 2

Missing the links in manufacturing strategy

Wickham Skinner

This book is focused on a set of ideas which its authors believe can change substantially how industry is managed. Enthusiasm for the concepts of 'manufacturing strategy' has never run higher. Academic papers, executive courses, and case studies of industry practice are bursting out all over.

But if we are realistic, certain facts dump some rain on our party and ought to dampen our high spirits. For example, since the original ideas are more than 25 years old, someone should be asking why have they taken so long to penetrate the academic world? And what is their influence in industry? What percentage of manufacturing actually employs these management approaches: 1%; 3%? I am afraid that the recent success we academics find so exciting is actually hollow in certain respects and more self comforting than real.

The conclusion of this paper is that industry adoption of the concepts and techniques I will call 'manufacturing in the corporate strategy' (MCS) has been held back by (1) strong instinctive premises and mind-sets cloned into generations of managers, (2) the conventional functional, departmental organization of business, and (3) missing conceptual links in the theory of MCS. Operation of this 25 year-old machine reveals some 'bugs' in it and these are hurting its performance.

The first two barriers describe a negative environment which we have to take as a 'given' for the time being. The third barrier is one we have built ourselves, in effect, by the incompleteness of our work. The paper concludes therefore that the theory, art and practice of manufacturing strategy must be improved if it is to be more influential.

So we must be self-critical and go back once more to the worn

Specially commissioned for *Manufacturing Strategy – Process and content.*
Edited by Christopher A. Voss.
Published in 1992 by Chapman & Hall, London ISBN 0 412 43660 4.

drawing board if we are to realize the enormous potential of this still unrealized, uniquely unconventional approach to the management of industry. To buttress our conviction that the effort is really worth while we begin on the positive side with a look at the recent good news.

2.1 ON THE PLUS SIDE

Even the most self-critical cannot deny that MCS is on some kind of a 'roll' and is now in product life cycle terms shooting upwards at a steep rate on the S curve. Backed by the sober though generally belated recognition in government, industry, and academia (trailing, as usual) that manufacturing matters, there is widespread interest in finding a better way to manage in industry in order to compete. This has sparked a frenetic though recent burst of energy in academia, a significant proportion finding its way into MCS research.

The astounding outpouring of research papers and doctoral theses is demonstrated by the very successful conference at the University of Michigan organized by John Ettlie in January of 1990 and by the conference at Warwick assembled by Chris Voss and Terry Hill. Over 100 papers in all were available for these conferences, all focused on MCS in one way or another. The literature of MCS has more than quadrupled in the last five years.

MCS was virtually untaught prior to 1984, except in about six schools. Now colleges and universities all over the world are introducing MCS into many required and elective courses. Short courses on MCS concepts and techniques for POM professors have sprung up in a number of schools such as, for example, North Carolina and Iowa. Professors describe students taking MCS courses as 'finally getting excited about operations courses'. The intellectual content combines with the strategic-importance/top-management flavour in these courses to develop classroom enthusiasm generally rare in POM.

Many universities now offer executive courses and seminars in MCS and these have been uniquely successful. Often attended by consultants eager to learn new tricks, the top consulting firms such as Arthur D. Little, Bain, BCG, Booz-Allen, A.T. Kearney, Management Analysis Center, McKinsey and United Research have developed an understanding and a degree of competence in MCS. Many of the large accounting firms which have consulting wings, such as Arthur Anderson and Coopers–Lybrand, have trained a cadre of staff in MCS by engaging professors to teach them what it is all about.

So there are a lot of pluses to tout. It must be observed, however, that most of the positives centre around the universities, professors, doctoral students and courses. And we must also confess ruefully that progress in the first 20 years was only about 20% of the progress in the last five years.

It took a long time for this art form to become popular with the critics. It was not criticized much; it just didn't seem to be noticed. It was business as usual in old academia. Apparently what brought MCS to academic attention was a clearly recognized need for a new approach to industrial management brought on by US and European competitive failures in industry. Plus, it should be added that, at least in some schools, there were serious failures in executive and MBA courses tied to teaching the conventional POM, OR, and Industrial Engineering curricula. When Western industry, managed by the use of the same concepts taught in the management schools for decades, was steadily failing in global competition, academics apparently began to recognize the need to change their product line.

Whatever the reason for the MCS takeoff in academia, it appears to have momentum and, indeed, acceleration. The same cannot be said for the rate of progress and penetration in industry. We turn now to the negative side of the balance sheet.

2.2 QUESTIONS, NEGATIVES, CRITICISMS

The negative side is so clear that it can be explained quickly. The adoption of MCS in industry has been quite modest. Very few firms have a manufacturing strategy. Manufacturing is generally perceived as a necessary but costly and high-risk investment. Manufacturing is seldom a competitive weapon. Manufacturing people seldom get to the top of the corporate ladder. Manufacturing is still the whipping boy, criticized for costs, delivery, service, inventories and inflexibility with no sense of the trade-offs at work. And so it goes on.

In spite of our failures in the market-place, manufacturing is managed pretty much the same as it has been for 50 years. How is it managed? Generally, this is by:

1. piecemeal decisions,
2. short-term performance objectives and measurement systems,
3. attempting to optimize on almost every dimension,
4. command and control workforce management,
5. a primary focus on reducing costs and increasing productivity, and
6. buying and installing the recommendations of academics and consultants willy nilly off the shelf, such as (beginning in the twenties): Gantt charts, time and motion study, stop-watch standards, predetermined standards, work sampling, statistical process control,

work simplification (this brings us up to about 1960), queuing theory, simulation, heuristic algorithms, economic order quantity, min./max. inventory control, line of balance expediting, the unit manager (upgrading the foreman's role), computerized inventory control, and in the 1980s, quality circles, MRP 1, MRP 2, TQC, JIT, FMS, CIM, CAD, CAM, and MFT (that is management for time!).... Show us a new technique and we buy it; we like sliced bread.
7. We are, nevertheless, chary of major process innovation and automation. That's dangerous and risky, not to be avoided of course, but do be careful and let your competitor go ahead and get into trouble first.

So why is it, we must ask, if production managers have demonstrated for 100 years that they are always hungry for the latest technique that might possibly make a difference to the bottom line, that they have not seized upon MCS long before now? And why does not course-induced exposure to MCS last long? Even executives who have taken courses in MCS which they rated as 'mind-blowing' generally snap back to their old modes of production management when they return to work. There seems to be a kind of biological rejection of the new approach even after a thorough implantation.

What is wrong? I would like to suggest four obstacles which are blocking wider, effective penetration of MCS in industry in spite of its growing acceptance in the schools of management.

2.3 FOUR LINKS ARE MISSING

Analysis of cases of the attempted introduction of MCS suggests that four common, linked prerequisites to radical change are missing typically. Only one condition, that of motivation, is typically present these days.

What is missing concerns leadership, managers capable of implementing MCS, ideas which are conceptually sound and complete, and favourable organizational structures.

Motivation for doing a better job in production and operations management is a ready commodity now. The loss of market shares in global competition, sick industries, declining profits, layoffs, recession, the ascendance of the Japanese and other Pacific Rim competitors, and problems with the level of the 'playing field' all have provided and continue to provide ample and evident motivation for seeking new ideas and approaches to managing industry. The feeding frenzy over JIT and TQM support the conclusion that POM managers are as hungry as always for easy to assimilate new techniques.

Analysis of the last four linkages reveals serious problems. Taken one by one they are enough to explain why MCS has been so slow to penetrate industrial practice.

2.3.1 Leadership which understands and accepts the new concepts

What causes the 'biological rejection' of MCS is simply that manufacturing and corporate leaders have been trained, rewarded, promoted and thoroughly conditioned all through their careers to think of a factory and its management in terms of efficiency, productivity and costs. The job of a production manager is to meet schedules while keeping costs and investment low. The heritage of the values of industrial management is mass production, economies of scale, level production, a limited product mix, and keeping each operation and operator busy and at full efficiency. These sacred ideas, born of engineers and efficiency experts, form the philosophical ground of the profession. We snap back obsessively to these premises of the right ways to manage an operation. We respond with guilty consciences to top management's entreaties that we do better on any element of financial and physical performance which causes the criticism of the weak.

MCS is a 'foreign language' and in a cognitive sense demands that the production manager adopt a strange new way of making decisions. In MCS the manager's thinking is conceptually different and requires an entirely different set of mental premises. MCS says the structure is the key, and a competent focus on daily operations is only a necessary housekeeping chore. To focus the structural decisions as to make *v.* buy, capacity, choice of equipment and process technology, and major systems — around the business' strategic task demands new skills which are wholly different from those engrained from a century of the 'cloning' of successful production managers.

The bulk of industrial leaders are conditioned to analyse operations in terms of productivity and/or quality and efficiency, not on the ability of a system to compete by providing a competitive weapon. Leaders trained at a good course in MCS return to the job determined to manage by developing focused and congruent structures but after a few weeks slip back into their habitual, conventional mode of managing. They polish and smooth the operation, cutting out waste and inefficiency, and push the team to exert more effort and care. These are all good steps but if the basic structure of that operation is wrong, it cannot compete in spite of such marginal improvement.

The champions needed for leading industrial management into thinking differently and setting examples of how the operations function can be structured to create a strong competitive resource have been few. Minds have been set into the efficiency ethic for a century. It should not be surprising that acceptance of so radical a change as asking the housekeeper to become architect and structural designer is proving to be a slow proposition.

2.3.2 Problems in middle management

Problems in middle management constitute a second major set of impediments to MCS. Essentially, the difficulty is that middle level managers running manufacturing and service operations are not surprisingly, prisoners of their training by the leaders just described. Inevitably cloned by their superiors, they have inherited the same premises and instincts.

In 1984 I did a study of about 60 production executives whose ages averaged 38. These men and women in twelve companies were selected by their superiors as 'comers', defined as being expected to be promoted within 5–7 years to positions of major responsibility. Each manager was interviewed for two hours in a comparable patterned interview. Among the conclusions reached were the following:

1. The 'comers' devoted their time almost exclusively to problems within their department, with an emphasis on the current month and quarter.
2. They showed relatively little knowledge of the business situation of their company or their division. They knew little, for example, about the competivive situation, the company's strategy, or the economics of the business unit.
3. They did not show a glimmer of understanding about manufacturing stategy. The term itself was unknown to nearly all the 'comers' and they showed little interest or knowledge of the structural decisions of manufacturing which literally surrounded them. Their attention to structure was only incidental to the particular function in which they worked, for example, as in quality control or production control. They did not recognize the options in front of them.
4. They were parochial in their interests. If the 'comer' specialized in production control, for example, they showed very little interest in or concern for other functions.
5. They were basically technology aversive. They strongly expressed the belief that it is not necessary to be proficient or knowledgeable in understanding their process technology. Almost to a body the view was expressed that technology is for engineers and that it is unnecessary to have much to do with it. This was particularly disturbing because process and equipment selection is at the very heart of the design and structuring of any operation. It was clear that few of these able men and women would ever be competent or confident about the very processes upon which the competitive success of their businesses depended.

This set of attitudes, skills and competences is not only disheartening relative to its impact on the competitiveness of US business, but it also tends to subvert any attempts to introduce new ways of managing in manufacturing. The environment for MCS is unhealthy because it is these middle-management 'comers' who must make it work. Without extensive changes in production managers' thinking and skills, the prospect is dim for developing manufacturing strategies and operating with a longer-term focus on the needs of the business.

What is to be done? The vicious cycle is obvious: the ideas inherited control the attitudes and behaviour of top management and, influenced by measurement and performance appraisals, the next layers of management develop the same weaknesses and limited vision. To break the log jam will require, I believe, some practical and tangible concepts and rules of thumb for MCS which do not yet exist.

The model of JIT and TQM is relevant. These techniques offer a straight 1, 2, 3, 4 set of procedures which MCS still lacks. This is our fault in academia. Once new tangible missing link ideas are supplied and begin to work, there will be more success models, and those successes may lead the way. Next, we must discuss some of these missing links.

2.3.3 Problems with the ideas of manufacturing strategy

The basic ideas of MCS, I think, are quite sound. An abbreviated summary of the basic ideas is included as Exhibit 1. Although these ideas are essentially correct, there are two problems with them. The first concerns the issue of **trade-offs**. The second is a serious gap in the syllogism of moving from a specified **manufacturing task** to making the actual choices of manufacturing policies which form the structure. In my opinion the trade-off problem is one of inadequacy of explanation or, possibly, of semantics and thus is of less importance. But it bothers many students of the art of operations management so let us take it up first.

The tradeoff problem

The notion of trade-offs in system design is basic to the MCS theory. The idea is that a manufacturing or service operation is limited in its performance by the technologies it employs. The conceptual parallels to a goods or service producing operation include a vehicle, or a boat, a plane, a building, a bridge or a railroad system. An airplane, for example, must be carefully designed to meet performance objectives and, clearly, cannot be the best airplane ever built in every factor such as speed, capacity, fuel economy, short takeoff and landing distances, initial cost, comfort, etc., etc. All technology-based systems are limited in what they can do by their technologies of equipment and processes, materials, information systems, and management systems. That there are trade-offs no one will deny.

It is also clear that trade-off curves or relationships between cruising speed and landing distances, for example, have been drastically revised over the years with developments in airfoils and changes in wing flap designs. Similarly, the jet engine totally shifted trade-offs of speed, fuel economy, and capacity. Changes in technologies change trade-offs and relationships between design objectives.

Now consider factories. They are also inevitably limited by the technologies employed. Trade-offs between costs, investments, quality, leadtimes, service levels, flexibility for product and output changes exist just as surely as they do in other technologically limited systems. This fact was readily accepted by MCS academics and practicioners several decades ago as trade-offs such as cost *v.* quality and inventory *v.* leadtimes were easily perceived. Then along came new ideas in quality (actually not so new, but newly merchandized and applied) and just-in-time scheduling and inventory procedures and suddenly 'quality is free' and inventory is bad. There were no more trade-offs!

This is the wrong conclusion. After the invention of the jet engine did aeronautical designers suddenly decide that there were no more trade-offs between cruising speed and fuel costs? Of course not. They simply had to draw up a whole new set of curves. The penalties assumed in the word 'trade-offs' may actually be algebraic; that is, the two functions may turn out to be positive concurrently over a certain span.

This may actually take place in the cost-quality 'trade-off' and it depends in many cases on how each objective is pursued. For example, if reduced cost is a primary objective and is vigorously undertaken, generally quality will suffer in the short run. But if quality is the prime objective, costs will usually rise for a year or two and then decline as quality substantially increases. These relationships are depicted in Fig. 2.1.

What do we conclude about trade-offs? They are just as real as ever but they are alive and dynamic. As technologies develop, and this includes management technologies as well as physical ones, the relationships will change. To ever assume that a plant or service operation can function without trade-offs is absurd. As technologies improve, so can operations, but there are and always will be relationships which are important, indeed critical, between different design and performance characteristics. What is important about trade-offs is not to generalize loosely about them but, as is essential in aircraft design, operations structure designers must be as specific as the state of the art will allow. This is another missing link. We hypothesize about trade-off relationships but have failed to do research and gather the data needed for better understandings.

It may be argued with good reason that the very word 'trade-off'

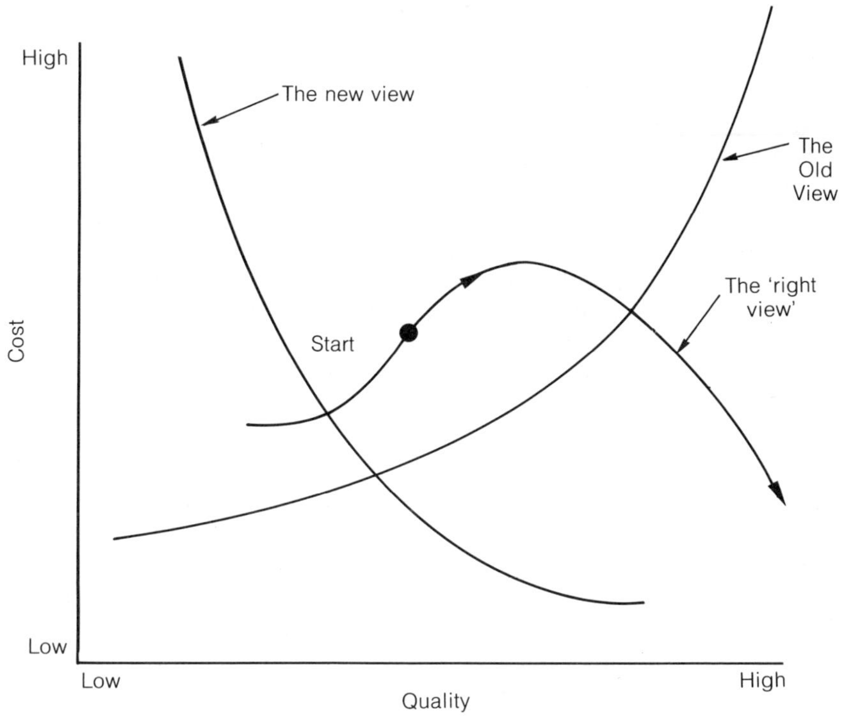

Fig. 2.1 The cost/quality relationship.

creates a problem. For if the relationship between *x* and *y* is such that both *x* and *y* increase or decrease together, is there in fact a trade-off? Trade-offs imply that to get more of *x*, one has to have less of *y*. But to me this is a simplistic understanding of the word because in many technologically based relationships between outcomes, generally called trade-offs, the variables may run in parallel or counter to each other at different levels or amplitudes.

So the basic notion of trade-offs turns away some academics and managers. This is a fact whether or not we perceive the criticism as reasonable or simplistic. The fix is not to stop using the idea (which is simply that operations systems need to be designed to maximize performance on the few success criteria of strategic importance because a system which is technologically constrained cannot perform superbly on every measure). The fix could be just to drop the word and substitute 'performance relationships' or something better. But what is really needed is research to develop more knowledge and the specifics of these constraints.

A big gap in MCS as a management tool

To illustrate the gap, let me offer up a problem in MCS as follows.

Suppose a management team seeking to develop a powerful new manufacturing strategy works out that the critical competitive and business challenge for the operations function is to get into production smoothly and rapidly on many new products, which over the next three years will comprise 75% of the sales volume. That is the 'manufacturing task'. Now, what specific structural decisions should be made, for example, concerning make or buy, capacity, equipment and processes, production control and quality management, workforce management and other manufacturing policies?

There is no textbook or article that helps managers make these decisions to design the structure to meet the manufacturing task. We say that manufacturing policy decisions must be made such that the system designed meets the manufacturing task. But the manager is left to figure out how to get from task to structure without much specific guidance.

In my two books on MCS I offer several menu lists to suggest the options available, but I have not suggested that for task *abc* and industry economics *def* and process technology *ghi*, the management should select backward integration *jkl*, capacity strategy *mno*, processes with descriptors *pqr*, production control system *stu*, and so forth (Skinner, 1978; and Skinner, 1985). The menu is set forth but missing is the probable impact on system performance of each menu listing. That in brief is the gap.

In my own and others' defence we have not been prescriptive because it is more art than science and there are 999 variables and 998 equations, so to speak. This is true and reasonable, but I would suggest that we could be a lot more helpful to managers and that we need to be. Until MCS can provide more links between tasks, objectives and specific manufacturing policies, only bright students in a well-led case-method classroom will do the synthesis needed and MCS will continue to miss its potential. This is the greatest research, writing, and consulting need and opportunity in the field.

Exhibit 1 The basic ideas of manufacturing strategy

1. The operations or manufacturing function can be a strong competitive resource if properly designed and operated.
2. Cost, efficiency and productivity, while commonly accepted as the most important performance objectives, are generally too narrow and limited and, paradoxically, self-defeating to create competitive advantage.

3. Seven performance objectives need to be considered:

   (i) Cost, efficiency, productivity,
   (ii) Delivery leadtimes,
   (iii) Quality,
   (iv) Service, reliability,
   (v) Flexibility for product change,
   (vi) Flexibility for volume change, and
   (vii) The investment required in the production system,

4. A focused strategic objective or 'manufacturing task' is based on one or two of the seven objectives and is derived from the firm's competitive strategy, economics and technological opportunities.
5. To meet the task, the operations system is designed and tailored to focus on that task, with limited ranges of products, markets, technologies, degrees of process demands and order quantities.
6. The key job of high-level managers is the design of the structure. A strategicly designed structure is the key to the operations function becoming a powerful competitive weapon.
7. The structure of an operating system is derived from decisions concerning make *v.* buy, capacity, equipment and processes, number, size and location of facilities, what products are made in which plants, and the management systems for production planning, scheduling and control, information systems, quality control, organizational structure, and work-force management.

2.3.4 Problems of the functional organization

Fifteen years ago analysis of problems of Western industry made it very clear that to compete we had to do a much better job on quality, costs, new product development, human resource management, and mechanization. We have made progress on all these fronts but in retrospect the progress has been surprisingly modest and slow. The Japanese and other Asian competitors have moved ahead on all these facets as fast or faster than we have. As a result in spite of over a decade of the most energetic, enthusiastic work in production management history we find ourselves in a worse competitive position than before.

Analysis of progress on these fronts reveals a strikingly similar problem on each. The problem is that progress has been delayed and frustrated by the functional departments surrounding manufacturing and operations. Managers of the different functions such as engineering, marketing, finance, personnel and accounting are seldom rewarded for progress in manufacturing.

On the contrary, they are rewarded for their own professional accomplishments. For example, at the risk of oversimplfying, in engineering the objective is products that work and do not fail. In marketing the goal is more sales and increased market shares. In finance it is increased cash flow and return on investment and to secure capital cheaply. In personnel it is to keep the peace, and to avoid disputes and possibly unfortunate precedents. In accounting it is to get out reports on time that meet the requirements of FASB and the firm's auditors.

The needs of manufacturing are quite different. Manufacturing needs products which are easily manufactured, long runs, fewer products, more innovative experiments to unleash the ideas and energies of employees, and a cost-accounting system which reveals sources of costs; these are the cost creating activities. Finance meanwhile turns down investments in mechanization and automation when they fail to achieve satisfactory hurdle rates. Engineering resists concurrently designing the process with the product.

These functional conflicts and internal barriers are entirely normal to the Western corporation. Managers in one function rarely if ever are transferred to manage in another. Rewards are given down the ladder of the organization for meeting the professional objectives of the department rather than for the competitive success of the business unit.

Is it any surprise that a management concept such as MCS which requires all-out cooperation and sharing of knowledge and authority across functional boundaries is 'biologically' rejected? MCS requires building a structure based on strategic objectives and it requires wholehearted support across all functions to transform a company into a true 'manufacturing enterprise'.

So the fact is that most Western corporations engaged in manufacturing are not true manufacturing enterprises. Manufacturing may require the bulk of the corporate assets, the largest number of employees and may totally limit the firm's competitive success, but the management traditions of line and staff, and decentralized bottom line judged operations focused around meeting the annual plan are condemning us to mediocrity.

2.4 CONCLUSION

Great progress in MCS academic research, writing of books and articles, seminars and executive courses contrast with modest penetration and adoption in industry. Our explanation has been that the ideas are as yet incomplete and that the inherited mind-set makes for a dearth of champions and middle managers who are prepared to approach their jobs with a radically different set of premises and objectives. Add to that hostile environment the functionally dominated management of the traditional Western corporation and the climate for change could hardly be worse.

But what about the great changes actually taking place in operations such as the wholesale adoption of off-the-shelf techniques such as JIT, TQM, and MRP? These certainly don't demonstrate that production management is still in the sleepy doldrums of the 1960s and 1970s. But that is just the point: there is a crisis of failing competition and in our desperation to survive and recover we are grabbing at anything new that offers a hope to do better on any single criterion of performance. As a result productivity and quality are up, inventories are down and we are still losing market shares in dozens of industries. Why?

Our competitors can make and in fact do make these single-criteria improvements better and faster than we. We chase after them with these off-the-shelf techniques and it should be no surprise that this does not restore a competitive edge. And why is this quick-fix approach so instinctively grabbed? It is the inherited conception of the factory as a kind of efficiency machine, measured on cost and efficiency and return on investment, and always improvable by slapping on new techniques. This is the mind-set of old-fashioned efficiency experts and industrial engineers.

Of course the problems are deeper than JIT or TQM can fix because the issue is not inventory or quality; it is how to compete with companies that are already way ahead in JIT and TQM. Copying competition won't do it. What is needed are new manufacturing strategies which innovate entirely new integrated structures and create a true manufacturing enterprise.

What is needed in MCS to break out of these vicious cycles is, I believe, the supplying of these links we are missing. The primary need is the research and conceptual development of concrete, explicit bridges between task and structural decisions. With these fundamental, prescriptive spans in place, tried out, improved and debugged, success examples will be added to the modest but spectacular successes already in the record books and the problems in leadership, middle-management, and excessive functionalism will moderate with the use of specific 1, 2, 3, 4 techniques.

The 100 papers published this year demonstrate the excitement many are finding in MCS. But before we can be honestly proud we have to show much more influence on business practice. The basic problem, I must conclude, is that our concepts are still incomplete. My urging as we finish this conference is that we close those gaps.

REFERENCES

Skinner, W. (1978) *Manufacturing in the Corporate Strategy*, John Wiley, New York.

Skinner, W. (1985) *Manufacturing: the Formidable Competitive Weapon*, John Wiley, New York.

PART TWO

The Process of Manufacturing Strategy Formulation and Implementation

Despite the growing literature on manufacturing strategy, it is widely recognized that there has been little attention paid to the **process** of manufacturing strategy development and implementation. Many frameworks have now been developed for manufacturing strategy, and one of the most widely used of these is that of Hill (1989). He views manufacturing strategy as a series of discrete and sequential steps. These are embodied in his manufacturing strategy framework, which consists of the following five stages:

1. Define corporate objectives,
2. Determine marketing strategies,
3. Assess how products win orders,
4. Establish the most appropriate mode of manufacture the sets of products – process choice, and
5. Provide the manufacturing infrastructure to support production.

As Hill points out, implementation in a particular company does not necessarily start at the beginning and finish at the end. Manufacturing strategy implementation is an iterative process. In the first article in the previous section, Terry Hill described his framework. One of the best known frameworks in which to think about implementation is that proposed by Hayes and Wheelwright (1984). They describe four 'stages' in the evolution of manufacturing strategy:

1. Internally neutral,
2. Externally neutral,

3. Internally supportive,
4. Externally supportive.

In their review of the literature on manufacturing strategy, Anderson *et al.* (1989) state:

> perhaps even more revealing in the literature is the lack of attention on the process of operations strategy. There seems to be an implicit assumption that once operations strategy elements are defined, the process of operations strategy is merely perfunctory or self fulfilling, both in an analytical and an organisational sense.

In a similar review, Adams and Swamidass (1989), propose a number of missing themes in operations strategy. Missing theme number one is 'Operations Management research needs distinct research streams investigating content and strategy processes.'

This section presents a series of articles examining the process of implementation from various perspectives. Ken Platts and Mike Gregory present a workbook-based audit approach that operationalizes the process of formulating manufacturing strategy. Tom Vollman, Bob Collins, Jinchiro Nakane and Mike Oliff present a framework for manufacturing strategy implementation. Two papers, one by Chris Voss, the other by Ann Marucheck, Ronald Pannesi and Carl Anderson examine how companies actually go about formulating manufacturing strategies. Craig Giffi, Aleda Roth and Gregory Seal develop a set of operating principles to guide companies in the 1990s. Finally, Francis Jouffroy and Jean-Claude Tarondeau put forward a framework for developing an industrial strategy.

REFERENCES

Adam, E. and Swamidass, P. (1989) Assessing operations management from a strategic perspective. *Journal of Management*, **15**(2), 181–203.

Anderson, J.C., Cleveland, G. and Schroeder, R. (1989) Operations strategy, a literature review. *Journal of Operations Management*, **8**(2), 133–57.

Hayes, R. and Wheelwright, S. (1984) *Restoring Our Competitive Edge*, Wiley, New York.

Hill, T.J. (1989) *Manufacturing Strategy, Text and Cases*, Irwin, Homewood.

CHAPTER 3

A manufacturing audit approach to strategy formulation

K.W. Platts and M.J. Gregory

3.1 ABSTRACT

The need for companies to develop manufacturing strategies which are consistent with their overall business strategies has been well established. However, many companies still do not have an explicit strategy for manufacturing and all the signs are that they do not know how to set about creating one, even if they appreciate the need. Formal approaches to strategy formulation appear in the business strategy literature but are rare in the manufacturing strategy literature.

This paper describes a process for the analytical stages of manufacturing strategy formulation based on an audit of market requirements, current performance and manufacturing practices. Essential elements of the process are the systematic involvement of key personnel, and the use of worksheets to allow both data and judgements to be recorded and revisited.

The experience of following the process in six companies is summarized and the main findings reported. The process has proved to be feasible and usable, and has provided a useful way of structuring the analytical stages of manufacturing strategy formulation. Some suggestions are made for steps which might be taken to further refine the process.

Specially commissioned for *Manufacturing Strategy – Process and content.*
Edited by Christopher A. Voss.
Published in 1992 by Chapman & Hall, London ISBN 0 412 43660 4.

3.2 INTRODUCTION

Traditionally, improvements to manufacturing systems have been initiated, specified and introduced by engineers with insufficient information about the business context within which they will operate. Often cost reduction has been the primary aim, with little awareness of, or attention paid to, non-cost factors such as quality and delivery performance. The result is that manufacturing systems have evolved which do not satisfactorily contribute to the competitive position of the company. In short, many companies have not had sound manufacturing strategies which are consistent with their overall business strategies.

This weakness has been widely recognized over recent years and many authors have described and reported the problems which arise from the lack of a manufacturing strategy and have indicated the issues which need to be considered. It remains the case however that many companies do not have an explicit strategy for manufacturing and all the signs are that they do not know how to set about creating one, even if they appreciate the need. Manufacturing strategies must not only be consciously created but they must also be subject to regular review so that the elements of the strategy develop as the business develops and internal and external consistency are maintained.

Although the concepts of manufacturing strategy have been around for many years (Skinner, 1969) most of the published work has concentrated on the content of strategy with little attention being paid to the processes by which strategies are formulated.

A review of research is the area of production/operations management (Voss, 1984) considered that the basic frameworks for manufacturing strategy had been developed but that research was required to help lead to effective practice. A key requirement was: 'Getting an organisation to define the manufacturing task. . . . The first element required is some form of manufacturing audit'. The aim of the work reported in this paper was to develop and test such an audit procedure as a mechanism to facilitate the analytical stages of the manufacturing strategy formulation process.

3.3 STRATEGY FORMULATION

The majority of the literature on strategy formulation applies to business strategy. The traditional view was developed in the 1960s by, among others, Ansoff (1965) and Andrews (1971). This approach is based on an analysis of the environment (opportunities and threats), and the internal strengths and weaknesses of the organization. The strategy formulation process seeks to identify action plans which will use the organization's strengths to exploit opportunities while minimizing its vulnerability to threats.

This traditional, prescriptive approach is essentially analytical and rational. It specifies how strategies should be formulated consciously. However, when the process is observed in practice other modes of the strategy-making process can be seen. Mintzberg (1973) identified two further modes, the entrepreneurial mode (a strong leader controlling the organization) and the adaptive mode (the organization adapting in small disjointed steps). Mintzberg (1978) observed the processes of strategies emerging rather than being planned deliberately. Strategies can form as well as be formulated, they can emerge in response to evolving situations. Mintzberg introduced the definition of realized strategy as a pattern in the stream of actions.

This descriptive view of strategy formation is in marked contrast to the previous formalized planning view.

These two extremes were reconciled to some extent by Quinn (1978). He proposed a process of **logical incrementalism**, proceeding flexibly and experimentally from broad concepts towards specific commitments making the latter as late as possible. His rationale was that strategy dealt more with the unknowable than the uncertain, therefore one could not predict events in a probabalistic way, and should delay decisions as long as possible to ensure that the best possible information was available. However within this process he acknowledged a role for the classical, formal planning techniques. They:

1. Provide a discipline forcing managers to take a careful look ahead periodically,
2. Require rigorous communications about goals, strategic issues and resource allocations,
3. Stimulate longer term analyses than would otherwise be made,
4. Generate a basis for evaluating and integrating short-term plans,
5. Lengthen time horizons, and
6. Create an information framework

Our work, whilst acknowledging the descriptive theories' contribution to an understanding of the more intangible elements of the strategy formulation process, sets out to provide a prescriptive model for the restricted functional area of manufacturing strategy. We have adopted a rational, analytical process framework within which to provide an audit methodology for companies to develop their manufacturing strategies.

In contrast to the literature on business strategy formulation, little has been written on the process of manufacturing strategy formulation. Adam and Swamidass (1989) stated:

> . . . although business strategy literature investigates strategic planning, planning processes, and planning models, manufacturing

strategy literature does not. There is an urgent need to develop a body of literature on the manufacturing planning process.

There is an implicit process in the strategy frameworks of Skinner (1969), Wheelwright (1978), Hayes and Wheelwright (1984) and their derivatives. This implicit process depends on breaking manufacturing down into a number of decision areas and making the goals of manufacturing explicit in terms of a number of performance criteria. The steps of identifying these criteria, prioritizing them and relating the decision areas to them form the process. Within this process the areas which have received the most attention have been the linking of marketing and manufacturing and the issue of process choice/process organization. See, for example, Fine and Hax (1985).

The linking of manufacturing strategy into the business strategy is achieved by using market requirements to establish the performance criteria against which manufacturing should be measured. It should be noted that a company is not expected to perform well against all the criteria. It is the prioritizing of the criteria and the achievement of a match between manufacturing and marketing that determines how the company will compete. Hill (1980) emphasized that this is not simply determining the manufacturing implications of marketing decisions, it is instead an iterative process involving marketing and manufacturing jointly developing a competitive position in which each supports the other. Hill provided a useful way of linking marketing and manufacturing by introducing the concept of order-winning and order-qualifying criteria (Hill, 1985).

3.4 THE MANUFACTURING AUDIT APPROACH

The audit procedure which we use guides the user through a logical process of identifying manufacturing objectives, measuring current manufacturing performance, determining the effects of current manufacturing practices and identifying where changes are required. The framework for the process is shown (Fig. 3.1). The audit procedure has sought to 'operationalize' the existing strategy frameworks of Skinner (1969), Wheelwright (1978), Hill (1985), etc., and make them accessible to the management of operating companies. This has been done by using worksheets at each stage of the process (Fig. 3.2) to facilitate the recording and progressing of the procedure. The use of worksheets is most important as they require the users to interact formally with the process rather than simply reviewing a 'checklist'. The worksheets also provide **traceability**; the logic and data of the analytical part of the process are recorded and can be revisited periodically to check that the bases for the strategy are still appropriate. The design and logic of

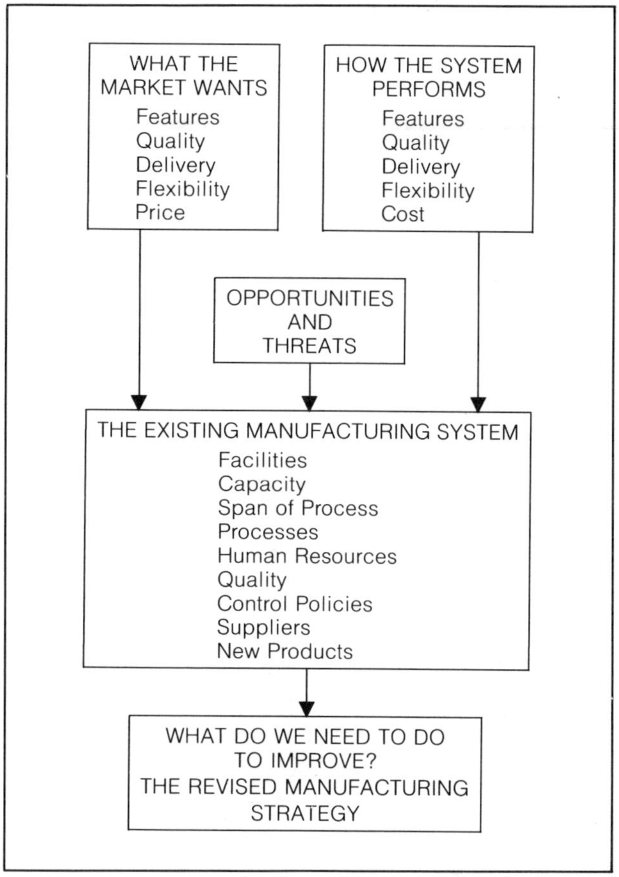

Fig. 3.1 A framework for manufacturing audit.

the audit procedure are reported more fully in Platts and Gregory (1990), and is also available in workbook form (DTI, 1988).

3.5 APPLYING THE APPROACH

We have conducted this process with six companies in various sectors. Table 3.1 summarizes the characteristics of these. Rather than act as external 'auditors', we have sought to act as 'facilitators'. Personnel within the company have been intimately involved in the process, our role being to provide the required structure, and to advise and assist when required. We have used multi-disciplinary workshops involving, in addition to manufacturing personnel, representatives from marketing,

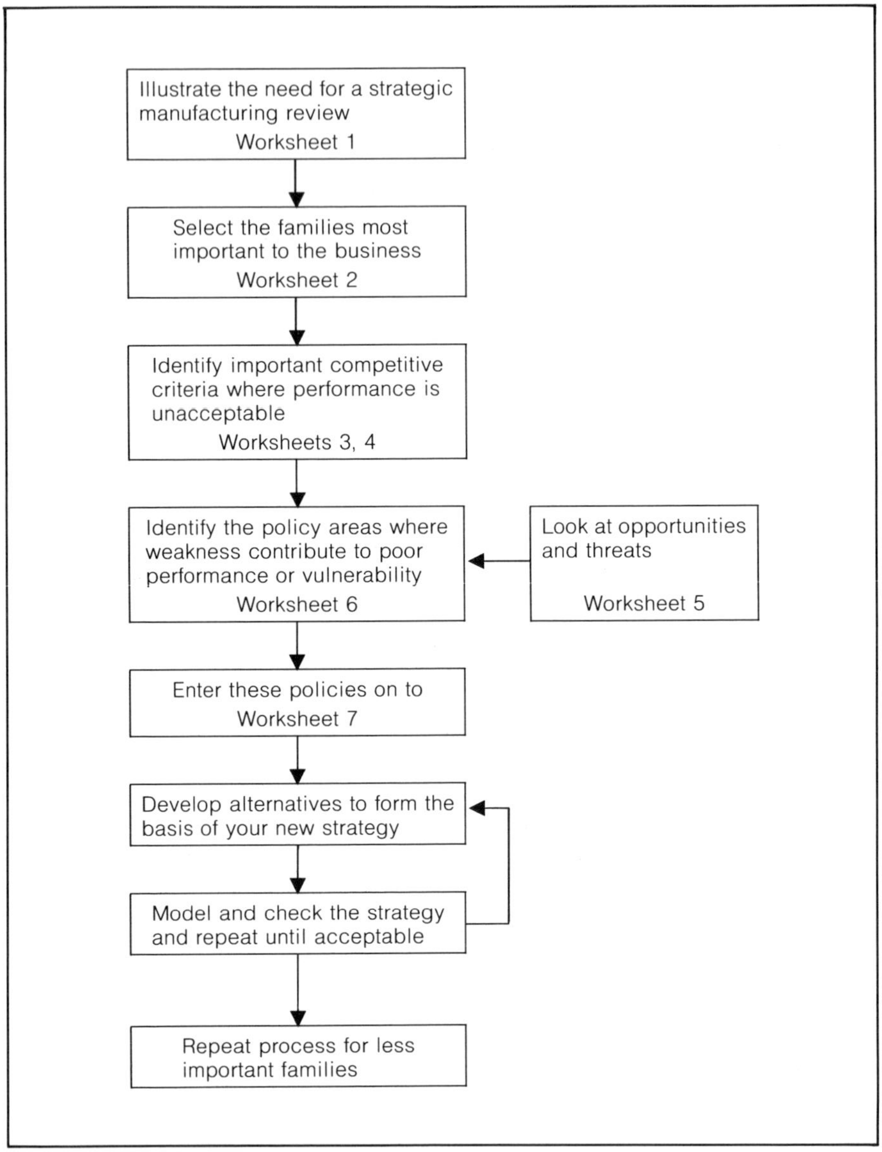

Fig. 3.2 The manufacturing audit process.

product engineering, finance, personnel, etc. Generally speaking the representatives have been at director level and the workshops have been chaired by the Managing Director. We structure the workshops by using the worksheets as an outline agenda and use them as the basis for discussion aiming to reach a consensus view at the completion of each

Table 3.1 Summary of case studies

	Company					
	P	*Q*	*R*	*S*	*T*	*U*
Product	Relays, solenoids	Mechanical seals	Ink-jet printers	Underground roadway supports	Aircraft power systems	Cranes
Size/turnover	336/£9 m	120/£5.5 m	300/£32 m	65/£20 m	1116/£51 m	300/£20 m
Sector	Electro/mech	Mechanical	Electro/mech/electronic	Mechanical	Electro/mech	Mechanical
Type of manufacture	Batch/line assembly	Batch m/c and assemble	Batch assembly	Line	Batch m/c and assemble	Project
Organization	Autocratic/hierarchical	Hierarchical	Young/flexible weak hierarchy	Flat hierarchy	Formal, strong hierarchy	Formal, strong
Business strategy	Written	Written	Written	Defined/unwritten	Written	Feel
Manufacturing strategy?	Written subset	Written subset	Feel	Defined/unwritten	Written	Feel
Access	MD	MD	MD to chairman	Non-executive director to MD	Group manufacturing executive to MD	MD to GM
Company project leader	None	MD for 2nd half	MD	PA to MD as a link	None	None
Chairman of process	MD	MD	Chairman	MD	MD/man. engineering Man	GM

Table 3.1 Continued

	Company					
	P	*Q*	*R*	*S*	*T*	*U*
Worksheet 1						
What type of data?	Opinion	Not used	Opinion	Not used	Opinion	Opinion
From whom?	6		6		20	6
What then?	Fa analysis		Fa analysis		Fa analysis	Fa analysis
Worksheet 2						
What type of data?	Mainly factual	F-factual Mkt-opinion	F-factual Mkt-opinion	Mainly factual	Business strategy	Opinion/ published
From whom?	FD. MD. MktM.	Mkt-SD from 3S	Mkt-SM. MktM	Various then 2. 4 man		Sales M
What then?	Workshop for 1 & 2	Workshop	Workshop for 1 & 2	Workshops		Fa analysis
Worksheet 3						
What type of data?	Opinion	Opinion	Opinion	Opinion	Opinions	Ind. opinions
From whom?	Marketing consensus	Sales consensus	12ind w/sheets	MD. his PA and Fa	9	5 (GM. DeM. ManM SM. CoM)
What then?			Fa collated, summary	in workshop session	Fa averaged, then presented	W/shop to consensus
Worksheet 4						
What type of data?	Factual	Factual	Opinion (some facts)	Opinion (some facts)	Mainly opinions	Opinions, tenders
From whom?	Fa collected analysed	Fa collected analysed	9	6 & Fa in	Executive	2 (SM. CoM)

What then?	Workshop for 3 & 4	Workshop for 3 & 4 facts, 2nd w/shop agree w/sheets		Workshop session		Fa analysis
Worksheet 5						
What type of data?	No response	Opinion	Opinion	Opinion	Business strategy and division. MktM	Industrial opinion
From whom?		MD. no other resp.	Same time as 3	Same time as 4		5 (as 3)
What then?						Group brainstorm and rank
Worksheet 6						
What type of data?	Opinions	Opinions	Opinions	Opinions	Industrial w/ sheet	W/sheet and interview
From whom?	TD. ManM. QM. MatM	ManD. ManM. CIMM	16 w/sheets	6 in workshop	9 in 2 workshops	5 (w/shop) and 4 interview
What then?	2 w/shops discuss consensus	Discuss, consensus Workshop	Fa summary, w/ shops (4), 9 people, consensus	2 sessions	Discuss, consensus	Discuss, consensus

Key to personnel

M	Manager	D	Director	Fa	Facilitator	PA	Personal assistant.
De	Design	F	Finance	G	General	M	Managing
Man	Manufacturing	Mkt	Marketing	Mat	Materials	Q	Quality
S	Sales	T	Technical	CIM	Computer Integrated Manufacturing		
Co	Commerical						

stage. Between the main workshops there are data gathering activities and mini-workshops usually involving lower levels of management.

The following sections review our main findings at each stage of the process.

3.6 POINT OF ENTRY (WORKSHEET 1, FIG. 3.4)

In all cases the point of entry was at a very senior level in the company, generally the most senior manager on site. This was seen to be important in obtaining the right perceived status for the exercise. In itself, however, this did not ensure commitment.

Worksheet 1 (Introductory Profiles) (Fig. 3.3) was introduced as a way of illustrating the need for a strategic review of manufacturing by graphically illustrating the differences between market requirements and manufacturing performance. It was hoped that this would then engender a commitment to the exercise. The profiles were used at four of the six companies and appeared to be useful in showing a need for a strategic review of manufacturing.

3.7 WORKSHEET 2 BASIC PRODUCT FAMILY DATA (FIG. 3.4)

The objective of this worksheet was to obtain on one sheet of paper a sufficiently detailed picture of a company's products and markets to enable the most important families to be identified as an initial focus for the exercise.

Few problems were experienced in the initial definition of product families; however, there was difficulty in obtaining some of the basic product family data. Financial data was generally available from the accounts but the obtaining of market data was much more difficult. Market data on share and growth was not explicitly known in most companies. Overall, there was considerable reliance on the use of the subjective opinions of managers in establishing the market position.

No problems were experienced in choosing the families which were the most important to the business. The choice was generally made by group consensus in a workshop at which Worksheet 2 was presented. The fact that the market data had a substantial subjective content did not appear to affect the decisions. Although all the companies identified the important families, most proceeded through the subsequent stages with several families only using the results of this stage to minimize future data collection by concentrating on the most important.

3.8 WORKSHEET 3 COMPETITIVE CRITERIA (FIG. 3.5)

The purpose of this worksheet was to identify for each family the market requirements in manufacturing terms. This was done by allocating 100 points among the competitive criteria to reflect their relative importance.

3.8.1 Obtaining the data

The methods of completion differed between sites but fell into two categories: individual views and consensus views.

When individual views were used an extremely large spread of responses was received. Merely averaging these lost the opportunity to explore the reasons for the differences which could have provided insights into the market. Discussion leading to consensus revealed more but the method which we found to be the most productive was individual assessment of the requirements followed by a group discussion in an attempt to reach a consensus. Here all views could be exposed for questioning whereas group discussions without prior individual assessment could result in some individuals' views being suppressed. However care needed to be taken in this process to avoid antagonizing or appearing to evaluate personnel who held views at the extremes of the distribution.

3.8.2 Defining the competitive criteria

Although all six companies started with the same set of criteria, three modified the set in some way usually by the addition of non-manufacturing criteria, e.g. field reliability (*R*), after-sales service (*U*). The commonly used set of manufacturing performance criteria were sufficient to identify all the manufacturing requirements imposed by the market; no company introduced further manufacturing related criteria. The addition of non-manufacturing criteria was useful as it gave an indication of the overall importance of manufacturing to the business.

Generally there was no problem in choosing and defining the criteria. The exception was with the 'flexibility' criteria especially 'volume flexibility'. The problem seemed to be caused by expressing the concept of order-winning criteria by the question, 'How do products win orders in the market?' This was interpreted by companies as, 'Why does a customer buy our product?' The argument then went, 'A customer doesn't buy a product because of "volume flexibility", he buys it because the supplier can offer the delivery that he wants. How the supplier achieves that delivery is an internal feature, not a market criterion.'

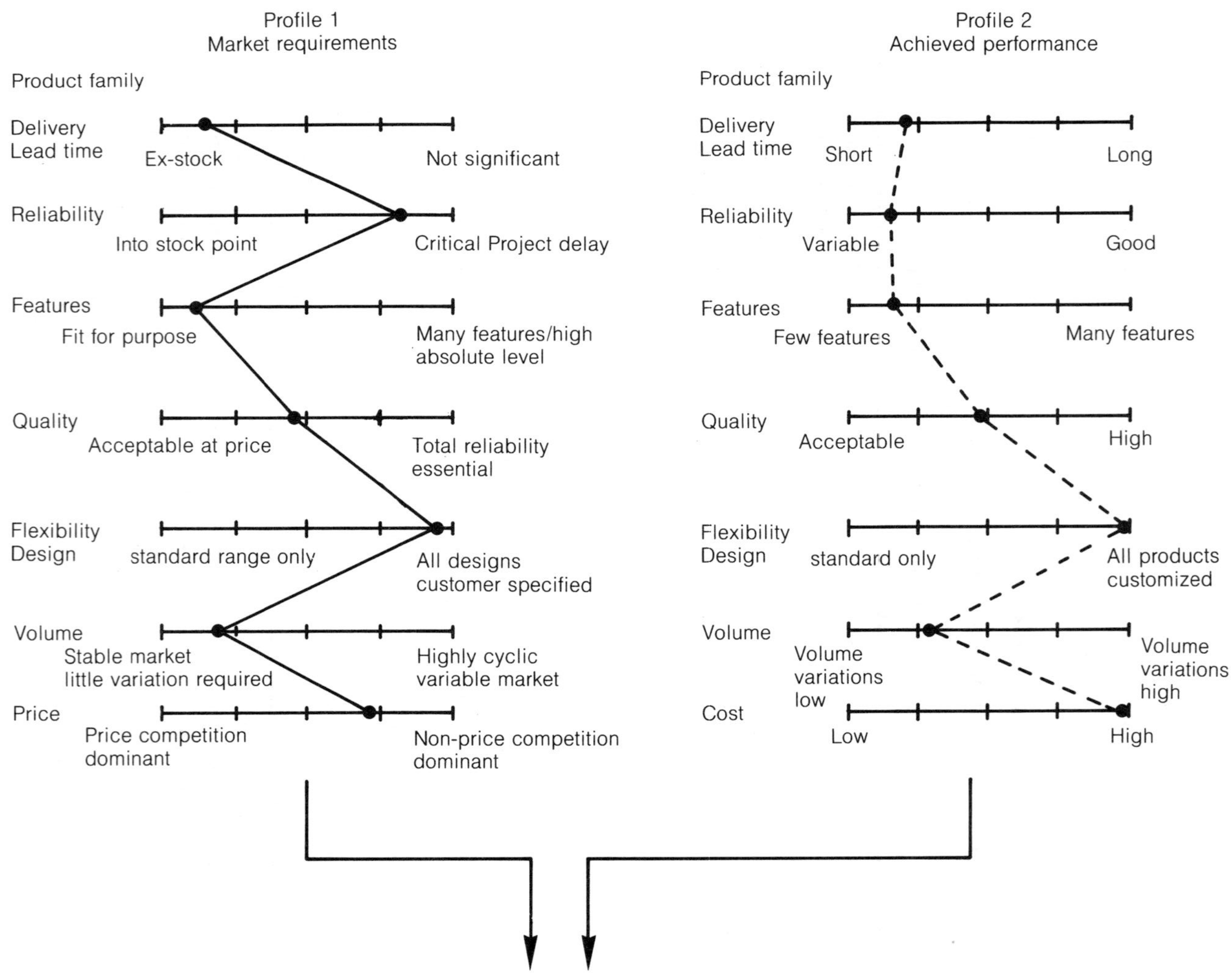

Profile 1
Market requirements
Product family
Delivery
Lead time
Ex-stock
Not significant
Reliability
Into stock point
Critical Project delay
Features
Fit for purpose
Many features/high absolute level
Quality
Acceptable at price
Total reliability essential
Flexibility
Design
standard range only
All designs customer specified
Volume
Stable market little variation required
Highly cyclic variable market
Price
Price competition dominant
Non-price competition dominant
Profile 2
Achieved performance
Product family
Delivery
Lead time
Short
Long
Reliability
Variable
Good
Features
Few features
Many features
Quality
Acceptable
High
Flexibility
Design
standard only
All products customized
Volume
Volume variations low
Volume variations high
Cost
Low
High

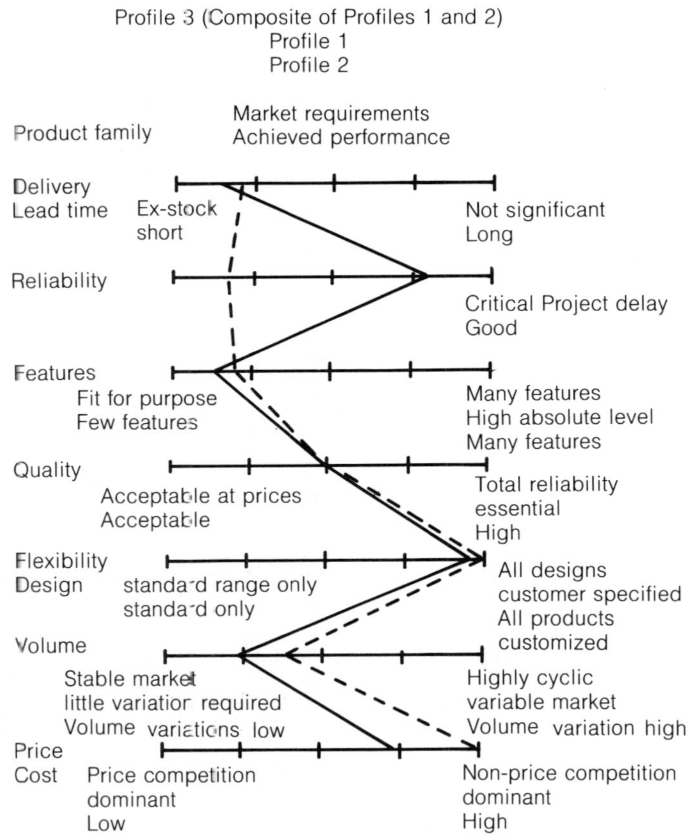

Fig. 3.3 Worksheet 1: Initial profiles.

Product families	Sales	% sales	% contrib.	Market share	Growth of mkt share	Market growth

Fig. 3.4 Worksheet 2: Basic product family data. For growth, if actual figures are not available, use scale of −2 (declining rapidly) to +2 (growing rapidly).

Product family				Delivery		Flexibility		
		Features	Quality	Lead time	Reliability	Design	Volume	Price

Fig. 3.5 Worksheet 3: Competitive criteria.

The underlying basis of the problem was the confusion of 'the customer' with 'the market'. Although the 'flexibility' criteria may be redundant from the viewpoint of a single customer, they are extremely important in specifying the manufacturing requirements imposed by the market. It is the latter which need to be established to define the objectives for manufacturing and it is the latter which the worksheet is designed to obtain. Thus a better question to ask is, 'What are the requirements imposed upon manufacturing by the way products win orders in the market?'

3.8.3 Existing business *v.* new orders

In several cases there was a difference in the competitive criteria for maintaining existing business and winning new business. Company T exhibited this feature strongly; it seems to be a characteristic of companies which have long-term renewable contracts. The maintenance of existing business generally reduces to achieving adequate performance in the qualifying factors. These qualifying factors however are not necessarily the same set as those considered at the gaining of the initial business. There will usually be additional 'qualifying' factors which have moved from the 'winning' category, i.e. factors which were sensitive in winning the business initially become less so in maintaining that business.

Product family	Features	Quality	Delivery		Flexibility		Cost
			Lead time	Reliability	Design	Volume	

Fig. 3.6 Worksheet 4: Achieved performance. Rate performance on a scale of −2 (performance gives us a strong disadvantage *v.* our competitors) to +2 (performance gives us a strong advantage *v.* our competitors).

3.8.4 Factual *v.* subjective assessments

A feature of this stage of the process was the complete reliance of the companies on the subjective opinions of their managers in determining the needs of the market. None actually went to the customers and asked for their views. It might be that such exercises had been done in the past by the marketing personnel within the company but this was not established. The impressions gained by the facilitators was that there was relatively little factual information on market requirements but that collectively the sales managers seemed to have a reasonable understanding of what was needed. The issue of the validity of subjective measures was not pursued in any depth and is a potential area for further work. The indications were that the companies were relatively satisfied that they understood enough about the market to proceed to the next stage.

3.9 WORKSHEET 4 ASSESSING PERFORMANCE (FIG. 3.6)

The objective of this worksheet was to obtain an assessment of the current performance of manufacturing against the criteria identified as important in the previous stage (Worksheet 3). Various issues arose during this stage of the process. These are considered under the following categories: internal *v.* external performance measures; factual *v.* subjective performance measures; and standards for comparison.

3.9.1 External *v.* internal performance measurement

The distinction made between external and internal performance measures is that external measures provide an assessment of the performance of the company as perceived by the market, while internal measures reveal how the various manufacturing sub-systems are performing.

3.9.2 Obtaining external measures

No major problems were encountered in the measurement of quality and delivery, although it was necessary on occasions to think around the measures required as they were not always available from the expected source. Minimizing the effort in data collection was perceived as an important factor if the approach was to be acceptable to the busy manager. Small samples were found adequate in so far as the results that they produced were consistent with the subjective views of managers. The concentration on one or two representative products as indicators of overall performance was also valid by the same criterion. Existing data were used wherever possible providing that they appeared to be reasonable measures contributing to the overall assessment of performance against a competitive criterion.

Price is an external measure according to the definition above; it is not, however, a useful indicator of manufacturing performance as it is affected by too many extraneous factors. It is fixed more by the market than by the performance of the company. The internal measure, cost, is a more useful assessment.

3.9.3 Obtaining internal measures

Internal measures of quality and lead times were readily obtained in a similar way to the external measures. Cost data, however, were not easily obtained in a form which could be used. The problems were with the costing systems which were designed primarily for the valuation of inventory (which is needed for the statutory accounts) and not for use in exercises such as these.

It was impracticable to attempt other forms of costing for these exercises due both to time constraints and the amount of managerial effort which would be required from the participating companies. Consequently it was necessary to accept the existing product costs and to look at alternative cost indicators such as utilizations, efficiencies, stock levels and staffing.

3.9.4 Factual *v.* subjective performance measures

This issue is closely related to that in the section addressing market requirements. To what extent is it necessary to measure performance in an objective way? The extent to which factual performance measures were used varied considerably among the companies. Where factual performance measurement was carried out it proved most useful, providing increased confidence in managers' views However when it came to filling in Worksheet 4, the requirement was to assess performance against the competition and as the companies had little factual data on their competitors this was again done by the subjective judgements of the management team.

3.9.5 Standards of comparison

Worksheet 4 was structured around comparisons with competitors but no attempt was made during the exercises to obtain comparative data from competitors of the participating companies. All the assessments made relied on existing data within the company (of which there was generally very little) or the subjective assessments of company personnel, typically from marketing and sales. Surprisingly companies seemed to be able to do this rating easily and confidently despite the lack of factual information. As there was no check on the rating, it was impossible to say whether they did it accurately in an absolute sense. It was clear, however, that they were able to do it to a level of satisfaction which enabled them to proceed to the next stage of the process.

3.10 WORKSHEET 5 OPPORTUNITIES AND THREATS (FIG 3.7)

The objective of this worksheet was to identify opportunities and threats which might affect manufacturing in the future. It was designed as a 'catch-all' to pick up items not revealed on the other worksheets. This worksheet had less structure than the other worksheets; a virtually free format was adopted to encourage the inclusion of any relevant observations. There were no major concerns in the development of this worksheet as it was half of the widely used SWOT (strengths, weakness, opportunities and threats) analysis; it was expected that individuals would be able to identify some items in each category fairly readily.

3.10.1 Obtaining the data

In practice this worksheet was the least well used in the whole audit process. The best responses were obtained when individual assessment was followed by group discussion as with Worksheet 3.

Product family	Opportunities	Threats

Fig. 3.7 Worksheet 5: Opportunities and threats.

Where the worksheet was completed, the opportunities and threats identified had more to do with products and markets than with manufacturing. This may have been partly due to suggesting that Porter's framework for the basis of industrial competition might provide useful guidelines or it may be that people still think of strategic opportunities and threats in terms of products and markets and not manufacturing. Manufacturing may still be the missing link in corporate strategy. In none of the cases were the results of this worksheet used in the subsequent stages. There are two possible explanations for this; either this type of analysis is of little value in the formulation of manufacturing strategy; or there needs to be a much more structured approach, as in the other worksheets, to guide people through the manufacturing issues which they need to address.

3.11 WORKSHEET 6 ASSESSING THE MANUFACTURING OPERATION (FIG. 3.8)

The objective of this stage was to identify the current manufacturing strategy and to assess what effects this had on the achievement of the objectives established in the previous stages.

The problems associated with the identification and specification of practices revolved around the questions of 'how many?' and 'how detailed?' In theory the complete specification of the practices at a company would be almost endless, in practice, over the six cases, the

Policy area			Delivery		Flexibility		
	Features	Quality	Lead time	Reliability	Design	Volume	Price
Facilities							
Capacity							
Span of process							
Processes							
Human resources							
Quality							
Control policies							
Suppliers							
New products							

Fig. 3.8 Worksheet 6: Existing practices. This worksheet is diagrammatic. Each of the policy areas would be expanded to allow the entry and rating of several practices.

number of specified practices ranged from 26 to 76 with a mean value of 50. These were distributed among the decision areas as follows (rounded means);

Facilities	4	Human resources	9
Capacity	4	Quality	5
Span of process	3	Control policies	8
Processes	6	Suppliers	6
		New products	6

The personnel completing the worksheets were generally able to identify those practices which they thought were important in the sense that they were the ones which shaped the way in which manufacturing performed. Obtaining the balance between defining the practices closely enough to enable meaningful assessments to be made, and keeping the practices sufficiently aggregated to encompass 'strategic' decisions was the key to this stage of the process. In the later studies, checklists or questions were circulated with the worksheet to prompt the user to consider the main areas of strategic choice.

In many of the cases, 'negative' practices were identified. These expressed the lack of something which the respondents thought desirable. For example, at company *U*, the 'Human Resources' area included the practices 'Little use of teams' and 'Little management training'. These were rated in the same way as other practices and although they did not define strategy, except by defining what it wasn't, they proved to be useful in identifying problems and when considering alternative courses of action.

The rating of the practices was generally carried out by individual assessment followed by group discussion leading, in most cases, to consensus. Occasionally a common view could not be reached and the practice was referred for further investigation.

The overall finding was that companies were able to identify and rate the current practices with relative ease.

3.12 WORKSHEET 7 DEVELOPING THE NEW STRATEGY (FIG. 3.9)

This stage is strictly speaking outside the scope of auditing, however in all of the cases company management proceeded to the stage of identifying the main manufacturing problems, and in three of the six cases went to the next stage of formulating action plans.

The problem identification stage was conducted by inspection of the Worksheets. Worksheets 3 and 4 together revealed where performance was unsatisfactory against an important criterion and Worksheet 6 was then used to identify the practices which contributed towards this.

Policy area	Weakness/required strength	Possible actions/strategic choices

Fig. 3.9 Worksheet 7: Action worksheet.

Generally a visual check of the worksheets was all that was required. If much more complex audits were conducted it might be necessary to adopt some of the techniques allied to decision analysis, however they would need to be used with care. One of the benefits of the approach adopted was that it was easily understood by practising managers, they were able to trace the logic leading to the conclusions. If more complex techniques were adopted this aspect of the process might be lost.

The results of this stage fell into two main categories which might be labelled operational improvements and strategic improvements. Operational improvements dropped out of the analysis of Worksheet 6, there was no need to look at competitive criteria or performance, simply looking at the effects of the practice itself were sufficient. Strategic improvements were obtained when the full analysis outlined above was followed and a programme of actions was embarked upon which related improvements directly to the needs of the market.

3.13 ORGANIZATION OF THE PROCESS

A key feature of the audit process was the intimate involvement of company personnel at all stages. The earlier worksheets were, as expected, completed mainly by the marketing personnel and the later worksheets by manufacturing personnel, however, the results were generally fed back to a group containing representatives of all functions.

The methods of worksheet completion differed slightly between companies but the approaches which were perceived to be the more successful were generally variations on the process of individual assessment followed by group discussion. For example, for Worksheet 6 the process was:

1. Blank worksheets were circulated to a number of personnel in manufacturing related roles. They were asked to identify the current practices.

2. The Facilitator collated the results into a master summary sheet. This often required selected interviewing of the respondents to clarify practices.
3. Copies of the summary master sheet were circulated to the manufacturing personnel who were asked to assess the effects of the practices.
4. A workshop was held involving the manufacturing personnel and the Facilitator at which the worksheet was discussed, often at great length, and a consensus view on the practices and their effects was (almost always) reached.

Most companies made substantial use of workshops, with the most popular arrangement being a full workshop at the completion of Worksheet 2, a marketing workshop for Worksheet 3, a full workshop at the completion of Worksheet 4, several manufacturing workshops for Worksheet 6 followed by a full workshop at the completion of Worksheet 6. The full workshops would include representatives of all functions, generally at director or senior manager level; the functional workshops would generally include the representative of that function from the full workshop and several members of his staff. This arrangement was considered to work well. Companies found the multi-functional representation throughout the process very worth while, considering that it improved understanding and fostered a feeling of common purpose. The functional workshops were considered to be more successful when they contained more rather than fewer people. The use of more people was thought to bring in a wider experience of the detail of the business and to promote discussion. It was also seen as a way of involving lower level managers in an area where they may not normally participate and was said to have had beneficial effects on their motivation and enthusiasm. The problem with fewer people is that there is a danger of one person dominating the workshop. There is also a danger of free discussion being inhibited by the presence of certain people. This becomes particularly pronounced if one of the members is of a perceived higher 'rank' and is of an autocratic nature. This was the situation in some companies where the presence of the Managing Director was felt to suppress discussion at some of the workshops.

Various methods of participation in the completion of the worksheets were tried from individual assessments followed by Facilitator analysis to group discussion and collective completion of the worksheet. The method that seemed to be the most successful both from the perception of the Facilitator and from the comments made by the participating managers was the use of individual assessments followed by group workshops. This was most often used at the Worksheet 6 stage. The advantages of this approach are that each individual gives careful consideration to the issues before the workshop and is more inclined to become involved, the downside is that it can make participants less inclined to accept views different to their own and may alienate people with extreme views. In practice, although consensus was not always reached, there was little evidence of intransigence and people appeared to be trying to reach a common view.

The overall view of the managers involved in the process was that participation involving several functions and the use of structured workshops were important features of the approach. Companies felt that full participation in the analysis improved the recognition and acceptance of problems and expected that this would lead to increased cooperation in the implementation of solutions.

At five of the six companies the exercise was sponsored by the most senior executive; the Chairman, Managing Director or General Manager. At the other company the Manufacturing Director sponsored the exercise. This top-level sponsorship was perceived to be useful in giving the exercise the necessary status to ensure that activities were carried out. At the company where the Manufacturing Director sponsored the exercise a reasonable response was obtained at Worksheet 6 (the manufacturing practices) but the Facilitator had more of a problem in chasing the responses to Worksheet 3 (the competitive criteria) than was experienced in other cases. Although there could be many explanations for this, it is consistent with a problem of sponsorship at the wrong level.

Although the Chief Executive might sponsor the exercise, this will not ensure that it progresses satisfactorily unless he has a close involvement on a day-to-day basis (which is unlikely to happen and is probably undesirable). There needs to be a senior company manager who is assigned the internal responsibility for the exercise. This was not recognized initially. In the early studies no internal leader was nominated with the result that the exercise became relatively low priority and was extended over several months. In the later studies each company was asked to appoint a project leader at the start of the exercise. Various internal project leaders were used but the most successful tended to be at manufacturing director level.

The conclusion is that the progressing of the exercise will be enhanced and its chances of success improved if it is sponsored by the Chief Executive and a senior manager (preferably the Manufacturing Director) is given internal responsibility for it and becomes involved on a day-to-day basis. The use of a multi-disciplinary steering group is also recommended.

3.13.1 The use of a facilitator

The Facilitator supplied the external expertise in the use of the audit methodology. In the early cases there were misconceptions of the

Facilitator role. This was initially perceived as that of an external consultant who would provide solutions to the company's manufacturing problems, rather than providing and expediting a process which would enable the company to solve its own problems. In these circumstances it was harder to obtain the participation of the management team. In the later studies great emphasis was given to ensuring that the Facilitator's role was clearly specified at the start of the process, even so it was often still difficult to obtain sufficient company management involvement and tasks fell to the Facilitator which would have been better handled by company personnel.

Most companies thought that an external Facilitator was necessary. Although they felt that they could have followed the process themselves, as it seemed quite straightforward, they did not think that they would have done so. There would always have been disruptions, operational pressures, month-end purges, etc., which would have diverted their attention. They needed to have someone who would catalyse their involvement and progress the exercise. Another identified role for a Facilitator was the questioning of accepted beliefs, this however is much more dependent on the knowledge and skills of the individual than is the simpler progressing function.

The overall conclusion was that a Facilitator to support the company leader was desirable and maybe essential. The company must be made fully aware at the outset of the role of the facilitator and of their own responsibilities for the outcome of the exercise.

3.13.2 Timescale

From a small sample it is difficult to be prescriptive about the length of time to be spent on the exercise. The first two studies were spread over several months and were considered in retrospect by all parties to have been too long. At the time, however, there were always 'good' reasons why things (data collection, workshops, etc.) needed to be delayed. Operational pressures, year-ends, etc., diverted attention. The result was that the exercise tended to lose focus and direction. The four later studies were constrained to a timescale of 6–7 weeks and this seemed to be satisfactory. There was sufficient time to collect data and conduct workshops but the timescale was tight enough to give a feeling of urgency and maintain momentum once it was running. A shorter timescale may be possible but this was not tried.

An alternative put forward by some managers was to conduct the exercise as a two-day workshop off-site. This may be a possibility, however there may be problems arising from:

1. Lack of time to collect and verify data (unless it could be prespecified);

2. Less opportunity to 'sleep on a problem', i.e. less time to reflect;
3. Less opportunity to vary working groups; and
4. Developing a mentality that strategy is developed by short intensive exercises rather than the mentality that strategic thinking should become more of a way of life.

These problems may or may not be real, further work would be required to investigate them. However two-day workshops for manufacturing strategy formulation have recently been developed and used (see, for example, Schroeder and Lahr, 1990).

3.14 ASSESSMENT OF THE PROCESS

It was extremely difficult to measure objectively whether the approach was successful and therefore we have relied on subjective assessments from the managers within the firms against the criteria of feasibility (could it be done), usability (how easily could it be done) and utility (how useful was it).

All stages of the process were clearly feasible. Although problems of various kinds arose, none was insurmountable. The major unexpected finding was that the majority of companies studied did not have clear views of their position in the market, did not appear to have adequate information on market size and growth and did not have much factual information on competitors. In short, their marketing functions were weak. Although this is not directly a manufacturing strategy issue it is obviously a critical business issue and will impact on manufacturing via the manufacturing objectives which can only be properly defined against known or predicted market needs.

That said the companies felt that following the process had supplied them with a clear enough picture of the market requirements to enable them to specify their manufacturing objectives at least in qualitative if not quantitative terms. It had enabled them to identify manufacturing priorities and to define the manufacturing task.

The process was usable. The worksheets were readily understood and were straightforward to complete when factual data was available. When there were difficulties with the collection of factual data, subjective views and the use of group workshops were found to provide sufficient data for the exercise to progress.

As for utility, all the companies found the approach useful. In addition to the output of action plans, several key features of the process were identified:

1. The use of worksheets provided an effective way of structuring both the process and the data.
2. The use of group working and full participation in the analysis was

perceived to improve the recognition and acceptance of problems and was expected to lead to increased cooperation in the implementation of solutions.

3. The use of an external Facilitator catalysed the process by promoting involvement of the management team, reducing interdepartmental stresses, progressing the exercise and stimulating the analysis by questioning.

3.15 CONCLUSION

This research has resulted in an audit process which has been shown to be feasible, usable and useful for structuring the analytical stages of manufacturing strategy formulation. However, because of the breadth of the approach, many issues could not be researched in depth and as a result many of the findings must be regarded as indicative rather than definitive. It was considered more desirable to introduce an overall structure which could be applied in companies before seeking to refine each element of the process in depth. Such refinement is however felt to be an important next step and the following list indicates some of the key areas.

3.15.1 The 'mechanics' of the audit

The acquisition of market data is one problem area which could be addressed in more detail. Typical questions for investigation are, 'How can the obtaining of market requirements be improved? Are subjective measures adequate? Is there close correlation between managers' opinions of market requirements and "external" measures obtained by, for example, market survey?'

Much of the performance measurement is fairly straightforward, however the measurement of costs and flexibility is still problematical. Can more usable measures of cost be developed? Can measures of flexibility be developed? Some companies made extensive use of subjective measures considering that the effort involved in obtaining objective measures was not worth while. Why was this? Are subjective measures adequate? Is there close correlation between subjective and objective performance measures?

The Opportunities and Threats worksheet is currently unstructured and tends to be the least used. It was seen very much as a 'catch-all' to pick up information which had not been obtained from the other worksheets. Is it possible to structure the worksheet in such a way that it stimulates the identification of *manufacturing* opportunities and threats? Is the worksheet needed at all for manufacturing strategy formulation?

The identification of existing practices was initially very broad and had only a loose structure. The introduction of checklists provided more structure and focus to this stage. Can the structure for the identification of existing practices be further improved? Could the worksheet be customized to suit more closely particular types of manufacture? The existing assessment of the effects of practices considers them in isolation; in practice there are interactions between practices. How can such interactions be incorporated into the process?

3.15.2 The process of audit

Further work should attempt to establish to what extent the following factors influence: the success of the process; the use and composition of teams; the presence of a designated leader for the process; the use of an external facilitator; the use of factual data (as opposed to subjective opinion).

There was a strong indication from the case studies that the sociological benefits of the exercise were substantial. Factors such as improved communications, increased motivation, the introduction of a sense of common purpose were frequently mentioned. These aspects should be more rigorously researched. How widely through the organization did these benefits permeate? How can the exercise be structured to maximize these benefits without losing sight of its prime objective?

3.15.3 Extensions to the process

The audit process has addressed the analytical stages of strategy formulation ending at the identification of manufacturing objectives and manufacturing practices requiring revision. There appears to be a need for a similar prescriptive methodology to be developed for the synthesis of revised strategy. This should cover the generation of alternative practices, the initial assessment of the expected effects and the selection of the most suitable set. There would need to be a way of considering the interactions of the alternatives to ensure that a coherent set was obtained. This would be a major research undertaking as currently very little process theory exists in this area. A research methodology similar to that used in the current work would provide a possible approach to the problem.

3.16 ACKNOWLEDGEMENT

The research upon which this paper was based was supported by the ACME directorate of the SERC, Grant No. GR/E 38719. This support is gratefully acknowledged.

3.17 REFERENCES

Adam, E.E. and Swamidass, P.M. (1989) Assessing operations management from a strategic perspective. *Journal of Management*, **15**(2), 181–203.

Andrews, K.R. (1971) (rev. ed. 1980) *The Concepts of Corporate Strategy*, Dow Jones-Irwin, Homewood, Illinois

Ansoff, H.I. (1965) *Corporate Strategy*, McGraw-Hill, New York

DTI (1988) *Competitive Manufacturing; A Practical Approach to the Development of a Manufacturing Strategy*, IFS, Bedford.

Fine, C.H. and Hax, A.C. (1985) Manufacturing Strategy: A Methodology and an Illustration. *Interfaces*, **15**:6, 28–46.

Hayes, R.H. and Wheel wright, S.C. (1984) *Restoring our Competitive Edge (Competing through Manufacturing)*, Wiley, New York.

Hill, T.J (1980) Manufacturing implications in determining corporate policy. *Int. Journal of Operations and Production Management*, **1**(1), 3–11.

Hill, T.J. (1985) *Manufacturing Strategy: The Strategic Management of the manufacturing function*, Macmillan, Basingstoke.

Hill T.J. (1987) Teaching and research directions in Production/Operations management: The manufacturing sector. *Int. Journal of Operations and Production Management*, **7**(4), 5–12.

Mintzberg, H. (1973) Strategy making in three modes. *California Management Review*, Winter.

Mintzberg, H, (1978) Patterns in strategy formulation. *Management Science*, **24**(9), 7–21.

Platts, K.W. and Gregory, M.J. (1990) Manufacturing audit in the process of strategy formulation. *Int. Journal of Operations and Production Management*, **10**(9).

Quinn, J.B. (1978) Strategic change: Logical incrementalism. *Sloan Management Review*, Fall, 7–21.

Schroeder, R. and Lahr, T.N. (1990) Development of manufacturing strategy, in *Manufacturing Strategy* (ed. J. Ettlie), Kluwer, Boston.

Skinner, W. (1969) Manufacturing – Missing link in corporate strategy. *Harvard Business Review*, May–June, 136–45.

Voss, C.A. (1984) Production/operations management – A key discipline and area for research. *Omega. Int. J. of Manufacturing Science*, **12**(3) 309–19.

Wheelwright, S.C. (1978) Reflecting corporate strategy in manufacturing decisions. *Business Horizons*, February, 57–66.

CHAPTER 4

A conceptual framework for manufacturing restructuring

Thomas E. Vollmann, Robert S. Collins, Jinchiro Nakane and Michael D. Oliff

This paper presents a framework for the *process* of manufacturing restructuring. Manufacturing restructuring encompasses significantly more than change in the manufacturing function. Manufacturing restructuring as used here is a fundamental alteration in a manufacturing company, affecting all functions and activities.

The word 'restructuring' connotes many things to different people. More often than not it is used synonymously with 'rationalization' – plant closures, redundancies, personnel reductions, and associated downsizing issues. The framework developed here includes this focus, but also encompasses a *proactive* approach to restructuring manufacturing where the appropriate response to the major forces for change is to achieve long-run enterprise well being, market dominance and even growth.

4.1 A RESTRUCTURING FRAMEWORK

Restructuring needs to be viewed – not as more of the same – (i.e. not simply JIT, TQC, MRPII, CIM), – or even merger/acquisition, – but as a process of fundamental rejuvenation throughout the company. The fundamental posture is that doing things better is necessary, but not sufficient. It is essential to do better things! Successful restructuring requires a critical understanding of:

Specially commissioned for *Manufacturing Strategy – Process and content.*
Edited by Christopher A. Voss.
Published in 1992 by Chapman & Hall, London ISBN 0 412 43660 4.

1. The forces and constraints for change – the WHY?
2. The paradigm shift required – the WHAT?
3. The implementation process – the HOW?
4. The problems awaiting solution – WHAT IS NEXT?

4.1.1 Forces and constraints for change

Forces, both internal to the organization and external, mandate fundamental changes in business operations. Constraints, on the other hand, act to either prohibit or limit the restructuring that is undertaken. Together, the forces and constraints constitute the WHY for manufacturing restructuring. From all fronts, the forces for change are expected to increase, not decrease, in intensity in the future. Companies need to understand the dynamic forces precipitating change, as well as how constraints are transformed over time, – for forces and constraints both external and internal to the organization.

1. *External forces and constraints* stem from the environment specific to an entire industry, or to a particular firm within an industry. They include economic, social, and political forces and constraints as well as technological thrusts which impact on all players in an industry. Also included are these same forces as they uniquely impact on a particular firm with its individual set of strengths and weaknesses, – both human and physical.
2. *Internal forces* for change include any mismatch between the strengths required to compete successfully and the existing inventory of company strengths, a restatement of the overall mission and concomitant strategy for a business unit, or an initiative stimulated by some perceived need for performance improvement.
3. *Internal constraints* include all the forms of resistance to change.

4.1.2 The Paradigm Shift

Successful restructuring requires a well thought out set of objectives. That is, WHAT is the purpose of restructuring and how will the company be different as a result? The paradigm shift can be conceptualized as encompassing three distinct (but related) dimensions (see Fig. 4.1):

1. *Culture* includes the broadest, most enduring (and most difficult to change) aspects of business. Existing culture directs and constrains restructuring efforts.
2. *Configuration* relates to both organizational designs and relationships, and to physical/geographical distributions of people, capital and equipment. Restructuring typically results in new designs and arrangements of these entities.
3. *Coordination* refers to management and control within the business

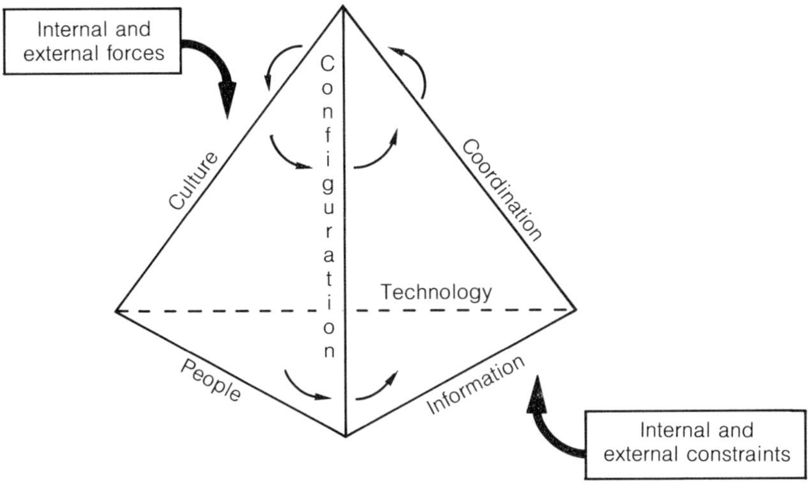

Fig. 4.1 Framework for manufacturing restructuring.

system itself. Restructuring normally requires new flows of information and materials – as well as new sets of managerial responsibilities (Oliff, Arpan and DuBois, 1989).

The foundation for restructuring rests on three fundamental resources: people, technology, and information. Restructuring manufacturing can then be defined as:

> The process (not a single activity) of changing significantly any one or more of the three dimensions (culture, configuration, coordination), through the deployment (or redeployment) of any of the three resources (people, technology, information).

The base of the pyramid in Fig. 4.1 depicts the three fundamental resources that are deployed, redeployed or substituted for each other to achieve the desired restructuring. In most cases, restructuring involves all three. People perform different jobs, utilizing new technology, coordinated with new information systems.

1. *People deployment* is at the heart of most restructuring projects, with the nature of jobs and responsibilities changing fundamentally. Examples include the elimination of managerial layers, replacement of pyramid organizations with flat or network organizations, mobilizing people to undertake new initiatives, and the continuing absorption of what has traditionally been 'staff' work into the basic manufacturing infrastructure.
2. *Technology deployment* for restructuring often involves a major shift in

fundamental manufacturing methods and equipment. Examples include flexible machining systems (FMS), robotics, and the transfer of technology from one firm to another. Introduction and successful implementation of these major advances in technology require significant change in the culture of the company.

3. *Information deployment* includes both the development/introduction of new management information systems (MIS) and the obsolescence of existing systems. Computer-integrated manufacturing (CIM) mandates increase integration among key manufacturing activities, including engineering design, manufacturing engineering, process planning and machine scheduling. At the same time, JIT moves materials through the factory so quickly that the need for systems to track materials is greatly reduced. Moreover, the rapid development of ever more powerful and inexpensive computers encourages decentralization of MIS and the better leveraging of all forms of information.

4.1.3 The implementation process

Figure 4.1 shows a circular arrow encircling the pyramid. This arrow depicts the process of restructuring, with the six elements of Fig. 4.1 intimately involved. Fundamental changes must be achieved in culture, configuration, and/or coordination by the redeployment of the three resources: people, technology and information. Defining the desired new configuration and the new coordination to be achieved in restructuring is more straightforward than defining the new culture that is to be achieved. But what is fundamentally more difficult is determining the set of steps required to achieve the desired result, i.e. the HOW of responding to the forces, recognizing the genuine constraints, and achieving the desired paradigm shift.

1. *Top-down* implementation efforts are typically directed by senior management. Included are: the decision to restructure, usually based on a careful analysis of the forces and constraints; the overall objectives and approach for the restructuring, perhaps based on scenarios of various actions and competitive responses; an inclusive plan which is *not* simply an overall goal statement or wish list without adequate attention given to the detailed action steps required; and personal leadership to achieve the desired results and timing.
2. *Bottom-up* implementation energies are typically required from a large number of people, in most cases the more the better. Also required is a set of key managers who will lead the implementation process. The changes required in restructuring are usually fundamental. If cultural change is necessary, the new culture can be adopted more quickly if everyone is a part of the implementation process.

Project teams play a key role in most restructuring undertakings when formal organizational barriers are broken. Education on the fundamental forces and constraints, as well as the paradigm shift (in clearly understandable terminology) is required in order to create a bandwagon effect. It is critical that the organization both understands and *believes* in its definition of the forces and constraints and its paradigm shift. Too often the paradigm shift is stated as some lofty goal associated with 'getting closer to the customer', but the organization sees it as only cost-reduction and head-count reduction.

Evaluation of a restructuring program is a major challenge. It is difficult to tell whether the results achieved are truly successful, mediocre, or poor, or if short-term results are being achieved at the expense of long-run health. Moreover, it is necessary to take into account the continuing impact of changing forces. Frequently, restructuring requires a new system for performance measurement in the company (Dixon, Nanni and Vollmann, 1990).

4.1.4 A view to the future

Restructuring needs to be clearly bounded. What is to be included or accomplished? It also needs to have a definition of what is and is not to be included. But the WHAT IS NEXT question must also be asked. Too often restructuring is viewed as a one-time adjustment rather than a continuing process. As a particular restructuring effort is taking place, it is important always to be evaluating the next steps. To the extent that this viewpoint can be proactive, it may well help avoid the painful reactive mode of restructuring.

4.2 FORCES AND CONSTRAINTS

In this section the WHY question of restructuring will be explored in more detail. The culture, configuration, and coordination dimensions of the pyramid in Fig. 4.1 depict a set of restructuring responses. Also depicted are the resources used to enact the restructuring. It is important for companies to articulate clearly why the changes are necessary; that is, to identify the underlying *forces* that drive restructuring, and the *constraints* that limit the response and preclude certain restructuring actions.

4.2.1 External forces and constraints

External forces and constraints include economic, social, political and technological events which impact all firms in an industry, as well as those influencing a particular firm more than others in the same industry. Differences between firms are a function of how each is structured and positioned relative to forces and constraints. For example, a political event might pose very different problems to one firm than another. Whether it is a force for change (or a constraint to change) depends upon the particular situation within each firm. Part of successful restructuring is to develop a culture, configuration and coordination that is less easily disturbed by, but more intelligently responsive to, external forces and constraints than competitors' culture, configuration and coordination.

Economic forces and constraints include actions by competitors, market place dictates for shorter product life cycles, world-wide competition, requirements for improved product features and prices, the demand for smaller lot sizes, the dictates imposed by the trade in consumer products and the competitive might of other firms who have undergone successful restructuring. On the more purely economic side are costs of the factors of production, the difficulty in acquiring these factors, the advent of new factors, new uses for existing factors and the inevitable changes in all of this. In some locations there are severe shortages of skilled personnel. In Japan there is concern over the traditional Japanese work ethic; the result will be a shortage of a 'factor' of production, and the need to 'substitute' other factors – either less well motivated employees or capital equipment.

Sometimes the need to restructure is clearly imposed by external economic forces and events largely outside the control of the company. Examples include mergers and acquisitions, as well as situations where the marketplace dictates an immediate response. The minicomputer industry is currently undergoing this sort of imposed restructuring. Wang, Siemans, Norsk Data, Data General and several other minicomputer companies are frequently in the news with major work-force reductions due to serious erosion of the market for minicomputers (to some extent based on the increasing capability of less expensive computers and workstations).

For many companies the costs for new generations of products are escalating rapidly. Other firms find themselves with a similar but less drastic economic problem: the need to spend more money on R&D and marketing, and the need to generate the funds through cost reductions in manufacturing. Sometimes the changes are relatively gradual, but at other times dramatic change mandates rapid response through restructuring. A current example of economic forces is the recent entry into the luxury segment of the automobile market by several Japanese manufacturers. The result has been a reduction in demand for some traditional sellers in this market. At the same time, these companies are facing market pressures for new models and features. The results are typically some form of restructuring (including merger/acquisition).

Continuing with this automotive example of economic constraints, some of the smaller auto companies are facing limits on their ability to raise funds for new product development, their ability to attract excellent employees and the availability of other production factors. Rapid change affecting the automobile industry was clearly demonstrated with the oil embargo of the 1970s and swift escalation in the price of gasoline. Some manufacturers were much better able to restructure to meet the challenges than others.

Social forces and constraints also drive restructuring and limit the use of people, technology and information to change significantly culture, configuration and coordination. Demographic trends impact the underlying demand for certain kinds of products, like diapers. Life-styles have a similar impact. Examples include postponement of marriage, family size and educational levels. Constraints include taboos associated with seven-day operations, women working night shifts and the number of engineering college graduates entering the labour force.

A manager from a large European food manufacturer expressed the belief that his company would not close factories in response to the forces associated with 1992, even though analysis of the situation clearly indicated significant over capacity. He said that plant closure would lead to political pressures, such as frequent tax audits. He attached a strong importance to the image of his company in each of its manufacturing locations. This view of forces and constraints is much broader than that of an individual factory operation, or indeed an entire company – it includes the society in which the firm will do business in the long term.

A concern over the social reaction to plant closures was echoed in several other cases. In one, a consumer products company was making several acquisitions, primarily to obtain branded products. In most of the situations, the product could only be purchased *with* the factory, and the buyer had to agree not to close the factory for a specified time period (in one case this was two years, and in another, ten!). The selling firms wanted any plant closure image problems to be associated with the buyers, not with them.

Growing environmental awareness is another social force which increasingly will drive restructuring. It will be felt differently and with more intensity in particular industries and companies. Current examples include soap powders, disposable diapers, garbage bags, other uses of plastics, and the more dramatic case of fur coats being now regarded as exploitive of animals.

Recent events in Eastern European nations have led to a massive set of social changes and pressures. Personal expectations and the range of acceptable business practices have changed dramatically. These forces and constraints will impact many manufacturing companies, since we increasingly live in a worldwide economy – and a worldwide environment.

Political forces and constraints tend to be more volatile than economic and social forces. Governments change laws frequently, and impose conditions that can have major impacts on restructuring. An example is the situation that is facing cigarette manufacturers in Germany. Fiscal incentives and subsidies have strongly encouraged manufacture of cigarettes in Berlin rather than in alternative locations; how this might change with German reunification is anyone's guess. Another example is a company that is locating a new high-technology plant in France, even though it would mesh nicely with their operations in the UK. The reason is an agreement with the French authorities regarding product prices, which are largely paid for by the French Government. Rapid change in these forces and constraints is easily seen in the 1992 agreements aimed at creating one market within Europe.

A particularly interesting political action from the point of view of restructuring can occur when environmental protection laws are changed. Examples include the nuclear power industry, and regulations for water discharge from industrial plants. A large manufacturer of transformers used PCBs in their construction, as did all of its competitors. The conclusion that PCBs were carcinogenic resulted in a ban on their use. Some firms using PCBs left the industry, others found alternative materials, and most faced massive clean-up operations and lawsuits. One executive at the transformer company expressed the opinion that a large change in external constraints like this represented an excellent opportunity: restructure to meet the challenges faster than the competition.

Technological forces and constraints are associated with the underlying forces of technology, and the impacts they can have on particular industries and companies. Included is the necessity to adopt technological innovation more quickly than one's competitors. In most cases, adopting manufacturing practices such as JIT, MRPII, TQC and CIM implies some form of restructuring. Each of these may lead to changes in culture, configuration, and coordination.

A basic technological force driving restructuring in many companies derives from a fundamental degree of productivity gain or 'learning' that is associated with particular kinds of manufacturing industries. An example that impacts almost all firms in the electronics industry is the growing power of semiconductors and the resultant decrease in the number of component parts required to make electronic items. The result is a continuing increase in the number of sales dollars produced with each hour of direct labour. Firms facing this problem either need to grow at significant rates (20% per year in many firms) or face up to a continuing need for downsizing.

Fundamentally new technology can create an industry, but well-known technological phenomena also play a key role. Examples include microwave energy for cooking, recombinant DNA for synthesis of organic compounds, new advances in microelectronics, and new generations of computers. In each of these cases technological advances resulted in new products, new uses and new markets. Those companies that have been able to integrate the new technologies, and restructure as required, have been able to reap the rewards.

One interesting constraint imposed by certain new technologies is the cost to develop the next generation of products. Examples include new telecommunications systems, aircraft and military hardware. Increasingly, firms are entering into joint ventures to develop these new systems, and to agreements and mergers for marketing the new products. In each case, significant restructuring is implied.

A *constraint* associated with technology is the underlying capability of a company to absorb new technology. This constraint is manifested in the operating functions of the company as well as in the financial area where capital budgets are approved. It is amazing how often a plant visit reveals no new capital equipment or other technology having been added during the last 10 years. Is it possible that *no* useful new ideas have been generated during this time period? It is also useful for managers of such plants to understand that in a restructuring it is *precisely* this kind of plant that will be the primary candidate for closure!

Industry structure and dynamics provide more forces for restructuring as well as constraints against achieving restructuring. In fragmented industries, it is difficult (but not impossible) for one player radically to alter the competitive balance that prevails. Lack of scale and small market shares typically preclude the emergence of price or technological leaders. In some cases firms break out of this stalemate by restructuring manufacturing in such a way so as to reduce costs significantly, gain market share and constrain the responses available to their competitors.

Not all players adopt the same approach, and it is possible to preempt competitors in certain situations. In Europe, one dominant manufacturer in pet food can dictate many competitive conditions, supported by its high volume focused factory operations. Seiko also preempted several competitors in the watch industry by building a factory that could produce a much larger market share than any manufacturer had before that time. Some companies are now considering taking a similar preemptive position in the Eastern European countries and in China, with a focus on the long term. In each of these cases, the industry structure has changed because of individual company actions, and the resultant forces and constraints for subsequent restructuring are also altered.

Industry structure and dynamics in the consumer products industry are quite interesting in Europe. The trade is much more heavily dominant than is the case in either North America or Japan. Powerful retailers and wholesalers increasingly dictate terms and conditions to manufacturers, even backward integrating into manufacturing or contract manufacturing with their own brands. Coupled with this phenomenon is growing pressure for equal prices throughout the EEC. The result is a series of forces – mostly cost pressures – that will require restructuring responses among the consumer products companies.

The proportion of value added by a manufacturer in the supply chain plays a key role in determining the relative power balance between the manufacturer and its customers. In the packaging industry, manufacturers of metal, paper, and plastic containers are increasingly caught in a squeeze between large upstream suppliers of raw materials wanting to pass along cost increases and large downstream packaged goods firms who are adept at resisting price increases for what to them are commodities.

4.2.2 Internal forces and constraints

Internal forces and constraints emanate from inside the company itself. Included are the mission, goals and strategy of the firm, the vision of its top executives, the value system in the company, its organizational structure, and other aspects of the company culture. Some firms have created environments that foster change more than others, but this typically refers to evolutionary change, not fundamental change. Restructuring is more profound, and the internal forces create a relative sense of urgency, or they constrain the necessary fundamental adjustments.

An example of a company where top management vision was clearly a dominant force for change is Olivetti. Mr Carlo De Benedetti transformed Olivetti from a production-driven office products company to a market-driven manufacturer of distributed data processing systems with recognized expertise as a supplier to the banking and insurance industries. In effecting this transformation, Mr De Benedetti was able to communicate his vision of where he ultimately expected Olivetti to be positioned within the industry, a sense of urgency for how the restructuring had to take place and a sense of practicality in how to go about the restructuring. In a nutshell, the company had to be restructured rapidly from producing and selling electronic typewriters to servicing the users of networked electronic computer systems.

Cost structure can be a primary internal force for restructuring. Many firms have great dissatisfaction with their cost structures. In order for the firm to be financially healthy, restructuring will be required. Cost structure can also be seen as a constraint, in that short-term financial objectives retard deployment of resources (people, information, and technology) to the achievement of long-term competitive capabilities. The proactive approach to restructuring should focus on adding value rather than reducing cost, and on insulating the firm from the actions of competitors.

A typical problem with cost structures is overhead. Restructuring often appears as a force for reducing overhead, perhaps as a mandate to reduce head counts. However, head-count reduction by itself will not solve the basic problems. In too many cases, restructuring results in fewer people doing the same work in the same manner, by working harder. This may treat the short-run symptoms of the problem, but it cannot be considered as an adequate long-run solution. Cost reduction needs to be seen as a necessary but not sufficient condition of restructuring. It will be necessary to define the new configuration and co-ordination (typically the result is simplification). Moreover, it will be necessary to determine the culture required both to carry out the restructuring and to operate effectively thereafter.

A proactive way for improving the cost structure is often manifested in a quest for performance excellence. Improvement is seen as a constant, and experimentation as a way of life. A good example here is the Northern Telecom drive significantly to grow the company while maintaining head counts, particularly in indirect labour personnel. This mission led to an ongoing search for improvement in design, procurement, quality and all facets of manufacturing. Restructuring has taken place several times at individual manufacturing locations and at the corporate headquarters. The ability of this company to take on the restructuring tasks, achieve the results, and go on to new horizons has been supported by a fundamental drive for excellence in all functions. The result of the restructuring is a basic change in the cost structure.

Inertia is a major internal constraint against restructuring. Overcoming inertia – that is, surmounting the inherent mismatch between the present deployment of the enterprise resources and the issues of most importance – is a significant constraint in most companies. Some research illustrates the problem. A study of the United States textile industry found productivity limited because:

1. 75% of manufacturing losses result from response delays to problems and disturbances.
2. Up to 66% of the labour force adds little or no value to the final products.
3. Up to 70% of the decisions made in manufacturing have predictable responses (Oliff and Collins, 1989).

This adds up to a gross misuse of the three primary resources in manufacturing: people, technology and information. Overcoming this inertia was an internal driving force for restructuring, and the inertia itself acted as a constraint to restructuring.

In many firms organizational inertia acts as an immovable constraint against restructuring to achieve faster new product introductions. In several cases, the only way the firm has been able to achieve a profound change has been to set up a new organization to accomplish a carefully defined mission. **Skunk works** groups are one alternative. Another is the IBM PC model – simply set up an entire new organization. Still another is the growing use of **internal venture capital groups**. The convenient scapegoats for unsuccessful new product introductions are political pressures, trade unions, etc. In far too many cases, the real culprit is the inertia of the company.

A related constraint comes from a fixation in many firms with avoiding takeover attempts. A preoccupation with combatting hostile takeovers in industries where acquisitions are prevalent means that management is less inclined to implement strategies which make the firm look attractive. Actions which impact short-term measures such as return on equity, sales growth rates, and share price related to book value are avoided by some firms. This is an inherent constraint to restructuring. Paradoxically, this constraint allows takeover firms to achieve their objectives, because they operate under no such constraints. Additionally, the preoccupation with avoiding a takeover has its own costs in terms of the culture, philosophies and overall direction that can be given to the firm.

The immobility of key personnel can also act as a form of inertia to restructuring. Multinational firms are increasingly in need of international managers who can thrive in multicultural/multi-national environments.

4.3 THE PARADIGM SHIFT

This section addresses the WHAT of manufacturing restructuring in more detail. Manufacturing restructuring needs to be seen as a *significant* response to the forces for change – as a paradigm shift. The result is a new pattern of culture, configuration and coordination, with a redeployment of people, technology and information, within some set of **manufacturing subunits**, or the company in general. Culture includes the strategy or purpose for the set of subunits, as well as the organizational beliefs that will guide subsequent practice. Configuration relates to organizational design – how many subunits and where are they located? Coordination is fundamentally management control – how are the subunits coordinated? People, technology and information are the resources utilized to accomplish the restructuring.

Subunits may include individual factories, functions (marketing, engineering), departments within manufacturing (work centres, cells), staff activities (scheduling, quality), territories (countries, regions) or other organizations (suppliers, customers). The definition of subunits includes all activities performed within a manufacturing company – and with partners in the value-added chain – not just those typically considered as production functions. The term 'subunits' also allows for the restructuring framework to be applied at different levels in the firm. That is, it can be used like a zoom lens: culture, configuration, and coordination – as well as people, technology and information – can be viewed in terms of an entire company, a division, a factory or a cell. Finally, it is important to note that major changes might occur in only one or two of the three dimensions of restructuring, but the more general case encompasses all three. A similar argument holds for the three resources.

4.3.1 Culture

Changes in culture are major shifts in the driving forces of a manufacturing company. Included are changes in manufacturing strategy, mission, fundamental objectives, values held, philosophy and basic policies. Examples encompass shifting from a cost-driven company to one where high quality, time-based competition, shorter product life cycles, more highly customized products, and/or flexibility are seen as keys to competitive performance. Another frequent culture shift occurs in mergers or acquisitions, in which two firms come together and develop new cultural underpinnings to support a combined organization.

Figure 4.2 provides a more detailed view of the cultural dimension of the pyramid presented in Fig. 4.1. Here, the intent is to depict increasing levels of intensity for cultural shifts associated with restructuring. At one end of the continuum is a shift in the policies or objectives of the company. At the other is a corporate merger, where an entirely new set of values and philosophies is required – and will evolve over time.

It is important to understand how culture can support restructuring, as well as the constraints it imposes. An example is the current situation at the NUMMI plant, a joint venture between General Motors and Toyota. The demand for the automobile produced there, the Nova, has been decreasing. The result is overcapacity at NUMMI and a force for restructuring. The response from General Motors is to restructure by reducing the work force. This response, while consistent with GM culture, is diametrically opposed to Toyota's. They want to retain all workers (except those leaving by attrition), and actively to search for a better solution – perhaps a new model to stimulate demand.

Post-restructuring culture is strongly influenced by the implementation process used (the HOW) to attain restructuring. An unfortunate

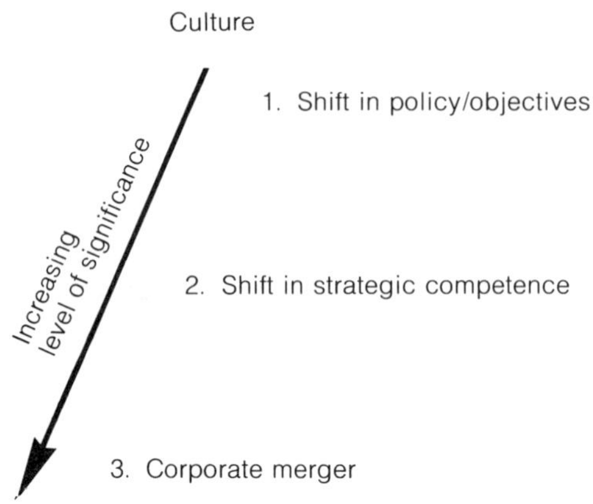

Fig. 4.2 Culture and manufacturing restructuring.

example is a US company with a long history of excellent employee relations that was a takeover. They fought and successfully resisted the takeover, but at the end of this battle the firm's financial situation was quite poor. One result was a need to restructure for cost reduction, cutting out some business activities. A consequence was to close one of the company's distribution centres. The manager in charge of the project insisted that planning for the closure be carried out in secret, and that a small group fly in by private plane one day and announce closure by noon of that day. The result was quite devastating to the culture of the rest of the company. By this one act, management had communicated a new culture where past performance was not valued, and destroyed a great deal of faith in the inherent fairness of the company.

A proactive culture change typically requires a well-articulated marketing and manufacturing strategy as well as a set of plans for implementation. Northern Telecom developed a **time-based competition** strategy, which was then translated into a carefully defined manufacturing strategy. This in turn led to a new set of action programmes in manufacturing, and to major changes in the set of performance measures used to evaluate manufacturing. As one part of the restructuring, an executive at Northern Telecom developed the chart shown in Fig. 4.3, clearly showing the change in culture required. In order to achieve this culture change, the company embarked on a massive education programme for all of its personnel.

Proactive cultural change is the focus of a study currently being conducted by Waseda University and a group of Japanese manufacturing companies. The study envisions the need for a major paradigm shift by many Japanese manufacturers in order to be world-class competitors in the 21st century (many of them are already world-class competitors

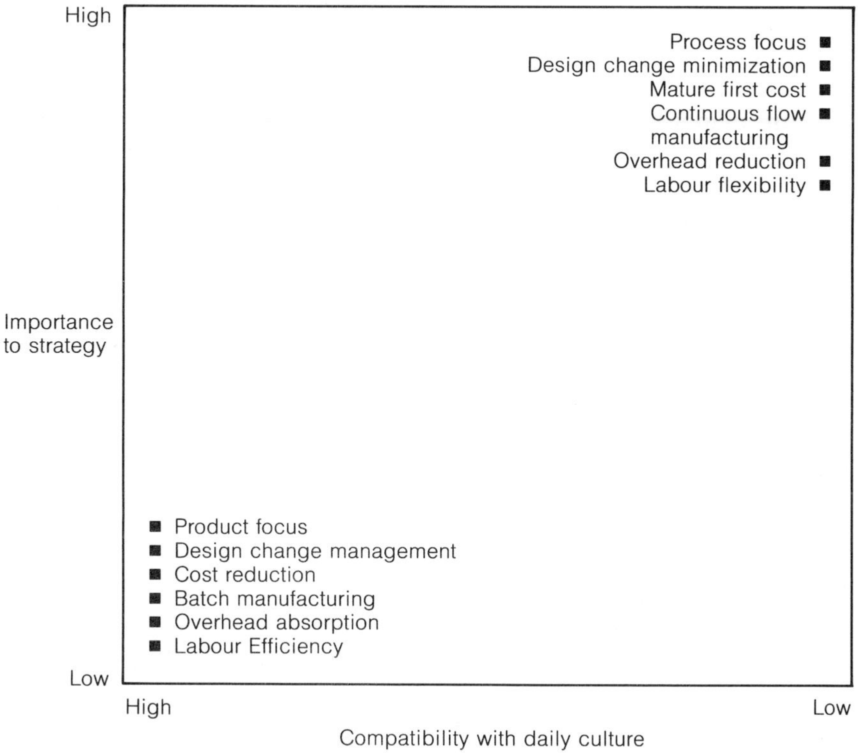

Fig. 4.3 NTI's analysis of new manufacturing strategy versus existing corporate culture. Adapted from Davis (1984).

and see the need for fundamental change in order to remain so). Some aspects of the anticipated cultural change are: a shift in the work ethic of Japanese workers; the need to embrace totally different work practices in non-domestic Japanese factories; the need to abandon the policy of advancement based on seniority in order to empower younger people with important new ideas; and the adoption of **strategic leap** philosophies for change, in addition to the current Japanese practice of **small steps**.

The proactive nature of the project at Waseda is illustrated by their development of scenarios for future success. Fundamentally, manufacturing companies need to adopt a strategic posture for becoming strong competitors in the future – say five or even 10 years from now. This vision has to be used to shape restructuring, and it needs particularly to be used in order to develop a proactive shift in company culture. This shift must address enduring values, new values that must be developed, plans for achieving the shift and the necessary leadership to see it through.

4.3.2 Configuration

Configuration changes include adding or deleting manufacturing subunits, such as building new factories, plant closures, downsizing, or removing managerial layers. Configuration change also includes transforming the definition of the basic tasks or charters of each of the manufacturing subunits. Inside the factory, configuration changes include regrouping of machines into cells, significant new methods of manufacturing, and major redeployments of overhead personnel. Another example of configuration change is a new organizational design where traditional activities of marketing, engineering design, and manufacturing are combined to focus on customer needs (or specific products) in a more integrated fashion. Figure 4.4 provides a continuum for this dimension of restructuring. At one end is the addition or deletion of some manufacturing subunits; at the other is a comprehensive vertical integration. The extent to which the other dimensions of restructuring need to change is strongly influenced by the scope of the configuration change envisioned.

A frequent cause for organizational redesign and configuration is mergers and acquisitions. It is necessary to define which organizational units are redundant under the new structure, what are the charters or task definitions for the subunits, their size, and where they are to be located physically.

Configuration change requires a definition of the set of subunits that is to be configured (and coordinated). Changes in this definition are occurring rapidly. One example is improved integration of subunits with other firms in the value added chain. Another is a redefinition of the subunits, either physically, such as cells made up of former machines in job shop configurations, or new organizational combinations of people to attack problems such as the provision of better field service.

One large European consumer products company is engaged in a major study of their configuration. The study is based on a simulation model that allows them to evaluate a large number of different com-

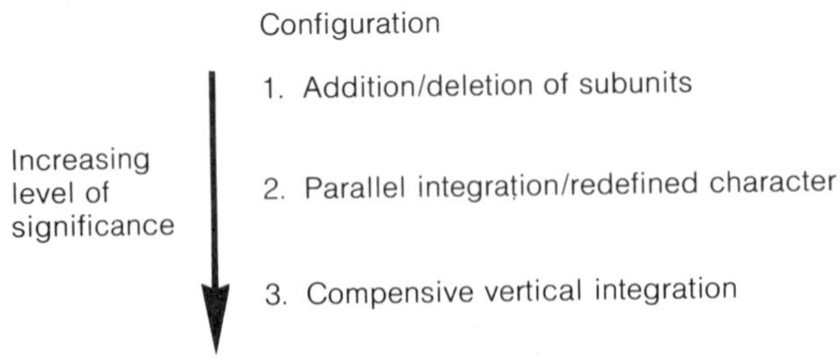

Fig. 4.4 Configuration and manufacturing restructuring.

binations of factories and distribution centres under various scenarios of demand, and to take into account realistic constraints imposed by political realities. The firm believes that the study is particularly useful because it allows many executives to examine the consequences of configuration alternatives on a rational basis. The hope is that a dominant pattern of configuration will emerge that will produce acceptable operating results under a wide variety of demand scenario assumptions.

4.3.3 Coordination

Coordination changes transform the interactions among a set of subunits in a manufacturing company, specifically in terms of the flows of material, information and responsibilities. Examples include new distribution systems, major changes in material planning/control systems, CIM or any new computer system that leads to a significant change in the relationships and management control of the subunits. Examples that cross traditional barriers are new approaches to support field service, better methods for new product introduction, electronic data interchange with customers and suppliers, and total quality-control approaches.

Figure 4.5 provides a continuum for the dimension of coordination restructuring. At one end is a redesigned set of performance measures, at the other is deployment of computer integrated manufacturing (CIM). The impact of significantly changing coordination will be felt in the other two dimensions of restructuring as well.

Coordination it many cases requires common manufacturing standards, packages, recipes and quality-control procedures. A large furniture manufacturer went through this problem in integrating the outputs from all of its factories. It was necessary to create a manufacturing infrastructure with similar processes and product specifications, where

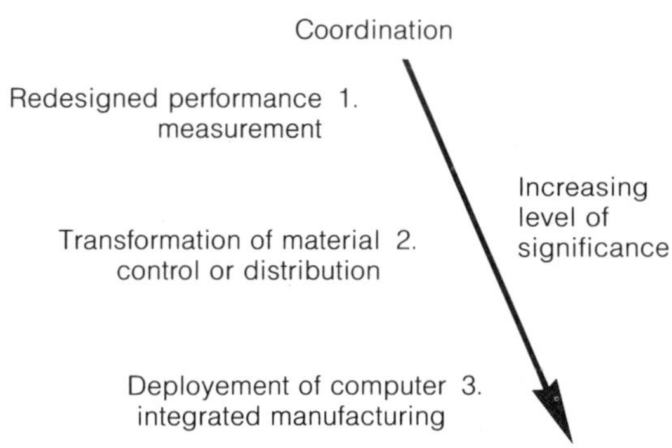

Fig. 4.5 Configuration and manufacturing restructuring.

changing the plant of manufacture was very straightforward and the products were uniform. To the extent that different customers have different expectations, or country-specific material-sourcing constraints exist, this problem is more difficult. Over time, it may be possible to get customers to accept greater uniformity – in the European case, perhaps with pan-European brands.

The advent of 1992 is a major driver for restructuring based on new forms of coordination. Consumer products companies have been among the first to change, particularly their distribution systems. As the importance of country borders decreases, the ability to satisfy demands from a region or other larger geographical area increases. A corresponding need is for overhaul of the approaches used to schedule manufacturing, and to rethink product branding and other issues. Pan-European marketing and pan-European manufacturing are important concepts, but significant restructuring is required to bring these concepts into practice.

Better coordination among subunits has been the design goal of many computer systems. Moreover, better coordination often leads to a redefinition of the subunits themselves (configuration). A good example is the increasing integration of computer-based systems to support engineering design, and numerically controlled (NC) programs for machine tools, a basic cornerstone for CIM in many companies. Major change in the jobs of design and manufacturing engineers, as well as those who actually operate equipment often results. Frequently, new work units emerge with manufacturing engineers on the shop floor working with NC programmers and machine operators.

A related example of coordination among a larger group of subunits is in firms who grow by acquisition with the intention to integrate the new products and services into an extant distribution network. In these cases, coordination mechanisms must be designed to be flexible enough to incorporate readily the new configuration of products and services.

Coordination change is also driven by growth. A US furniture manufacturer doubled its business from approximately $40 million to $80 million by everyone running faster. When business doubled again, the entire approach to coordinating their business activities had to be changed. Centralized control over decentralized factory operations was achieved through new manufacturing planning/control systems. Distribution was also overhauled. Products coming from different factories required coordinated shipping. Regional warehouses were built to replace individual store warehouses, and new information systems were developed to manage these warehouses. New marketing approaches were implemented which required reorganization of field sales people, marketing managers, information systems and mechanisms for evaluating and rewarding performance.

Coordination changes clearly include changing performance measurements. Many products now have such a small direct labour content that differences among plants are often more a function of the abilities of the overhead employees than any standard measures of productivity based on direct labour content. Traditional measures of product cost are increasingly irrelevant. If it takes 15 minutes of direct labour to make a television or less than 10 minutes to make a personal computer, it really does not matter a great deal if the labour is paid $3.00 per hour or $15.00 per hour. The maximum difference in terms of labour content is much less important than the number of overhead support personnel. Of even more importance is the *deployment* of these people, i.e. what do they actually *do*? How should they be evaluated? In the best factories, very few overhead support personnel will be devoted to the typical coordination activities associated with quality control, production scheduling, material planning, and other manufacturing staff activities. The major part of these jobs will be integrated into the basic infrastructure of manufacturing; that is coordination will be done by the direct labourers and their immediate supervisors (who are more facilitators than supervisors in the usual sense).

Control as a concept, needs to be seen as not producing value. Increasingly, world-class manufacturing companies are eliminating (or at least minimizing) quality control, production control and many financial controls. These activities are being made a basic part of manufacturing – quality products are guaranteed by the manufacturing processes. The work force produces production schedules as well as meets production schedules. Financial controls are more highly aggregated and not used for detailed performance measurement.

4.3.4 Resource deployment

The skilful deployment (or redeployment) of people, information and technology is the key to enterprise well being. Significant redeployment is an integral part of restructuring. As the extent of the change envisioned for restructuring increases, there is a concomitant increase in the extent to which redeployment of the three fundamental resource types takes place, and there is also an increasing interaction among the three resources. That is, the more fundamental the restructuring, the greater the change in resource deployment and the greater the change in the ways in which the three resource types are related.

A good example is found in the experience of the National Cash Register (NCR) corporation. Historically, their business was based on cash registers and mechanical parts. This business was abandoned in the 1960s, and the firm entered into the computer/electronics business. The result was a completely new set of requirements for people, for information and for technology. The changes in culture, configuration and coordination were extraordinary. The work-force in Dayton, Ohio, was reduced by over 50%. The entire company was reorganized to support new products, a new technological base and the emerging computer markets. Similar stories can be told for the telecommunications companies that changed from mechanical switching devices to electronics for telephone exchange systems. In each case the resource needs were fundamentally different and the restructuring changes were major.

4.3.5 People deployment

The redeployment of people is the most difficult part of restructuring, particularly when the restructuring comes as a reaction instead of as the result of a proactive adjustment based on growth. In the case of downsizing, redeployment almost always includes personnel reductions. But even when the approach is proactive, the desired result is a work force reduction in most cases – at least in terms of the number of workers per monetary unit of sale.

Examples include the replacement of hierarchical organizations, flatter organizations and the consequent elimination of managerial layers. Ford Europe, after studying Japanese auto manufacturing in the early 1980s, formulated a delayering plan called AJ (i.e. After Japan). A more recent European example is a Swissair reorganization that removed two entire levels of management. A variant that is usually more easy to implement is restructuring accompanied by growth. Northern Telecom plans have included a doubling of sales with no increase in personnel, and the assumption of ever more work that traditionally had been the work of staff employees by the direct labour force.

A measure of the magnitude of the problem (and opportunity) is shown in Fig. 4.6, which provides relative data on overhead deployment for the United States and Japan in 1984. As can be seen, in the United States the number of direct labourers per manager (not including first-line supervisors) is about 28; the companion number for Japan is 57. The approximate numbers for first-line supervisors are 14 and 8, for accountants 18 and 57, for human resource managers 58 and 28, for industrial engineers 19 and 12, and materials managers 6 and 9 (Boston University Manufacturing Roundtable and Waseda University, 1984).

The overall result is that US firms have about twice as many managers, three times as many accountants, and 50% more materials managers. Japanese manufacturing companies have 50% more first-line supervisors, twice as many human resource managers, and 50% more industrial engineers.

The Japanese manufacturing firms have their overhead personnel deployed closer to the actual operations, and in ways that facilitate continuous improvement, whereas US companies seem to focus on a top-down control driven approach to management.

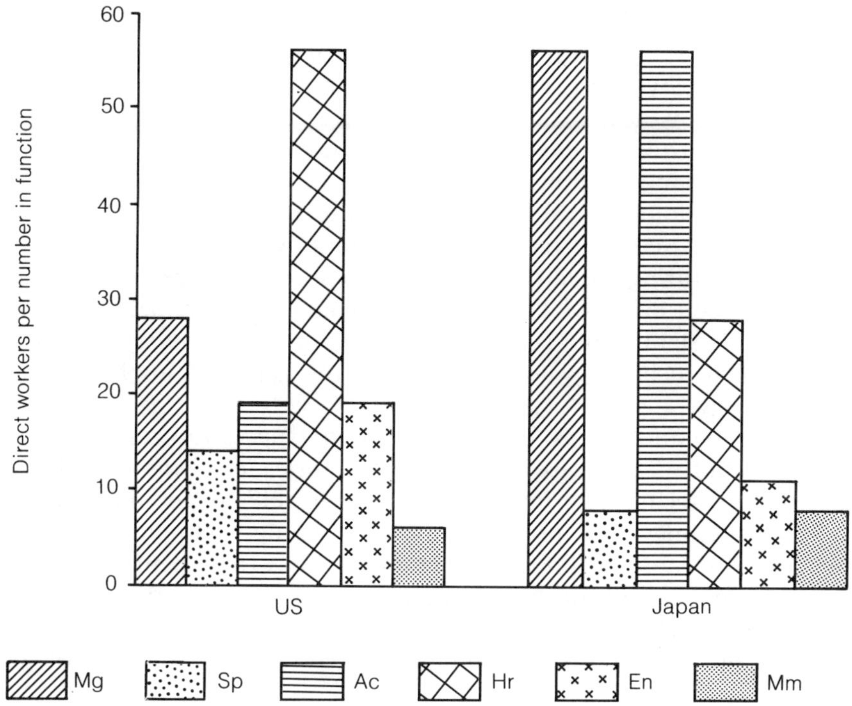

Fig. 4.6 Staffing patterns: USA and Japan.

In the years since 1984, significant changes in overhead deployment have taken place in all parts of the world. In the United States, for example, the 1988 Boston University Manufacturing Roundtable survey of North American companies respondents indicated that there had been a 35% decrease in the number of overhead personnel deployed in production planning and inventory control. At the same time the employment of people in both material handling and purchasing increased by roughly 20%. This reflects an overall decrease in 'materials managers' and perhaps indicates an increasing emphasis on purchasing and rapid movement of materials through the factory. The data also indicate that the number of managers per direct labour employee went up, while the similar number for first-line supervisors went down. This movement was opposite to that indicated by the Japanese approach, leading to questions about job definitions for direct workers. Perhaps restructuring has made comparisons over time much more difficult. Regardless, significant restructuring requires carefully crafted strategies and programs to mobilize people – both direct and indirect – in new ways that are focused on adding value to products and services (Miller and Roth, 1988).

4.3.6 Technology deployment

Technology, in the broadest sense, is the firm's second resource to be deployed (redeployed) in restructuring. Technology can be viewed both as 'what we know how to do' and as 'what we *need* to know how to do'. 'What we *need* to know how to do' is determined by both competitive pressures for change and by the availability of new technology. Whenever new technology becomes available, the firm needs to make an assessment of whether it represents something 'we need to know how to do'. Restructuring based on technology deployment is required when the gap between 'what we know how to do' and 'what we *need* to know how to do' is significant. Technology deployment involves both determining which technology to deploy to fill the gap, and implementing the technology.

Adopting technology is often the key to successful restructuring implementation. There are many horror stories of firms that have adopted new technology but have not been able to get the desired results. Sometimes, the problems are the wrong technology. Often, however, the cause is underdeployment or incorrect deployment of people. Companies all too frequently believe that signing the purchase order for major new hardware is the main task, rather than overall restructuring to create the capabilities required to utilize the technology fully.

Significant deployment of new technology almost always consists of a strategic leap, as opposed to an incremental small step. The kind of culture change involved makes this type of change more difficult, and dictates change in the people deployment. Often, entirely new kinds of employees are required to run the new equipment; this is habitually the case with flexible machining systems (FMS). When existing employees are to be a part of the new systems, they require significant education and training. Moreover, the culture and reward systems in the new situation must almost always be very different to be effective. This raises the issue of the process of implementing restructuring, which is addressed in the next major section of this paper.

4.3.7 Information deployment

The vast majority of improvement projects in manufacturing companies involve better deployment of information – the usual examples of MRPII, SPC, CIM, etc. More is not always better, as illustrated by JIT and the elimination of hidden factory transactions. The degree of foment in information redeployment is very high. Centralized computing departments have been decentralized, giving way to networks for linking users and sharing information, workstations as powerful as yesterday's mainframes, and ever more user-friendly computer software. A fundamental driver for the increasing use of new forms of information resources is the continuing reduction in the cost of computation. Computer systems such as MRP that required large mainframe computers 10 years ago now are capable of being run on minicomputers and even some personal computers.

These changes are having a significant impact in reducing the delays in responding to process problems and disturbances. In the case of real-time computers the lag time can be reduced to zero. Jaikumar (1989) describes a wire-drawing operation where measurements are made for a host of variables in real time as the wire is being drawn. The result is an ability to anticipate wire breakages, and adjust elements of the machine and related systems, so that the breaks do not in fact happen. He reports that the frequency of breaks has been reduced by over 90%.

The quality of information is also impacted by information deployment on the shop floor. In centralized systems, changes in computer systems are discouraged actively. In decentralized approaches, the systems can and *should* evolve as problems are brought under control and new issues raised. The presentation format of the data can be based on ease of user understanding. In a similar vein, the quantity and scope of the information can be problem driven. As evolution in problem statements and measurements takes place, the proper deployment of information needs to evolve similarly.

Most applications of computer resources have been problem specific, addressing a particular issue of the day such as materials planning, logistics or scheduling. Lately, the CIM philosophy is attempting to define a more global orientation for the deployment of information resources. In actuality, there are several quite different kinds of problems in manufacturing to which information resources are being deployed (Marchand and Oliff, 1989).

Machine control of manufacturing processes in the factory is based on the deployment of information for feedback and control, based on the principles of cybernetics.

The deployment of people in machine-controlled manufacturing processes is primarily to design the programs that provide the detailed instructions to the machines, and that monitor the actual machine operations, making adjustments as necessary.

Supervisory decision making is characterized by work situations where it is people who make the corrective adjustments to process problems and disturbances, rather than machines.

The kinds of information requirements for supervisory decision making are typically much wider in scope than those of machine control, because the problems and disturbances are likewise wider. That is, supervisory decision making is concerned with situations where machines are less likely to be able to control the environment routinely. These situations, however, are getting less common as robotics and other forms of machine control expand their applications.

Production planning and scheduling is concerned with the routine planning and execution of decisions that support the material flow through the factory.

Information deployment for production planning and scheduling tends to be quite specific and repetitive. These systems are usually complex because of the size of the database and the large number of data elements, rather than because of analytical complexity. These systems continually replan, developing new information as actual conditions differ from planned conditions. In the case of MRPII the database can be very large. In implementing JIT, many of the inherent uncertainties in manufacturing are eliminated, so that production planning and scheduling is less complex, and the deployment of information resources can be much more modest.

Coordination with suppliers is conceptually similar to production planning and control, but the deployment of information resources is quite different.

Electronic data interchange is now being adopted widely for coordination with suppliers, but there are fundamental problems to be solved. Essentially, it is necessary for the scheduling systems of the firm and its vendors to be compatible, and mutually supportive. This means that deployment of information resources is required in both firms and that they need to work together on the problem. This issue is compounded when the firm has many suppliers and/or the vendor has many customers.

Coordination of manufacturing processes is based on the deployment of information to better integrate the operations of separate manufacturing subunits.

Coordinated manufacturing processes in the auto industry use computer control systems to schedule the exact deliveries of component parts to the exact place needed on the line, just as they are required. This kind of coordinated planning goes beyond the routine production planning and scheduling done by MRPII or JIT. Coordinated planning has also been adopted widely between engineering and manufacturing so that new products can be introduced more effectively.

Improved integration of the manufacturing function with research and development, engineering, marketing, sales and distribution is obviously supported by production planning and scheduling systems such as MRPII.

But beyond this routine planning based on material flows are: the need for better coordination of the engineering pass-off to manufacturing; integration of manufacturing with marketing and the desires of customers; and the ability to better implement new manufacturing process technologies. This is the purview of what many people now call CIM.

Strategic management and manufacturing addresses boundary-spanning coordination. The coordination issues may involve bargaining, one-time specific problems and linkages with external players such as potential customers and sources of supply for both materials and knowledge.

Also required in strategic management and manufacturing are intelligence gathering from many sources and the support of analyses that will be driven by the particular predispositions of particular executives. The information required and the deployment of resources to obtain the information are not well structured.

The importance of this categorization scheme is that the type of information deployment envisioned needs to be driven by the goals for restructuring. Moreover, the kinds of information resources required are very different, as will be the needs for people and technology, and the resultant impacts on the three dimensions of culture, configuration, and coordination.

To sum up this section on the WHAT of restructuring, restructuring can be seen as a process of changing the culture, configuration, and/or the coordination of a manufacturing company through the deployment or redeployment of people, technology and information. There will almost never be an exclusive match between a dimension and a resource. Deployment will inevitably be in all three resources, although the primacy of one will often result. Similarly, the three dimensions of restructuring will almost always be impacted. Any change in configuration or coordination is surely going to influence culture, and the desired culture will play a role in defining the other two. This suggests that it is important to think carefully through all of the intersections among the resources, the dimensions and the fundamental goals for the restructuring.

4.4 IMPLEMENTATION

In this section the HOW of manufacturing restructuring will be examined in more detail. Restructuring is a process of responding to the forces, limited by the constraints. It is achieved by changing significantly culture, configuration and/or coordination through the deployment of people, information and technology. The forces are the cause, and restructuring is the effect. The change in each of the three dimensions (culture, configuration and coordination) is dependent upon the extent to which the forces dictate a change, and on the constraints that oppose change in each of the dimensions. It is critical for management to define objectives for restructuring carefully, and manage the process to achieve desired results.

4.4.1 Approaches

Approaches to restructuring vary widely. Some firms utilize consultants essentially to serve as 'hatchet men'. Others do the entire job internally. More frequently, some middle ground is found where advice is given by consultants based on seeing other companies undertake similar responses to similar forces.

It is important not to frame the restructuring project minimally. Too often this is the case – the project is defined to have the smallest possible impact on the labour force, culture, configuration and coordination. The frequent result is that it becomes necessary to undertake still another major restructuring project, with a combined impact that is far more serious for morale and culture. The approach needs to be based on the long-term competitive posture required, and the changes required to make it a reality.

Invariably, restructuring will require the learning of new values, skills and practices, and the unlearning of old beliefs. For this reason, virtually every restructuring project has to include education and training. In many successful restructuring projects, education and training have played a major role for the company in the adoption of the new culture. The emphasis is often formally on the issues of configuration and coordination, but the bottom line is culture.

Although restructuring involves a major change process, it is important not to see it as *simply* involving one big change or strategic leap. The Japanese approach of small steps needs to be coupled with strategic leaps in order truly to achieve the desired results. An example of the problems that can result from not coupling both approaches is seen in the implementation of flexible manufacturing systems (FMS). The forces dictating this approach include the need for more flexibility, shorter lead times, faster new product introductions, greater product variety and the availability of the technology itself. The resources required are a major deployment of new technology, supported by new information systems and new training of personnel. The configuration of the equipment is in one integrated unit. The coordination includes not only those entities inside the FMS, but the links to design engineering and programming. Culture changes impact the entire group of people who are in direct and indirect contact with the FMS, as well as less directly those who are in other areas and those who are displaced by the new technology.

In the same way it is dangerous to believe that an FMS is a one-time strategic leap. Many US firms have treated it as such. The result is a major resource commitment to put the FMS in place, almost no resource commitment to improve and integrate the operations subsequently, and a disappointing series of unsuccessful FMS projects.

The opposite kind of problem occurs with a 'salami-slicing' approach to restructuring. Restructuring needs to incorporate *major* change, change in fundamental relationships in culture, configuration and control, and the integration of these dimensions. This can not be accomplished well with small steps. As a Polish politician recently remarked, 'You cannot leap a chasm in two jumps.'

A similar problem arises when the need for restructuring is well understood, but grossly inadequate resources are available to accomplish the restructuring. (It is important to note that resources for restructuring will almost *always* be inadequate – but there are degrees of inadequacy.) The Greek subsidiary of a large European consumer products company has this situation. Business has grown significantly, but present conditions have required all of the executives to adopt a work week of five 12-hour days plus Saturday. A new approach to running their business is required. How are they to find the time and energy to restructure? Who will do it? Must it be imposed from headquarters?

The approach to restructuring is usually different when the restructuring is a proactive organic response to the fundamental forces for change rather than a reactive response imposed by a merger/acquisition, takeover attempt or turnaround circumstance. In the former case, even though hard choices often have to be made, the changes in the cultural dimension are usually less severe. If the firm has an ongoing business strategy and philosophy that are moving the firm in the right directions, the major changes will be in configuration and coordination.

On the other hand, mergers, acquisitions, takeover attempts and turnaround situations almost always require major changes in culture. Complicating many of these situations is a paucity of resources. Severe financial constraints make significant deployment of new technology less likely. Work force levels are frequently reduced, so that people resources are also under pressure. The same is true for information. The approach that many firms have taken in these situations is to employ a new top management team that specializes in these kinds of situations. A recent example is the hiring by Wang Laboratories of Richard Miller, who in turn brought in a turnaround team of consultants and managers. The intent was to change the culture *radically* – and quickly.

4.4.2 Leadership

The issue of managerial leadership for restructuring has not received the attention it deserves. One fundamental problem is in motivating good managers to take on a restructuring job as a career assignment. Many mangers do not want to be viewed as responsible for job reductions and other painful adjustments associated with downsizing.

Implementation with a lack of leadership occurred in a UK consumer products firm. It decided to merge two of its divisional companies, traditionally competitors, with a downsized headquarters group. The forces were largely associated with decreasing markets, but the firm was still quite profitable. Very little detailed direction or leadership for the merger was given from top management. Many problems ensued. A fundamental problem was the lack of a defined culture in the new entity. The three prior units each had a culture, but they were not the same. The configuration approach was to take the accounting group from one firm, the purchasing group from another, etc., integrating employees from the other groups as possible. The result was chaos. Coordination seemed to require that *all* computer systems from the three units be maintained – even when they produced conflicting results. And the problems were magnified by the lack of a consistent culture. There simply were no guiding philosophies, values or direction. Several suicides resulted, the firm lost market share, another downsizing was required, and eventually the entire division was sold off. As the restructuring was occurring one manager said, 'A takeover would have been so much easier.'

An issue associated with leadership is the question of the mandate that needs to be given to key managers who are to implement the restructuring, and the corresponding kind of contract they should request when taking on a restructuring assignment. To some extent the answer must begin with a clear understanding of the forces and constraints that are driving the restructuring: as good a definition as possible of the required changes in culture, configuration and coordination; the availability of resources; and the latitude available to change people, technology and information.

The experience needed for a restructuring assignment is also an interesting question. What set of prior assignments would be good preparation for restructuring? In several cases firms have chosen individuals who have had line management experience, as well as significant project experience (e.g. building a new factory in a foreign country).

One example was a manager who had many project assignments over his career. He started in England, with a technical education, working in the research department. After a few years he was transferred from Europe to the United States to work in industrial engineering. From there he was put in charge of building a factory in South America, then another in a different country in South America, then back to England as a plant manager, where downsizing driven by automation was a large part of his job. The next assignment came when the company bought another company and his job was to merge the two operations. This required three years to complete, and he has now been given still another restructuring assignment. This time he is to restructure the entire European management information systems (MIS), moving definitively to decentralization of most of the computer systems, while maintaining a central core of information and systems development experts.

4.4.3 Evaluation

Restructuring projects bring to mind one of the famous good news/bad news jokes. The bad news is that a restructuring project is not being implemented as well as it should be. The good news is that there is no way to find this out.

The measures of effectiveness for restructuring are often not focused on what is truly important: the ability of the new business entity to compete. All too frequently, the goal is to shed *X* people or *Y* dollars of cost by the end of some period *Z*, or to create a better set of financial statements by next year. While recognizing the primacy of financial measures under certain conditions, the short-run orientation of many companies comes at their long-run expense. It is not too difficult to reduce cost dramatically and improve financial performance in the short run. Cutting all expenses for marketing, R&D, maintenance, education and training will accomplish the result in most companies. An even better trick is to take such an enormous one-time write off (blamed on the past sins of others) that future results *have* to be good.

In order to evaluate a restructuring project effectively, it is necessary to return once more to cultural issues. The company needs to look deeply inside itself to understand the strategy, philosophy and underlying values that presently exist, and those that are desired after restructuring. From this focus, ground rules and measures of effectiveness can be developed. There are of course the usual timing issues for personnel changes as well as changes in configuration and coordination. Perhaps it would also be useful to perform attitude surveys of all employees, before and after, including those that remain with the company and those who do not. The attitude survey could focus on whether the desired set of fundamental beliefs are truly shared throughout the company. To the extent that they are not, it might be concluded that restructuring has not been completed.

Evaluation also needs to recognize the inherent complexity of restructuring. Figure 4.7 provides some insights into the underlying issues. When restructuring is focused on coordination, change efforts focus more highly on factual issues as opposed to relational or emotional issues. Also, the comfort and commitment level can be high. On the other end of the spectrum is a restructuring with culture as the primary focus. Here, the emotional aspects are high, and the degree of comfort is

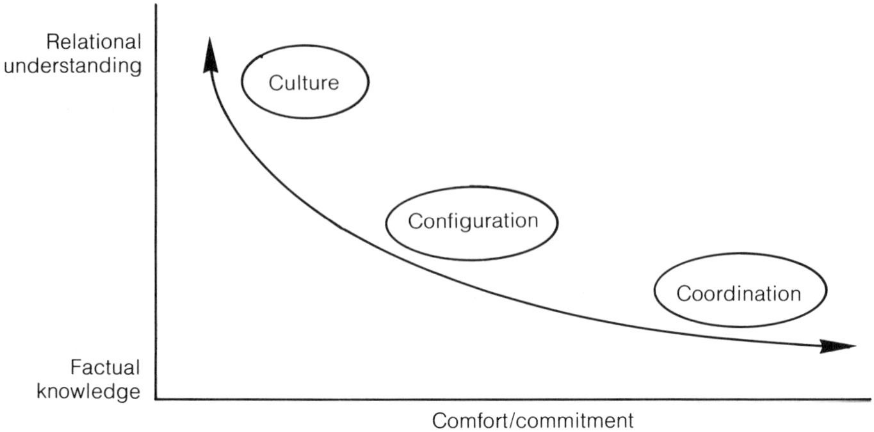

Fig. 4.7 Forms of restructuring.

low. The ease in implementing restructuring projects in these two areas is very different, and evaluation of progress must take these differences into account.

A focus on changing performance measurements has also been useful in several companies. The fundamental concept is to use a questionnaire to solicit opinions about what is important versus what is currently measured. Anomalies can be identified between beliefs and stated strategy, as well as between levels of management and across functional organizations. The result of this investigation is a new consensus on what is important, and the measures that need to be adopted, *and* eliminated, in order to achieve the desired results (Dixon, Nanni and Vollmann, 1990).

4.5 CONCLUDING REMARKS

This paper has attempted to define a framework for manufacturing restructuring to enable managers to better understand the intricacies of the restructuring process. It can act as a checklist as they attempt to plan, implement and evaluate restructuring in their companies. The framework is based on a review of related literature, plus interviews with executives who are presently implementing, or have implemented, restructuring programs. The paper should not be viewed as definitive, since restructuring is expected to grow in frequency and importance in the years ahead. It is hoped that the framework and related concepts can be of practical use as this growth occurs.

4.6 REFERENCES

Boston University Manufacturing Roundtable and Waseda University (1984) *Manufacturing Futures Survey*, Boston University Manufacturing Roundtable and Waseda University.

Davis, Stanley M. (1984) *Managing Corporate Culture*, Ballinger, Cambridge, Mass.

Dixon, J.R., Nanni, A.J. and Vollmann, T.E. (1990) *The New Performance Challenge: Measuring Operations for World-class Competitiveness*, Dow-Jones Irwin.

Jaikumar, R. (1989) Industrial transformation: the long view. Paper delivered at the Accounting and Manufacturing: Allies or Rivals conference, PICS Belgium, Ghent, September 1.

Marchand, D.A. and Oliff, M.D. (1989) Strategic information management in manufacturing. Working paper of the International Institute for Management Development.

Miller, J.G. and Roth, A.V. (1988) *Manufacturing Strategies*, Boston University Manufacturing Roundtable.

Oliff, M.D. and Collins, R.S. (1989) Expert systems: the bridge to intelligent manufacturing. *IMEDE Perspectives*, **3**.

Oliff, M.D., Arpan, J.S. and DuBois, F.L. (1989) Global manufacturing rationalization: the design and management of international factory networks, in *Managing International Manufacturing* (ed. K. Ferdows), North Holland, Amsterdam, 41–67.

CHAPTER 5

An exploratory study of the manufacturing strategy process in practice

Ann Marucheck, Ronald Pannesi and Carl Anderson

5.1 EXECUTIVE SUMMARY

This work presents an exploratory empirical study of the process of formulating and implementing manufacturing strategy within the framework of overall corporate strategy, as practised by a cross-sectional representation of leading-edge firms.

A conference was held where executives representing six firms in the following industries – computer equipment, pharmaceuticals, valves, telecommunications, furniture, and electrical submersible pumps – described the process of manufacturing strategy formulation and implementation as practised within their firms. Although the firms represented a diversity in terms of sales volume, longevity, and geographic location, they shared several commonalities including Class A use of MRP, excellent customer service records, and maintenance of accurate manufacturing databases. In addition, despite increasingly competitive conditions within their respective industries, none of the firms considered itself to be the low cost producer in the industry. Competitive strength was sustained through innovation, quality, and service. The process of strategy formulation varied among the firms in terms of participants, complexity, and degree of formalization.

Several findings emerged from the conference. First, all firms followed a traditional hierarchical top-down approach in formulating manufacturing strategy under the umbrella of corporate strategy. In general,

Journal of Operations Management: Marucheck, Ann, Pannesi, Ronald and Anderson, Carl (1990)

the process was reactive to corporate strategy although the executives reported greater involvement on the part of manufacturing in corporate strategic planning through the identification of competitive strengths and weaknesses. Manufacturing strategy was also viewed as reactive to marketing strategy with marketing playing the role of boundary spanner in assessing competitors and customers to support the manufacturing strategy effort. Finally, some of the executives stated that several of the traditional manufacturing resource decision categories were constrained by corporate philosophy. Examples of constraining policies included facilities, capacity, vertical integration, and organization. On the other hand, all the firms were able to develop strategies for process choice/ technology, quality, manufacturing planning and control, and people.

The executives described the implementation process as one of gaining employee acceptance of the strategy through lower level involvement and teamwork. Although communication was a major concern in implementing the strategy, the executives all stated that the most difficult task was changing corporate culture, which was often entrenched in obsolete cost accounting systems, to make it supportive of the new strategy. Other problems confronted by the firms included maintaining consistency among managers across all levels in the organization, gaining top management support, and developing appropriate styles of leadership.

The study shows that these firms' processes of formulating manufacturing strategy seem to follow the general conceptual models developed in the academic literature. However, the executives indicated that the real benefits of strategy come from implementation, which is a less structured and behaviourally oriented process. Future research must address infrastructural issues including culture, performance measurement, and managerial style.

5.2 INTRODUCTION

Ever since Wickham Skinner (1969) identified the absence of manufacturing in the corporate strategic planning process, the idea of formulating a manufacturing strategy tied to overall corporate strategy has gained momentum. Although numerous articles dealing with manufacturing strategy have been published over the last 20 years, the topic is still under considerable development and debate. Research in manufacturing strategy has focused on one of two topics: (1) the *definition and content* of manufacturing strategy; and (2) the use of conceptual tools to *formulate and implement* a hypothetical manufacturing strategy. In contrast, the purpose of this paper is to provide an exploratory examination of the process of formulating and implementing manufacturing strategy *within the framework of overall corporate strategy*, as represented by practice in six leading-edge firms.

Although there is no generally accepted definition of manufacturing strategy, several definitions have appeared which specify what it does or what it addresses. There seems to be agreement that manufacturing strategy should act to *support corporate objectives* (see Skinner, 1978; and Wheelwright, 1984), provide manufacturing objectives which offer the firm a *competitive advantage* (Hayes and Wheelwright, 1979b), and *focus on a consistent pattern of decision making* within key manufacturing resource categories (see Hayes and Wheelwright, 1984; Buffa, 1984; Edmondson and Wheelwright, 1989; and Hill, 1989).

Within these broad categories, manufacturing strategy focuses on the following competitive variables (Buffa, 1984; and Porter, 1980): cost/price, quality, delivery, flexibility, product design, and after market service. Hayes and Wheelwright (1984) categorized manufacturing strategy decision variables into the following eight groups: capacity, facilities, technology, vertical integration, workforce, quality, manufacturing planning and control systems, and organization. (This categorization is not unique with both Hill (1989) and Buffa (1984) proposing alternatives.) These categories suggest a clear link between manufacturing and corporate strategy.

Only recently has manufacturing strategy been empirically analysed (Miller and Roth, 1987; and Schroeder, Anderson and Cleveland, 1986). Despite Buffa's (1980) suggestion that researchers in operations management should consider developing 'a rational decision process' for strategic manufacturing decisions, there has been a lack of theoretical attention to the process of formulating and implementing a manufacturing strategy once an overall corporate strategy has been defined (Anderson, Cleveland and Schroeder, 1989).

The few empirical studies that have been done suggest that a stronger role for manufacturing within the hierarchy of corporate strategy formulation should improve perfomance. Richardson, Taylor and Gordon (1985) showed that improving the focus in both the corporate mission and the manufacturing tasks and increasing congruence between the corporate and manufacturing objectives each produced better corporate performance within the Canadian electronics industry. Swamidass and Newell (1987) demonstrated that an enhanced role for manufacturing managers in strategic decision making resulted in better business performance. Fine and Hax (1985) developed a framework for formulating manufacturing strategy in an electronics firm based on the product-process life cycle (Hayes and Wheelwright, 1979a and 1979b), the product-process innovation curve, and the concept of factory focus (Skinner, 1974). Recently, Edmondson and Wheelwright (1989) described the role that manufacturing plays in sustaining a competitive advantage over time. Overall, this research shows that strong ties between manufacturing and corporate strategy have positive effects on performance.

This study extends earlier research by examining the linkages between manufacturing and corporate strategy formulation and implementation as practised by a cross-sectional representation of firms. Although the results are exploratory, they provide a preliminary comparative analysis of how strategy is formulated and implemented in practice, as well as some hypotheses for future research directions. The overall question we examine is: In practice, how is manufacturing strategy formulated and implemented within a corporate strategy framework?

5.2.1 Methodology

Since this research is exploratory in nature, a qualitative research methodology was used. Burgelman (1983) states that such an approach is appropriate when the research is attempting to gain new insights into '. . . as yet incompletely documented phenomena.'

Data were collected during a practitioner conference in which industry representatives described how manufacturing strategy was actually practised in their firms. The conference setting allowed for a greater amount of unstructured participant input than would have been possible using questionnaires, and further, it allowed for dialogue between the researchers and participants to clarify issues as well as to develop points of interest that emerged during discussions. The cost of this higher quality of information was that a lower number of firms could be sampled.

Twenty firms who had made a conscious attempt at implementing manufacturing strategy were invited to the conference. Some declined because of schedule conflicts, while others declined because they felt their firm's strategy development was not far enough along to subject it to scrutiny. Thus, there was some bias and self-selection among the participant firms. The final list of participants consisted of six firms representing the following industries: computer equipment and electronic instruments, telecommunications, furniture, petroleum (electrical submersible pumps), valves, and pharmaceuticals. Each firm was represented by at least one person at the level of Vice President of Manufacturing or higher, and in three cases, corporate officials with titles such as Director of Strategic Manufacturing or Director of Corporate Development and Planning also participated in the presentation. Five of the presenters had been involved in the process of formulating manufacturing strategy for the corporation, while one of the presenters had been involved in the development of manufacturing strategy at the SBU level.

Each firm was given up to 90 minutes to make a presentation about its experience with manufacturing strategy formulation followed by a question and answer period. After the conclusion of the presentations, a half day was set aside for continued questions and unstructured discussion of high interest topics that had emerged from the presentations. The discussion period was initiated by an agenda of questions presented to the participants in round robin style. This agenda consisted of questions identified by the researchers as recurring problems in the formulation and implementation of manufacturing strategy after all the presentations had been completed.

To provide a common background and format for the presentations, the participants were provided with a definition of manufacturing strategy and a list of issues to address. Since the literature indicates that many different interpretations of 'manufacturing strategy' exist, particularly among practitioners (see Swamidass, 1986; and Schroeder Anderson and Cleveland, 1986), it was necessary to provide all participants with common terminology. The following definition was provided:

> Manufacturing strategy is a collective pattern of coordinated decisions that act upon the formulation, reformulation and deployment of manufacturing resources and provide a competitive advantage in support of the overall strategic initiative of the firm or the SBU.

In addition, the participants were asked to consider the following questions when preparing their presentations:

Manufacturing strategy issues

1. What are the objectives of manufacturing strategy?
2. How is manufacturing strategy formulated?
3. How is manufacturing strategy related to marketing strategy?
4. Who is responsible for the formulation of the strategy?
5. How formally and frequently is the strategy formulated or reviewed?
6. How is the manufacturing strategy related to corporate strategy?
7. What are the elements or the major categories of strategy?

Manufacturing strategy experience

1. What were the considerations in implementation?
2. How is evaluation and feedback done?
3. What problems were encountered in implementation?
4. How long has your firm been involved in manufacturing strategy?
5. What lessons have you learned?
6. How do you evaluate your manufacturing strategy effectiveness?

The conference was videotaped in its entirety to enable the researchers to provide an accurate representation of the processes described by the participants. In addition, this freed the researchers from the burden of note taking and enabled them to participate to a greater extent. All of the participants' presentation materials were also accumulated to help assure accurate depiction of the strategy process. Finally, the process descriptions that emerged from the conference were written by the researchers and then carefully reviewed by the corporate participants for accuracy. In each case, the participants provided final approval of the data and descriptions used in the research study.

5.2.2 Description of the participants

Each of the participating firms was selected on the basis of its known experiences with manufacturing strategy. Not surprisingly, each firm was performing very well within its industry as evidenced by its profitability and market share. The cross section of firms represented not only a diversity of products, but also diversity in terms of size, complexity and geographic location. Three of the firms were classified as 'large' firms with sales varying from $1 billion to $8 billion annually and employed from 18 000 to 82 000 people worldwide. The other three firms had sales which varied from $100 million to $310 million annually and employed from 600 to 6000 people. Many of the firms were multi-plant and multi-divisional, while others, including one of the larger firms, did all manufacturing in a single plant. Some of the firms had very long histories, having been in existence for over 100 years, while others had been in existence for only 30 years. Some of the firms, like the furniture manufacturer, competed in very mature industries, while the two electronics firms had significant activity within newly emerging technologies.

Despite these diversities, the firms had several things in common. All of the firms had established excellent manufacturing planning and control systems. Some firms and some divisions of the larger firms had attained Class A status in MRP use. Each firm had a manufacturing database which was accurate and up-to-date. Each of the firms was involved in implementing Just-in-Time although the actual stage of implementation varied widely among the firms. Most importantly, all these firms had achieved credibility within the manufacturing function by demonstrating an ability to produce and ship products on time through the successful use of a manufacturing planning and control system. This result may imply that manufacturing's 'getting its house in order' is a prerequisite to the initiation of the strategy process.

The participants reached a general consensus that conditions within their respective industries were growing increasingly complex. All of the firms mentioned the new challenges created by growing international competition, shorter product life cycles, and customer demands for high quality products. In addition, the two electronics firms mentioned problems created by rapid technological obsolescence. Two of the firms which supplied products to the petroleum industry had considered basic survival to be a key challenge as the price of oil dropped during the early eighties. The pharmaceutical firm focused on problems with government regulation of drugs and pharmaceuticals, while the telecommunications firm discussed the paradox of deregulation which had created both problems and opportunities for the firm. No firm considered itself to be the low cost producer in the industry, although many of the firms mentioned the increasing importance of costs as the industry grew more competitive or was economically depressed. All firms considered their competencies to be in innovation, quality, and service. These characteristics are consistent with the characteristics of the participants reported in an empirical study of business strategy conducted by Schroeder, Anderson and Cleveland (1986).

5.3 DESCRIPTIONS OF MANUFACTURING STRATEGY

The following sections describe the manufacturing strategy process for each of the six firms. At the end of the sections are two tables which provide a detailed comparison of the six firm's experiences.

5.3.1 Firm 1: The manufacturer of computer equipment

With annual sales of over eight billion dollars and 82 000 employees worldwide, this highly decentralized firm uses a hierarchical approach to corporate and manufacturing strategy formulation as shown in Fig. 5.1 Corporate strategy represents the starting point for manufacturing strategy and for the business plans of the operating units. At the corporate level, broad corporate strategic objectives are formulated which provide a set of expectations for lower level manufacturing strategies.

The responsibility for formulating group-level manufacturing strategy rests on the group manufacturing managers, who represent the manufacturing departments of the various related divisions within the group. A fundamental goal in formulating group-level strategy is to identify and integrate all strategic elements at the group level into the plans and strategies developed at other levels within the corporation. Thus, in addition to the corporate strategy and business plan, the group must respond to its own business plan, the plans of the Corporate Manufacturing Council (in which each group participates) and any ongoing *major* R&D programme activities which will involve the group. Initial efforts at a group-level manufacturing strategy focus on how the group's manufacturing resources can make a contribution to the higher level plans, and in particular, how they can provide a competitive advantage.

After potential competitive contributions are defined, each division within the group is polled twice in order to determine what manufactur-

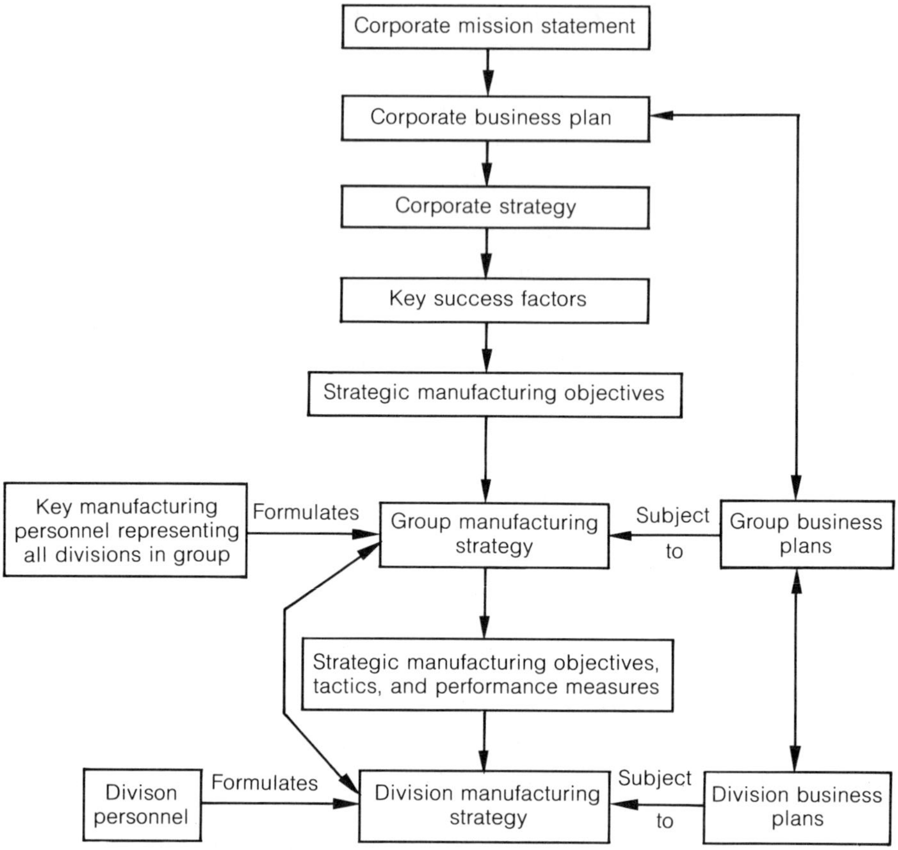

Fig. 5.1 The manufacturing strategy process for the computer equipment manufacturer.

ing strategy is *actually* being followed. The approach involves prioritizing the major bases of competition (e.g. quality, cost, availability, design) for each division. Each respondent is first asked to rank the competitive variables in importance to the customer, and at a later point in time, is asked how the variables would be ranked based on their perceived importance in day-to-day operations. The discrepancies in these results provide the basis for evaluating how patterns of decisions are being made. The group looks at categories of resources (using the Hayes–Wheelwright classification) to decide what opportunities for change in manufacturing are possible. Specific strategies are formulated for each of the categories of manufacturing resources and, in addition, performance measures and tactics for accomplishing the strategy are selected.

A division manufacturing strategy is formulated by establishing strategic objectives that are consistent with the group manufacturing strategy. Yet at the same time, there is an effort to provide an appropriate level of detail to specifically focus the strategy to a division's strategic objective. The process involves defining the strategic objective, creating a specific tactical plan for each strategic objective and then defining specific performance measures for each tactical objective. The process is further refined by identifying appropriate departmental tactical objectives for each department within the division. With the support of the tactical plans, performance measures, activity plans for implementation of the tactical plans, and supporting organizational plans, implementation can be easily monitored and strategies can be executed.

5.3.2 Firm 2: The manufacturer of pharmaceuticals

This firm currently has annual revenues in excess of $1 billion and employs 18 000 people. The firm is organized into functional units. A Unit Managers Committee, made up of the functional managers, is the corporate strategic planning entity for the firm. The responsibility for manufacturing strategy rests with the Production and Engineering functional unit of the firm. Manufacturing strategy is tied to the overall corporate strategies, goals and objectives of the firm as shown in Fig. 5.2.

In preparing their strategies, each of the functional units follows a method similar to the corporate strategy process. The corporate strategy is contained within three statements: a mission statement, a values statement, and an objectives and goals statement. Corporate objectives and goals are not only centred around the usual issues of new product introductions and profitability, but also speak to the importance of the various competitive variables for the firm. With these priorities in place, strategy statements are developed which serve as a guide and set priorities for each of the individual operating units.

With this background, manufacturing strategy is formulated by developing a mission statement for the manufacturing unit. The strategy addresses the key competitive issues identified in the corporate statements, i.e. cost, quality, and availability. The strategy process begins with an environmental analysis which is usually produced by a corporate strategic planning task force appointed by the president of the firm. To supplement the environmental analysis, the Vice President of Production and Engineering appoints a planning team to perform a strength and weakness analysis on the internal manufacturing environment.

As a result of the analyses, a small number of critical issues are identified for strategic action. Each of the critical issues is addressed by a strategic action item which includes: the specific action(s) recom-

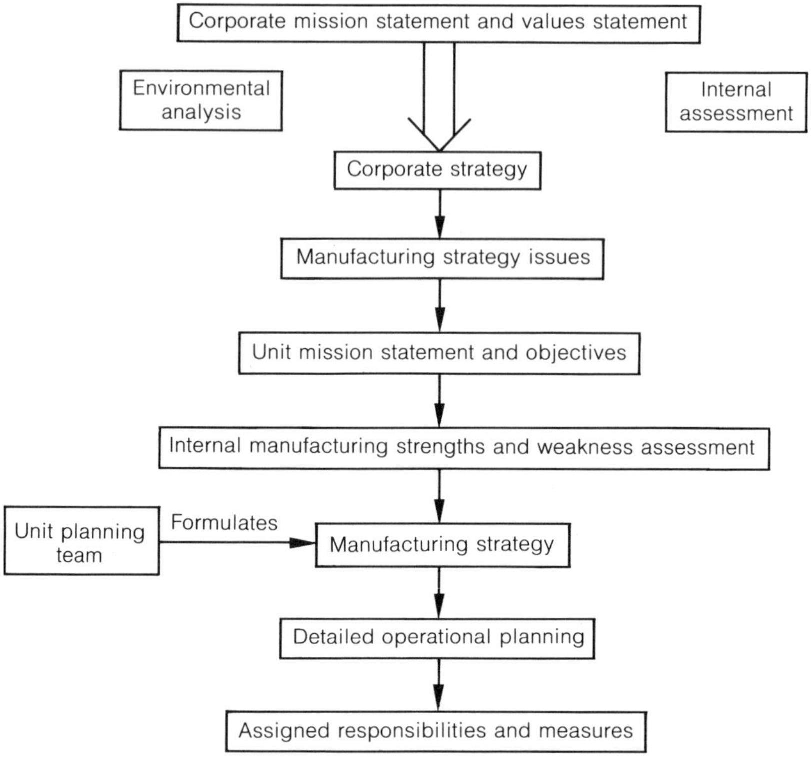

Fig. 5.2 The manufacturing strategy process for the pharmaceutical manufacturer.

mended, a rationale for the action, an identification of resource needs, and a matching of the action to both the specific corporate and unit goals. In addition, each year, a three-year operating plan is developed. The first year of the plan is highly detailed with specific operational objectives set and line manager responsibility assigned. Each objective is identified with one of the overall manufacturing strategy issues. The operational plans and objectives become part of the evaluation system for line managers and are reviewed regularly.

5.3.3 Firm 3: The manufacturer of valves

Firm 3, the smallest firm in the study, is a $100 million manufacturer of four basic types of valves and employs about 600 people. Because of the firm's small size, the strategy process is not highly formalized, yet the firm has the advantage of being able to maintain a highly dynamic process of strategy formulation because of the close communication

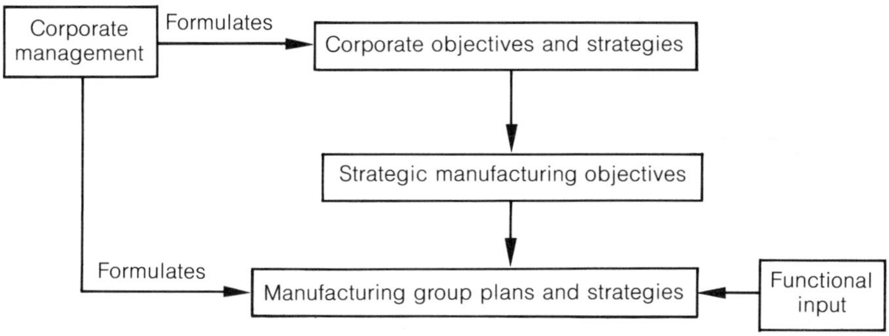

Fig. 5.3 The manufacturing strategy process for the valve manufacturer.

between top and lower management. This process is illustrated in Fig. 5.3.

In spite of the lack of formalization, the firm has identified and prioritized the competitive variables critical to its success. Manufacturing feels that it is proactive and involved in all strategic decisions; it is an equal partner in the formulation of other functional strategies.

Based on business conditions, top management produces a small set of strategic manufacturing objectives. The manufacturing group responds to these objectives by developing specific plans for each objective. The plans consist of a series of operational steps which are required to execute the strategic objective. Frequently, cross-functional task forces are assigned to identify and prioritize the resources needed to accomplish a specific plan. Resources are then assigned by top managers who regularly evaluate the progress of the implementation with a series of weekly operational committee reviews.

The small size of the firm gives it the flexibility to revise strategy on an ongoing basis. Top management develops strategic directions and objectives, while both middle and lower management indicate what can be realistically accomplished given current resource levels. This information is fed back to top managers who may reformulate the strategic objectives. This process may occur anytime that business conditions dictate the need for a change.

5.3.4 Firm 4: The manufacturer of telecommunications equipment

This manufacturer has annual sales of over $5 billion and employs 20 000 people domestically. The operations strategy is highly comprehensive and covers the total product delivery process, including product design and development, manufacturing, procurement, order entry,

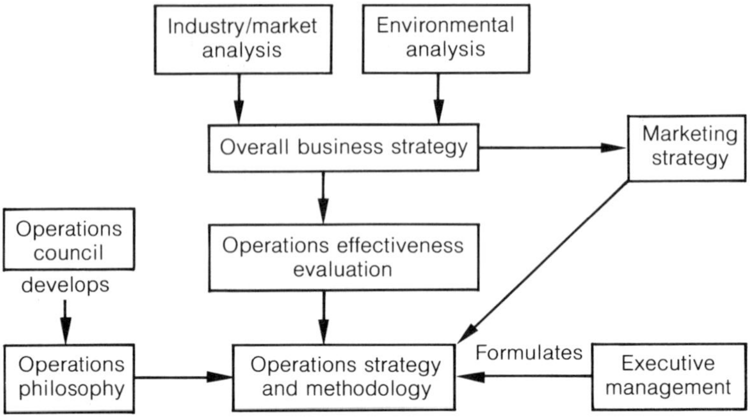

Fig. 5.4 The operations strategy process for the telecommunications manufacturer.

installation, and service. In contrast, the marketing strategy determines *which* products are to be delivered. Together the two strategies form the business strategy. The process of operations strategy is illustrated in Fig. 5.4.

The Operations Council, which consists of the directors of operations and/or general managers from all operating groups and divisions as well as all the operations directors at the corporate level, is responsible for developing the manufacturing strategy. The Council uses the marketing strategy and an environmental analysis as input to the strategy development process. This process is highly structured and consists of five steps.

The first step is an evaluation of the effectiveness of operations. Strengths and weaknesses of the manufacturing resources as categorized by Hayes and Wheelwright are analysed. An additional category for analysis is 'orgware', which includes the organizational metrics, control systems, human resource planning and development, organizational structure, capital investment decisions, plant charters, and corporate culture.

The second step is an industry/market analysis which seeks to describe what needs to be included in the strategy for a firm to be successful within the telecommunications industry. The third step is the development of an 'Operations Philosophy' which consists of the firm's own basic assumptions and beliefs that would pre-condition a successful manufacturing strategy.

The fourth step is the actual development of the manufacturing strategy as a specific strategic objective. The objective serves as a framework for planning throughout the organization and under which each

SBU configures its own manufacturing plans. The strategy is then encapsulated in a short document which is passed on to senior executives for additional input. The revised strategy is ultimately approved by the president. The final step is the development of an operations methodology which is a series of steps designed to execute the strategy and make the product delivery process more efficient.

Responsibility for strategy implementation also resides with the Operations Council, and specifically with the directors of operations within the operating units. Inherent problems with the strategy are identified by using a strategy–culture compatibility grid, which seeks to identify discrepancies between the behaviours and performance measures required for strategy implementation and those that seem to be currently practised within the firm. The strategy is documented in the form of a small booklet and a videotape to assist in implementation.

The Council continually reassesses the manufacturing strategy and monitors the status of its implementation with quarterly meetings. Although the strategy is broad enough that it only needs to be revised every three to five years or as business conditions dictate, the three-year operating plan is revised annually. The operating plan also contains financial projections, so that once it is approved, the first year of the operating plan automatically becomes the operating budget for the next year. In addition, each business unit also develops a strategic plan from the manufacturing strategy.

5.3.5 Firm 5: The manufacturer of furniture

The furniture manufacturer was one of the older firms to participate in the conference. It employs over 6000 people and has annual sales of about $310 million. About 20 years ago, the firm was acquired by a larger firm which had centralized all strategic planning; thus, the development of a manufacturing strategy was a fairly new experience. The purpose of the strategy formulation process was to integrate the efforts of marketing, manufacturing, materials, and other related functional areas within the firm under a single umbrella policy to achieve a 'decisive competitive edge' among furniture manufacturers. This process is illustrated in Fig. 5.5.

For this firm, strategy formulation is a four-part process. The first step is an evaluation of each of the manufacturing plants – its strengths, its problems, and how it should be focused on the basis of product mix and/or technology. The second step is a retreat for the entire management group where the problems that the firm and the industry face are discussed and possible solutions are generated. The third stage is a market analysis of what both retailers and consumers wanted from furniture manufacturers that they were not currently getting from any firm in the industry.

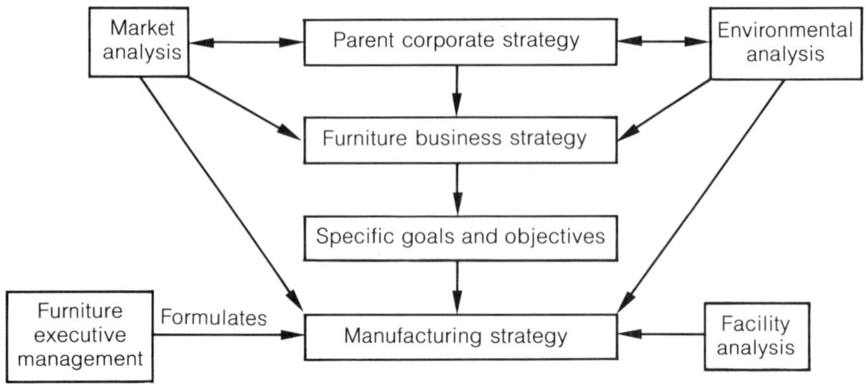

Fig. 5.5 The manufacturing strategy process for the furniture manufacturer.

The fourth stage, which was performed with the help of a consultant, is the development of a four-point manufacturing strategy. The four points are: product orientation relative to manufacturing capabilities, delivery, relationships with distributors, and the development of customer service goals. This strategy would serve as a planning framework over a five-year planning period.

In this firm, the entire executive management team, with input from key functional managers, is responsible for formulating and implementing the manufacturing strategy. The strategy is informally assessed on a continuous basis at many functional and cross-functional meetings and reviews. Formal review occurs on an annual basis where the strategic plan is reviewed and updated. The annual operating budget is prepared as a result of this annual meeting, and it is reviewed quarterly.

5.3.6 Firm 6: The manufacturer of electrical submersible pumps

This firm is only 30 years old and employs a total of 750 people. Its products support oil field operations, and its profits are directly correlated to the worldwide price of oil. Several years ago the company was acquired by a firm which placed a strong emphasis on strategic planning. Its strategy formulation process is illustrated in Fig. 5.6.

SBU teams are at the core of strategy formulation. There is one team for each of the four business units representing each of the four major product groups. Each SBU team consists of middle managers representing both manufacturing and staff support; the teams range in size from seven to 15 people.

The two key inputs at the SBU level are an analysis of market attractiveness factors and an analysis of the firm's current strengths and

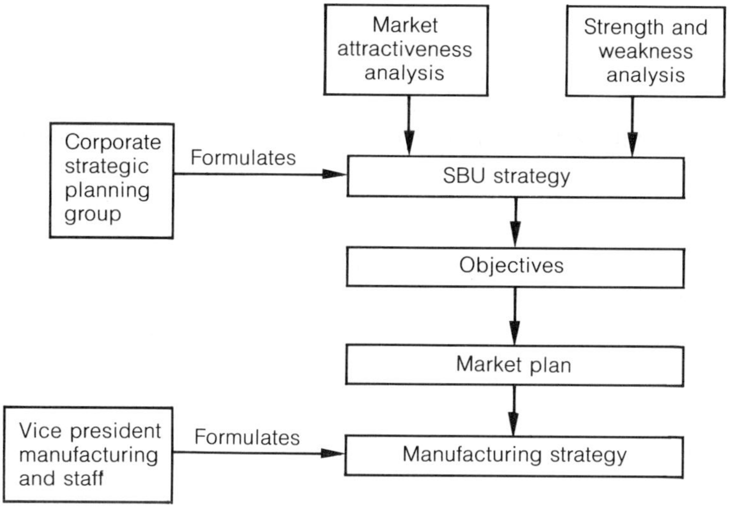

Fig. 5.6 The manufacturing strategy process for the pump manufacturer.

weaknesses. The market analysis is conducted along three different dimensions. First, the business potential of a geographic region is assessed; second, the potential size of a marketing region is estimated; and third, the relative strengths and weaknesses of each competitor firm are determined.

The second input is an assessment of the firm's strengths and weaknesses. A checklist consisting of categories (that closely match the Hayes and Wheelwright classification of manufacturing resources) was developed in order to categorize each business strength as high, medium or low.

The results of these two analyses are then plotted on a specially designed grid which relates business strengths to market attractiveness factors. The purpose of the grid is to get the SBU team to begin thinking strategically and to develop creative solutions for achieving strategic goals.

The SBU plan serves as the basis for manufacturing strategy. These plans are passed on to the Vice President of Manufacturing and his staff, who along with the corporate Strategic Planning Group, have the ultimate responsibility for formulating manufacturing strategy. The Vice President may provide additional input to the SBU plan and send it back to the group for further revision. In addition to the SBU plans, additional input to the manufacturing strategy is provided by externally developed sales forecasts, customer surveys, and technological forecasts. Currently, the purpose of the manufacturing strategy is to support the marketing plan in providing the high level of customer service which is required to be competitive within the industry.

Although five-year strategic plans were once common, the volatility within the oil industry has made that planning horizon unrealistic. Consequently, formal strategic plans are prepared annually and updated quarterly. Informally, the strategic plan is continuously assessed; monthly planning meetings are held to review changes within the strategic plan. In addition, a monthly review board meeting with all managers is held to analyse how well manufacturing is performing relative to the strategic plan.

5.4 DISCUSSION

Tables 5.1 and 5.2 provide an overview of the responses of the six firms to the set of common questions concerning manufacturing strategy formulation and implementation which the executives were asked to address in preparing their presentations. Other questions were posed during discussion periods at the conference, and the participants responded to the questions in round robin fashion. Some preliminary observations can be made about the practice of manufacturing strategy based on the results summarized in these tables.

5.4.1 Strategy formulation and strategy implementation

The findings are discussed in two sections, strategy formulation and strategy implementation. Strategy formulation is a rational activity involving identification of opportunities and risks, and determining material, technical, financial, and managerial resources, while acknowledging personal values and aspirations of senior management (Andrews, 1971). Strategy implementation focuses on achieving results through organization structure, relationships, processes, and bahaviour. Leadership is a key component (Hax and Majluf, 1984; Hrebiniak and Joyce, 1984).

5.4.2 Strategy formulation findings

Definition of the term. All of the firms reported that attempts at manufacturing strategy in its present form were a relatively new activity for the firm; most had started the present strategy process only within the past five years. (See 'Length of Strategy Experience' in Table 5.2.) The participants consistently portrayed strategy formulation as a planning mechanism to focus manufacturing activities in order to provide support for strategic business objectives and to achieve a competitive advantage (as summarized in 'Objectives of Manufacturing Strategy' in Table 5.1).

Size effects. In the smaller firms, the planning process was not as structured. Corporate objectives frequently served as the manufacturing objectives; and in some cases, the articulated manufacturing objectives were really tactical in nature since they were directed toward achieving a certain strategy as opposed to broad corporate objectives.

The larger firms had greater staff support and more corporate resources to assist in the strategy formulation process. Consequently, all of the smaller firms had used outside consultants in guiding their initial experiences with strategy.

Precipitating events. Although all of the firms mentioned that business conditions had become more complex in recent years, in each firm there seemed to be a single critical event that precipitated the manufacturing strategy initiative, ranging from a critical business event (e.g. an economically devastating year) to a new beginning with a 'slate that had been basically wiped clean' (e.g. moving into a new production facility). These are listed as 'Predisposing Conditions' in Table 5.2. All the firms explained the significance of this finding by stating that when things are going well, it is difficult to build a consensus for manufacturing strategy, since people aren't convinced that any changes need to be made.

This finding is consistent with other studies which show that an increase in environmental uncertainty leads to an increase in strategic planning efforts (Eisenhardt and Bourgeois, 1988). Generally, an appropriate planning system matches the complexity of the firm's environment and its internal culture (Hax and Majluf, 1984). The results from the six companies clearly demonstrate this.

Hierarchy of objectives. Strategic planning and management control follow a hierarchical approach. The process begins with strategic objectives at the corporate level, which subsequently serve as the basis for divisional and business objectives, action plans, functional unit operations, and monitoring activities. For control purposes the objectives may be used for analysis, performance, evaluation, and diagnosis. (Hrebiniak and Joyce, 1984). The hierarchy also produces activities at each level that contribute to the overall strategy and competitive advantage (Edmondson and Wheelwright, 1989).

The firms were remarkably consistent in describing the prerequisite hierarchical planning process for formulating manufacturing strategy under the umbrella of corporate strategy. (This hierarchy of activity has also been summarized by Ward, Leong, and Snyder (1990).) The process consisted of first developing a 'Manufacturing Mission', which was a statement of what was required to be successful within the industry and how manufacturing could best support the business. Manufacturing objectives were then developed which were described as 'measurable

Table 5.1 Summary of firms' experiences of defining and formulating manufacturing strategy

Firm	*Definition of manufacturing strategy*	*Objectives of manufacturing strategy*	*Responsibility for formulation*	*Relation to corporate strategy*	*Relation to marketing*	*Elements of strategy*	*Areas of analysis*
1	A shared vision and common vocabulary which unites the organization and determines the organizational, managerial and measurement systems	Provide a competitive advantage and support to higher level corporate plans	Corporate Manufacturing Council	Reactive to strategic business objectives with feedback	Increasingly market oriented. Plant charters change from product focus to customer/ market focus.	Process Capacity Technology Vertical integration Quality Mfg planning and control Human resources Information	Identify opportunities for change. Analysis of consistency in decision-making patterns at division level and on shop floor. How can change be effected?
2	A plan to adapt successfully to environmental changes	Match manufacturing resources to the environment	Vice President of Production and Engineering	Reactive to corporate objectives and goals with feedback	Definitely product driven. Also driven by R & D.	Facility size, inventory *v.* service, technology. vertical integration	Communication patterns
3	A plan to execule strategic objectives for long-term success	To realize new opportunities for success	Corporate management with cross functional input	Reactive to broad corporate objectives with continuous feedback	Supports marketing through innovation and solution of customers' problems.	New product analysis, costs, delivery goals, culture development	Identify opportunities for reorganization and consolidation
4	A framework for planning	Guide the total product delivery process	Operations Council	Together with marketing strategy constitutes business strategy	Complements marketing strategy to determine how products will be delivered.	Plant and equipment Capacity Facilities Technology Metrics Control systems	Strategy *v.* culture Compatibility analysis

Table 5.1 Continued

Firm	*Definition of manufacturing strategy*	*Objectives of manufacturing strategy*	*Responsibility for formulation*	*Relation to corporate strategy*	*Relation to marketing*	*Elements of strategy*	*Areas of analysis*
						Human resources Communication Plant charters	
5	An overall umbrella which coordinates planning and provides consistent functional interfaces	Develop a decisive competitive edge utilizing existing resources	Executive Management Group	Reactive to corporate SBU objectives. Consistency cheeks.	Identification of logistics/ service as a competitive weapon. Marketing and operations share common philosophy.	Product analysis, factory focus, product engineering, capacity utilization, cost structures, systems (mfg planning and control/ information)	Retailer and distributor analysis Quality analysis
6	A focus on what is needed for survival	To provide support for marketing and customer service	Vice President of Manufacturing with Strategic Planning Unit (cross functional)	Reactive with feedback to corporate strategic planning group	Reactive to marketing strategy. Provides marketing support in new product development serving new markets and maintaining showpiece manufacturing facility.	Forecast *v.* capacity, new products, technology needs, SBU plan, culture (MRP, Quality and Human Resources)	Market attractiveness/ internal strength analysis. What resources are needed to build strength in attractive markets?

Table 5.2 Summary of firms' experiences in implemeneting manufacturing strategy

Firm	Implementation plan	Evaluation/ feedback	Measurement categories	Length of strategy experiences	Problems encountered	Lesson learned	Strategy length reformulation	Cultural changes	Predisposing conditions
1	Specific strategies for each resource. Develop tactical plants, procedures and requirement plans.	Performance measures. Adherence to action plants.	Quality, Lead times, procurement prices, inventory reductions.	7 yrs	Focus on cost reduction. Collection of data for new cost accounting system.	Recognizing the difference between top-down conditions and bottom-up opportunities.	–	Communications. Cost acctounting system. First pass design.	New CEO
2	Plan and performance discussed at two-day retreat. Syndicates of senior managers develop statement and objectives for specific issues.	Review of operational plan. Semiannual formal meetings.	Inventory turns. Quality costs.	3 yrs	Developing commitment. Converting middle management. Building teamwork.	Plans must be concise.	Annual/quarterly	–	New CEO
3	Manufacturing group plans developed with middle and lower levels of managers.	Weekly committee reviews of group plans	Cost. Vendor performance. Quality. MRP/schedule performance. Inventory. Accounting.	4 yrs	Expectations set too low. Managerial inertia.	Involvement. Regular review. Objective performance measures.	Continuous	New performance evaluation. Organizational change.	Several poor economic years, new president
4	Development of an operations methodology specifying communica-	Quarterly meetings of Operations Council	Cost of quality. Delivery leadtime. New product introduction.	15 yrs	Building new culture to match new organization and	Must be a demonstrated need. Communication. Cultural change.	3–5 yrs/quarterly	New metrics. Teamwork. Change agenda: education	New president. flat revenue growth, revelation that certain

Table 5.2 Continued

Firm	*Implementation plan*	*Evaluation/ feedback*	*Measurement categories*	*Length of strategy experiences*	*Problems encountered*	*Lesson learned*	*Strategy length reformulation*	*Cultural changes*	*Predisposing conditions*
	tions, metrics and management of intangibles.				environment. Sustaining successes. Managerial lipservice.	Teamwork. It takes time.		communication procedures reward/ evaluation.	procedures weren't adding value
5	Strategy sessions, task force formation. quality training and implementation plans at cross-functional retreats.	Formal monthly meeting	Cost of quality. Cost reduction. Manufacturing cycle time. Service/delivery goals.	5 yrs	Changing plant profiles. Establishing new families of products. Matching marketing to manufacturing capabilities. Realistic service goals.	Pervasiveness. Participation/ involvement needed. Open lines of communication. Integrate functions.	5 yrs/quarterly	Design for manufacturing communication new vocabulary ownership of process	Poor economic year. new president
6	Three-day meeting of cross-functional management to develop implementation plan.	Monthly meetings with management Quarterly plant meetings.	MRP performance, Variances, cost of quality, inventory *v.* service, technology obsolescence.	5 yrs	Building teamwork and trust. Commitment means not changing priorities. Do it right the first time.	Settle for evolutionary change. Total involvement needed. Consistency. Breed success.	1 yr/quarterly	New layout. New labour classification. Decision-making style. Cost accig system. Egalitarian rules.	New facility. new president

requirements for success'. The firms were fairly consistent in using the competitive variables of cost, quality, delivery, and flexibility in formulating the manufacturing objectives. The firms either rank ordered these competitive variables or listed them without differentiating their relative importance.

Although none of the firms separated the competitive bases into order winning and qualifier factors as suggested by Hill (1989), there seemed to be intuitive understanding of the order winner concept in the formulation of the manufacturing objectives. In addition, many of the firms discussed the need for consistency between corporate and manufacturing objectives by reconciling manufacturing objectives to corporate objectives and relating both sets back to the competitive variables.

Analysis of strengths and weaknesses. The next step was an analysis of internal strengths and weaknesses to assess how readily the manufacturing objectives could be achieved given current industry conditions, the current manufacturing situation, and the current deployment of manufacturing resources. This analysis was used to determine additional resource allocations. The column in Table 5.1 titled 'Elements of Strategy' indicates the various categorizations the firms used to analyse the set of manufacturing resources at this stage.

The result of these analyses was the manufacturing strategy, which was a plan that communicated how manufacturing resources were to be allocated in some coordinated fashion in order to achieve the manufacturing objectives. Note that all the firms' definitions of manufacturing strategy (summarized in the first column of Table 5.1) describe the strategy as a broad plan which focuses, directs, and provides consistency for all manufacturing activities in an effort to achieve broad corporate objectives unique to each firm. These results are consistent with Miller and Hayslip's (1989) description of the process of manufacturing strategy formulation and implementation at Sonoco Products.

Tactical planning. Manufacturing strategy was accompanied by a tactical plan which contained a number of detailed objectives specifying how the strategy was to be carried out, what organizational resources were required to achieve a specific objective, and what performance measures were required to evaluate the objective. In the larger firms which were either functionally organized, or in one case, organized along product lines, objectives were set at several levels: corporate, group, divisional, and departmental. The tactical plan at one level became the strategic plan at the next lower level and so on through the hierarchy. This process assured consistency in manufacturing objectives among the various levels in the organization. Some of the firms also developed an action plan to correspond to each tactical plan.

Involvement of manufacturing in the overall process. Although the role of manufacturing has long been viewed as reactive with respect to corporate strategy, many of the firms reported that the interaction between corporate and manufacturing strategy was becoming more iterative. (See the column labelled 'Relation to Corporate Strategy' in Table 5.1.) Specifically, manufacturing played a major role in the assessment of corporate strengths and weaknesses which served as a basis for corporate strategy. Yet the firms felt that having a corporate strategy already in place was critical to formulating manufacturing strategy because the corporate strategy served as the 'acid test' in making resource allocation decisions. The consensus was that manufacturing decisions made in isolation could result in suboptimization of corporate strategy such as the case of deciding to invest in a new inventory control system when corporate initiatives were oriented toward reductions in WIP through Just-in-Time.

In addition, many of the firms reported that manufacturing was taking a more proactive role in identifying manufacturing strengths that could provide a competitive advantage to the firm. (This process is similar to strategic capability development described by Miller and Hayslip (1989).) Examples included time-based competition, manufacturing expertise that could be used to provide better customer service, and manufacturing capabilities that could be used in developing new products or in serving new markets. Two of the firms had developed formal mechanisms for providing this information as input to higher level corporate decisions.

Ties to marketing strategy. In general, it was felt that manufacturing strategy was, and should be, reactive to marketing strategy since an external analysis of both competitors and customers, as provided by marketing, was a prerequisite to the formulation of manufacturing strategy. As the column 'Relation to Marketing' in Table 5.1 illustrates, in all cases manufacturing strategy was closely tied to marketing strategy either formally (through marketing representation) or informally through shared corporate goals and objectives. Although all firms agreed that in the short term manufacturing could provide opportunities for marketing, long term they hoped that the manufacturing strategy process would assist the firm in developing the manufacturing resources needed to carry out the marketing strategy.

Constrained resources. The framework for classifying the elements of manufacturing strategy used by the firms was highly consistent with the classification of manufacturing variables proposed by Hayes and Wheelwright. (Refer to 'Elements of Strategy' in Table 5.1.) However, in each of the firms one or more of the manufacturing resources was constrained by decisions previously made at the corporate level and therefore was a 'given' in formulating manufacturing strategy. These constrained resources were often described as part of the operations philosophy in the sense that the decision had evolved over time and was now a guiding belief for the manufacturing organization. In each firm, the facilities decision was constrained although an evaluation of the 'focus' of the existing facilities was frequently included in the analysis. Other constrained resources in one or more firms included vertical integration, organization, and capacity strategies.

On the other hand, all firms were able to develop strategies for process choice/technology, quality, manufacturing planning and control systems, and people. The firms indicated that the categorization schemes served as checklists to be used in the process of strategy formulation, but that strategy was viewed as more encompassing than a series of sequential decisions made across a categorization of manufacturing variables.

5.4.3 Strategy implementation results

Table 5.2 summarizes strategy implementation issues. Each of the six firms was facing dramatic change in its environment and competitive position. As a result, we hypothesized a more dynamic implementation process than would be expected in stable situations (Pennings, 1985). Thus, we predicted more decentralization of decisions and involvement of lower levels, more teamwork, an informal culture, and a higher level of top management commitment to and involvement in the process in order to enhance information flows (Galbraith, 1977; Miller and Hayslip, 1989). The following sections develop these ideas in more detail.

Definition of the term. All of the participants mentioned that implementation meant 'getting the employees to buy into the strategy' or 'building a consensus among employees for the strategy'. Acceptance by the employees of the strategy and building the necessary teamwork were seen as the most critical tasks in strategy implementation.

However, from the beginning many of the participants felt that the definition of manufacturing strategy provided by the researchers did not go far enough in operationalizing the concept. The respondents felt that implementation characteristics of a manufacturing strategy were missing from the definition. These characteristics included the following: the strategy should be easy to articulate and understand; it should provide a basis for improvement; it should be difficult for competitors to match; it should focus only on a few key success factors; it should be durable; and it should imply internal performance criteria for measurement. These characteristics emphasize issues related to the implementation of strategy (such as communicating strategy to employees or evaluating the results of the strategy) as opposed to the top-down planning process which has been widely discussed in the literature. One firm called the traditional top-down planning process a 'necessary condition' for manufacturing strategy yet hypothesized that it was the bottom-up implementation process that would provide the firm with the opportunities and competitive advantages it was seeking by following a manufacturing strategy. Thus, the characteristics of manufacturing strategy which focus on implementation issues complement many of the currently used definitions which focus on strategy formulation mechanisms.

Involvement of lower levels. All of the participants demonstrated that although formulation is a top-down process, implementation involves lower levels of the organization. Frequently, the manufacturing strategy was summarized in a document that was to be distributed to all employees of the firm. Employee meetings were held where the new strategy was communicated and discussed. In addition, some of the firms held periodic meetings with all employees (frequently on a quarterly basis) where the status of the strategy was updated and further discussed.

One of the firms characterized the involvement of lower levels in strategy as the mechanism that identifies strategic opportunities for the firm by pinpointing the internal strengths of the firm that give it advantages that are difficult or impossible for competitors to emulate. One of the internal strengths frequently mentioned was the training and experience of the workforce which also facilitated the strategy implementation process.

Team approach. In these firms the formulation of manufacturing strategy was always a team effort (as summarized in 'Responsibility for Formulation' in Table 5.1). Although the head of manufacturing operations had the ultimate responsibility for manufacturing strategy formulation and implementation, in each case the process involved other people and, in some cases, other functions. The smaller firms involved more functions which is consistent with the open lines of communication and flexibility that a small firm typically enjoys. Conversely, the larger firms tended to formulate strategy in strictly functional teams consisting of the unit manager, his peers, and his direct subordinates.

Culture. All of the firms spoke of a corporate culture which was defined to be the norms, values, and informal beliefs held by people in an organization. Culture gives rise to the specific decisions, policies, and activities that govern daily activities. Culture is highly informal and perceptual, but usually dictates what activities/behaviours are necessary to become successful in a particular firm. Each of these firms had corporate cultures, developed under an old set of business conditions, which were slow, and even resistant to change under new competitive conditions. A major stumbling block was that people were often unwilling to modify their decision patterns and activities in a manner that would support the new strategy.

The problems of trying to communicate new cultural norms appear prominently in the columns labelled 'Problems Encountered' and 'Lessons Learned' in Table 5.2. All of the firms mentioned using various methods of effecting cultural change including education and training, employee participation programs, enhanced communication programs, revision of procedures and policies, and modifying the reward and evaluation systems. However, the consensus of the conference was that one of the biggest obstacles to cultural change were antiquated cost accounting systems which reinforced the building of huge inventories to absorb overhead expenses and often sabotaged strategic goals. All of the firms were beginning to examine their cost accounting systems with several of the participants already in the process of setting up activity based accounting systems, where costs are allocated on the basis of the activities which caused them to be incurred. The new accounting systems would focus on functional expenditures and provide simplified accounting information for better understanding. One firm even mentioned that it was revising its capital justification procedure to make it more compatible with the new strategy.

Consistency in implementation. Consistency among managers was mentioned as important to the implementation process in effecting the infrastructural changes needed to support the manufacturing strategy. To that end, each firm held an annual retreat for management to reevaluate the strategy and to update implementation plans. Regular meetings were held on a more frequent basis with lower level managers and employees to communicate how well strategic goals were being achieved and to enhance ownership of the process.

Top management commitment. As 'Predisposing Conditions' in Table 5.2 illustrate, each of the participants mentioned that there was some person in top management (the president or some other executive) who was convinced of the importance of manufacturing strategy and provided the impetus to begin the process. The common characteristics of this person were that he was relatively new to the firm, was perceived as willing to take risks and provided top management support.

The concept of top management support as a necessary prerequisite for the successful implementation of a new programme, technology, or system is very common within the implementation literature (Galbraith, 1977). Indeed, the participants reported that a major problem in strategy implementation was senior people who gave lipservice to the importance of the strategy but failed to change their actions and behaviour in a manner which was compatible with the new strategy. Although the concept of a manufacturing strategy 'champion' at the executive level is consistent with the 'product champion' widely discussed in the R&D literature, it should be noted that, unlike the product champion, the strategy champion was a top level executive who was not part of the manufacturing function. Thus, corporate management, as opposed to manufacturing executives, had provided the impetus for a successful formulation of manufacturing strategy.

Furthermore, the firms hypothesized that there are some types of managerial style which would cause any efforts at strategy to fail. They cited an autocratic manager who was unwilling to share power, and middle and low level managers who wished to be told what to do (as opposed to middle managers who showed a willingness to act upon strategic initiatives) as being inappropriate managerial styles for successful strategy implementation.

Project orientation. While strategy implementation was commonly viewed as a people-oriented task, the firms also indicated that implementation was project-oriented. One firm stated that there was a hierarchy of projects which comprised the implementation plan. The hierarchy consisted of core projects which were strategic in nature and were 'legislated' to all divisions; common projects which were implemented to all divisions at the subsidiary level; and applications projects which were implemented at the divisional level. The hierarchy might be envisioned as a pyramid with very few core projects, several common projects, and many applications projects. In the implementational effort, as many projects as possible were pushed to the applications level.

Strategy evaluation. Since all the firms were relatively new at manufacturing strategy, executives were not comfortable in discussing ways to evaluate manufacturing strategy. Although each of the firms was evaluating its manufacturing performance along a variety of measurement categories determined at the time the strategy was formulated (see Table 5.2), these measurements focused on internal manufacturing performance and did not address improvements in business performance relative to competitors. Most of the firms indicated that it was premature to determine whether the strategy had been effective in producing the desired results.

However, Firm 1 reported that many of its divisions had uniformly reported considerable improvements in a wide variety of performance measures related to programmes developed under the manufacturing strategy. They included significant reductions in inventory, space requirements, scrap/rework costs, labour costs and most importantly, leadtime. Beyond these general measures, there were other positive results experienced by the firm which may have emanated from its strategic direction.

All of the firms expected the strategy to remain viable over a time horizon of three to five years (or until business conditions changed). In all cases, the strategy process was periodic with strategy assessment performed annually by the group in a retreat setting. Operating plans and the annual operating budget were also updated at the same time. As the column labelled 'Strategy Length/Reformulation' in Table 5.2 indicates, the manufacturing strategy was usually formally updated and reviewed on a quarterly basis although all the firms mentioned that the strategy was informally assessed on a continuous basis.

The only firm that didn't follow this process was Firm 3 which happened to be the smallest participating firm. The small size and top management's close contact with daily operations in this firm allowed the process of strategy formulation to be ongoing.

5.5 IMPLICATIONS FOR FUTURE RESEARCH

Although the results of this research are exploratory, they pose some interesting questions for future research. First, what are the necessary and common prerequisites for implementation of a manufacturing strategy programme? This research identified several preliminary conditions including a dramatic change in business conditions, a single critical event, and/or the presence of a 'strategy champion'. Although the size of our sample was small and there may have been bias in its selection, the range of industries and the diversity of firms represented suggest that these commonalities may be more than a coincidence.

A second question is directly related. Is it reasonable to expect that implementation of new manufacturing technologies should vary with different business strategies? For example using the Miles and Snow (1978) categories, prospectors (organizations continually searching for new market opportunities) should be particularly anxious to improve manufacturing flexibility and quality. On the other hand, defenders (efficiency-focused organizations with narrow product-market domains) would continually strive to maintain a low cost producer status. Thus, prospectors would be expected to adopt new manufacturing technologies and strategy based planning techniques at a faster rate than the risk averse defenders (Ruekert and Walker, 1987).

We found some evidence for this relationship in our sample. Although the firms were asked not to report on the content of manufacturing strategy, all of the firms mentioned quality as being at or near the top of their lists of competitive priorities. While this is consistent both with the literature (Miller and Roth, 1987; and Schroeder, Anderson and

Cleveland, 1986) and with statements made concerning the competitive conditions within their respective industries, it may also point to an underlying integration of both content and process in the formulation and implementation of strategy. One participant commented that the organizational structure of his firm was about to be changed from organization along product lines to customer segments, and that he expected the process of strategy formulation and implementation to also change. On the other hand, another firm reported that under the umbrella of manufacturing strategy, a firm had developed programmes like TQC, JIT and CIM which caused a reorganization of the engineering department around these programmes. Thus, the implementation of manufacturing strategy had caused fundamental changes which *in turn* caused organizational change. Thus, these observations may indicate that the process of strategy formulation and implementation is not static; instead there may be an iterative adaptive process of strategy formulation and implementation with organizational changes.

Third, does the formulation of a manufacturing strategy condition the firm into accepting a more proactive role of manufacturing in strategic planning? Perhaps in the long run, only firms who have been sensitized to the competitive support that manufacturing can provide will be successful in the formulation of manufacturing strategy. To that end, the participants speculated that there may be some industries where manufacturing strategy would not be important. Such an industry might be characterized as having a stable business environment, being highly market-driven, containing relatively simple manufacturing processes, and possessing very short manufacturing leadtimes. (However, no one could point to a specific industry that possessed all these characteristics.)

Fourth, how does a firm evaluate the effectiveness of a strategy? Most of the firms felt that since the manufacturing strategy process was so new, it was premature to evaluate the effectiveness of the strategy. What are the characteristics of a 'good' manufacturing strategy? Is a firm better off with an ineffective manufacturing strategy than none at all?

A final question relates to the great emphasis the participants placed on the role of people within the organization and the importance of cultural change. How significant must the cultural change be to allow strategy implementation? The literature indicates that truly effecting cultural change is a long evolutionary process. Is it possible to develop a climate for effective manufacturing strategy as a distinct subsystem of the organizational culture as Schneider and Reichers (1983) suggest rather than trying to change overall corporate culture? If so, what are the critical systems that must be in place to support the manufacturing strategy? Although all the participants mentioned changes in the cost accounting systems as being important, other supporting structures such as information systems, decision-making processes, personnel performance evaluation, and compensation programmes were also seen as important. These questions point to the need for continued empirical research in the area of manufacturing strategy.

5.6 SUMMARY

This paper has documented the experiences of six firms in implementing a manufacturing strategy process. Although the specific form that the process follows varies among firms, in general, the formulation of manufacturing strategy seemed to follow general conceptual models that have been developed in the academic literature. These firms provide some evidence that manufacturing is slowly shedding its image as the 'missing link' in corporate strategy with manufacturing now participating as an active partner. Further, managers in these firms recognize the competitive strengths that manufacturing can provide. These firms have also installed mechanisms to assure the consistency of manufacturing strategy with both corporate and marketing strategies.

However, the firms indicated that the real benefits of manufacturing strategy seem to come from *implementing* the manufacturing strategy. Implementation is a less structured and more behaviourally oriented process than strategy formulation. Further research on strategy implementation must focus on the infrastructural issues including corporate culture, performance measurements, and decision-making/managerial styles. While these results are exploratory, they do provide topics for further study.

5.7 ACKNOWLEDGMENT

This research was supported by a grant from the Oliver W. Wight Educational Trust and the University of North Carolina Business Foundation. The authors wish to thank the following individuals for their assistance: Richard Ling, Walter Goddard, John Schorr, Thomas Wallace, and Mrs Joan Wight.

5.8 REFERENCES

Anderson, J.C., Cleveland, G. and Schroeder, R.G. (1989) Operations strategy: a literature review, *Journal of Operations Management*, **8**(2), 1–26.

Andrews, K.R. (1971) *The Concept of Corporate Strategy*, Dow Jones-Irwin, Homewood, IL.

Buffa, E.S. (1980) Research in operations management. *Journal of Operations Management*, **1**(1), 1–14.

Buffa, E.S. (1984) *Meeting the Competitive Challenge*, R.D. Irwin, Homewood, IL.

Burgleman, R.A. (1983) A process model of internal corporate venturing in the diversified major firm. *Administrative Sciences Quarterly*, **28**(2), 223–44.

Edmondson, H.E. and Wheelwright, S.C. (1989) Outstanding manufacturing in the coming decade. *California Management Review*, Summer, 70–90.

Eisenhardt, K. and Bourgeois, L.J. (1988) Politics of strategic decision making in high velocity environments: Toward a midrange theory. *Academy of Management Journal*, **31**, 737–70.

Fine, C.H. and Hax, A.C. (1985) Manufacturing strategy: A methodology and an illustration. *Interfaces*, **15**(8), 28–46.

Galbraith, J.R. (1977) *Organization Design*, Addison-Wesley, Reading, MA.

Hayes, R.H. and Wheelwright, S.C. (1979a) Link manufacturing process and product life cycles. *Harvard Business Review*, January–February, 133–40.

Hayes, R.H. and Wheelwright, S.C. (1979b) The dynamics of process-product life cycles. *Harvard Business Review*, March–April, 127–36.

Hayes, R.H. and Wheelwright, S.C. (1984) *Restoring Our Competitive Edge*, Wiley, New York.

Hax, A.C. and Majluf, N.S. (1984) *Strategic Management: An Integrated Perspective*, Prentice-Hall, Englewood Cliffs, NJ.

Hill, T. (1989) *Manufacturing Strategy: Text and Cases*, R.D. Irwin, Homewood, IL.

Hrebiniak, L.G. and Joyce, W.F. (1984) *Implementing Strategy*, Macmillan, New York.

Miles, R.E. and Snow, C.C. (1978) *Organizational Strategy, Structure, and Process*, McGraw-Hill, New York.

Miller, J.G. and Hayslip, W. (1989) Implementing manufacturing strategic planning. *Planning Review*, **17**(4), 22–7.

Miller, J.G. and Roth, A.V. (1987) *Manufacturing Strategies: Executive Summary of the 1987 North American Manufacturing Futures Survey*, Report of the Boston University School of Management Manufacturing Roundtable.

Pennings, J.M. (ed.) (1985) *Organizational Strategy and Change*. Jossey-Bass, San Francisco.

Porter, M.E. (1980) *Competitive Strategy*, The Free Press, New York.

Richardson, P.R., Taylor, A.J. and Gordon, J.R.M. (1985) A strategic approach to evaluating manufacturing performance. *Interfaces*, **15**(6), 15–27.

Ruekert, R.W. and Walker, O.C. (1987) Interactions between marketing and R&D departments in implementing different business strategies. *Strategic Management journal*, **8**(3), 233–48.

Schneider, B. and Reichers, A. (1983) On the etiology of climates. *Personnel Psychology*, **36**, 19–39.

Schroeder, R.G., Anderson, J.C. and Cleveland, G. (1986) The content of manufacturing strategy: An empirical study. *Journal of Operations Management* **6**(4), 405–15.

Skinner, W. (1969) Manufacturing – Missing link in corporate strategy. *Harvard Business Review*, May–June.

Skinner, W. (1974) The focused factory. *Harvard Business Review.*, May–June.

Skinner, W. (1978) *Manufacturing in Corporate Strategy*. Wiley, New York.

Swamidass, P.M. (1986) Manufacturing strategy: Its assessment and practice. *Journal of Operations Management*, **6**(4), 471–84.

Swamidass, P.M. and Newell, W.T. (1987) Manufacturing strategy, environmental uncertainty and performance: A path analytic model. *Management Science*, **33**(4), 509–24.

Ward, P.T., Leong, G.K. and Snyder, D.L. (1990) Manufacturing strategy: An overview of current process and content models. In *Manufacturing Strategy*,

The Research Agenda for the Next Decade. Proceedings of the Joint Industry University Conference in Manufacturing Strategy, Ann Arbor, MI.
Wheelwright, S.C. (1984) Manufacturing strategy: Defining the missing link. *Strategic Management Journal*, **5**(1), 77–91.

CHAPTER 6

Manufacturing strategy formulation as a process

C.A. Voss

6.1 BACKGROUND

In a recent presentation on developing manufacturing strategy in a major US company, Schroeder and Lahr (1990) described it as a 'proven process'. The process described by Schroder certainly seemed proven in the context of the company, but in general, the process of manufacturing strategy has been somewhat neglected. Despite the growing literature on manufacturing strategy, it is widely recognized that there has been little attention paid to the *process* of manufacturing strategy development and implementation. In their review of the literature on manufacturing strategy, Anderson, Cleveland and Schroeder (1989) state:

> perhaps even more revealing in the literature is the lack of attention on the process of operations strategy. There seems to be an implicit assumption that once operations strategy elements are defined, the process of operations strategy is merely perfunctory or self fulfilling, both in an analytical and an organisational sense.

In a similar review, Adam and Swamidass (1989), propose a number of missing themes in operations strategy. Missing theme number one is 'Operations Management research needs distinct research streams investigating content and strategy process.'

6.1.1 Is manufacturing strategy formulation a process?

Hill (1989) states that the reason for poor manufacturing effectiveness lies in the lack of an effective process for linking manufacturing

Specially commissioned for *Manufacturing Strategy – Process and content.*
Edited by Christopher A. Voss.
Published in 1992 by Chapman & Hall, London ISBN 0 412 43660 4.

strategies to corporate strategies. He states that manufacturing tends to have a reactive role in the corporate strategy due to: the production manager's view of himself; the company's view of the production manager's role; the fact that production managers are too late in the corporate debate; the 'can't say no' syndrome; the lack of language, functional goals and measures; functional support for manufacturing being weak; tenure; and top management's view of strategy. He proposes a process of manufacturing strategy formulation that is intended to overcome these weaknesses. He views the formulation of manufacturing strategy as a series of discrete and sequential steps. These steps are embodied in his manufacturing strategy framework, which consists of the following five stages:

1. Defining corporate objectives;
2. Determining marketing strategies;
3. Identifying how products win orders;
4. Establishing the most appropriate mode of manufacture for the sets of products (process choice); and
5. Determining the appropriate manufacturing infrastructure to support production.

Hill argues that implementation in a particular company does not necessarily start at the beginning and finish at the end. Manufacturing strategy implementation is an iterative process which he shows as a series of loops in some of his diagrams.

One of the main contributions to the area has been his concept of 'order winning criteria'. He sets out a process whereby order winning criteria are identified for each product line. He recommends that marketing perform this analysis, then that this is fed into 'the strategic debate'. Marketing's opinion will then be contested by the opinion of other functions and by the provision of information relating to the issues highlighted by the selection and the importance attached to relevant order winners. Hill proposes two outputs of manufacturing strategy. The first 'concerns a manufacturing review of the implications for manufacturing processes and infrastructure support for current and future product volumes'. This is done by manufacturing. The second output is:

> that having determined the manufacturing strategy position and the necessary investments and the time period for change involved, these now form part of the corporate strategy debate. The company as a whole is now forced to review the business in both marketing and manufacturing terms. . . . The corporate resolution therefore leads to an agreed understanding of what are the business objectives, marketing strategy adopted and manufacturing strategy required.

An interesting approach building on the framework developed by Hill is that of Platts (1988). Platts developed a 'manufacturing audit' in which he endeavoured to develop the approach into a workbook that could be used by companies on a stand-alone basis. (This work is described elsewhere in the book, so it will not be detailed here.) In doing this he found that there had to be a considerable number of changes in order to make it work on a stand-alone basis. For example, he found the concept of order winners difficult to apply in a workbook form. His work raises the question as to whether the process needs a facilitator to make it happen.

Miller (1988) proposes a process, in many ways similar to that of Hill, leading to debate, that starts with a market-based review by marketing, followed by manufacturing analysis. He also proposes a process based around workshops: 'a disciplined procedure designed to help middle managers in the different functions work together to help implement the business strategy.' This comprises six stages:

1. The business unit is defined, the key players are identified and the business strategy and expectations are determined from senior management.
2. Field interviews are conducted with the key players to determine the critical issues that affect their ability to carry out the business strategy.
3. A plan with agreed principal issues and a focus on key areas of concern is developed.
4. A limited number of key managers are selected to make presentations during a workshop.
5. A two-day workshop attended by managers and staff connected with the business is conducted. Both presenters and non-presenters participate actively and issues are debated and problem areas are identified.
6. Interfunctional task forces are then organized to take action on a limited number of the most critical function areas.

Central to Miller's process are the two-day workshops. The first day focuses on the view from the outside in. This means a summary of the strategy, an assessment of the market, an evaluation of competitors and a financial review of the business, and what commitments are being placed on the organization. The second day focuses on the view from the inside out in order to understand how the business is going to meet the commitment imposed by the outsiders. This means an understanding of the technology and its ability to keep up with competition and the state of the art; manufacturing's capacity, abilities to respond to change, and ability to meet quality and performance commitments, etc.

One way of looking at the implementation of manufacturing strategy is that proposed by Hayes and Wheelwright (1984). They describe four stages in the evolution of manufacturing strategy:

1. Internally neutral,
2. Externally neutral,
3. Internally supportive, and
4. Externally supportive.

Implicitly, this approach sees the implementation of manufacturing strategy as an increasing, more proactive, involvement by manufacturing in the linking of manufacturing to the firm's strategic needs. This is a useful and revealing categorization of manufacturing's strategic role. What Hayes and Wheelwright do not do, though, is to describe how a company should go about reaching these states.

Voss (1989), argues that the responsibility for implementing a manufacturing strategy lies with the board of directors of a company. He argues that the board has four responsibilities:

1. To ensure that there is a process of manufacturing strategy development;
2. To support the development of a broad based vision of manufacturing;
3. To name, lead and support a unified programme of change and improvement; and
4. To put in appropriate measures and controls.

6.2 METHODOLOGY

The research reported in this paper is field-based case research in four companies that had been through a process of manufacturing strategy development. The prime data was collected through interviews with executives involved in the development of manufacturing strategy. In addition the researcher was a participant observer in some cases. Other sources such as documentation and public domain articles were also used.

6.2.1 Case studies

The companies involved were diverse, ranging from a small diversified company to a large motor vehicle manufacturer.

Case 1 is a specialist manufacturer of medical supplies. It had a wide range of product lines produced in a single factory. These product lines varied greatly in success, nature of the market demand and manufacturing processes. The common factor was that they were all plastic.

Case 2 is a medium-sized manufacturer of specialist vehicles. It is profitable and successfully competes in world markets. It has a highly vertically integrated operation with all facilities on a single site. It manufactures two main product ranges each of which is characterized by high levels of variety.

Case 3 is a divisionalized company, with a large number of plants producing products in a range of markets. The development of manufacturing strategies is led by a central staff group with a wide range of expertise.

Case 4 is a major manufacturer of motor vehicles. The division studied has two main product lines and its manufacturing takes place on three major sites: body and paint, power train assembly and vehicle assembly.

6.3 ANALYSIS

The extensive data collected from the case data was reviewed. It was categorized into decisions and activities in the sequence in which a manufacturing strategy is developed. It was found that there were four elements in formulating a manufacturing strategy that all companies went through: triggering the process, determining the scope, managing the process and the analysis. The data from the research has been reviewed in the context of these elements and of the outcome.

6.3.1 Trigger

Issues. The issues triggering the development of a manufacturing strategy were different in each company. In Cases 2 and 3 they were triggered by competitive threats in the market place. In the other two the triggers were internal. In Case 1, there was a perceived lack of focus; in Case 4 the company wished to develop a coherent strategy for one of its sites.

Person who initiated. In all four cases, the process was initiated by senior management. In Case 3, the multi-business company, the process was promoted from the centre, but it was expected that it would be triggered by the business' managing director. In Case 1 it was triggered by the managing director who had previously been the manufacturing director. In the other cases it was triggered by the manufacturing director.

6.3.2 Scope

The scope varied considerably. In two organizations it covered the whole manufacturing strategy, and included a marketing as well as a manufacturing review of every product line. In Case 2 the scope was much more specific, the development of a CIM strategy; in Case 4 it was the development of a site strategy for one of the three plants of an integrated manufacturer.

6.3.3 Process

Having triggered the process there were a number of common dimensions examined in each company. The first was the process leader.

Process leader. This varied from company to company. In Case 1, the process was led by the managing director supported by the manufacturing director. (Although once the process was well established, the manufacturing director took over.) In Case 2, the leadership was by senior engineer in the plant's manufacturing systems group. In Case 3 the leadership was again in a manufacturing systems group, this time a central group. In Case 4, the leadership was by a senior engineer in the company's manufacturing engineering group. It is interesting to note that the leadership for the process came from staff rather than line manufacturing functions. This is probably due to the volume of analysis needed in developing a manufacturing strategy. The volume may prevent line managers becoming too involved.

Facilitator. It was found that in each case there was a person or department who acted as a facilitator for the process. In three cases this was an outsider, in one case facilitation was provided by the corporate manufacturing systems group. The role of the facilitator was to advise on the process, to counsel senior management and the process leader, to facilitate interfunctional participation and to ensure that the process took place. In all four cases the individuals had technical rather than process based skills. In one case this lack of process based skills led to problems with getting commitment from management.

Functions involved. In all cases there was an implementation team. In Cases 1 and 3 the teams were drawn from a wide range of functions. In Case 1, which was a multi-product company, a team was drawn from each product line. In Case 2 the team was drawn from manufacturing, engineering and systems. This caused problems as data was required from marketing in order to develop an effective strategy. This was refused initially on the grounds that marketing data was too confidential to reveal to manufacturing! This problem eventually was solved by the manufacturing systems team doing its own market research, developing an assumed market strategy, and then (when presenting the manufacturing strategy) pointing out the assumptions to the board. This led to a more open relationship. Case 4 had a very explicit marketing strategy known to all in the company, and the strategy exercise had a narrower focus than in the other companies. As a result, marketing was not involved in the team.

Pattern of interaction. In three of the cases, a broadly similar pattern was followed, with a cross-functional seminar workshop followed by a period of analysis, and the cycle repeated until the process was complete. At the end of the process, the strategy was presented to the board. This either triggered a further cycle if rejected or led to acceptance and the development of an implementation plan. In Case 2, the process was somewhat less interactive: it followed a similar pattern but without the workshops. This resulted in initial rejection by the board and a longer repeat cycle.

6.3.4 Analysis

Analysis can be considered in two dimensions: outside-in analysis, that is analysis of marketing strategies; and inside-out analysis of manufacturing capability and current manufacturing performance.

Marketing analysis. This was performed in three of the four cases. In Case 4 it was already available.

Manufacturing analysis. Analysis of current manufacturing performance capability and performance was done in all cases except Case 2.

Competitor analysis. Analysis of the competition was done in all four cases.

Paradigm used. There was considerable variation between the analysis paradigms used in each of the cases. In Case 1, an analysis based on Hill's approach, including the use of order-winning criteria, was used. In Case 2, the paradigm was that of Computer Integrated Manufacturing. This reflected the scope of the analysis. In Case 3 a systems engineering approach was used, with use of other paradigms, particularly Porter's strategic frameworks and some of Hill's. In Case 4, a framework developed in-house was used. The framework was very similar to one previously published by Wheelwright. Its approach had a strong industrial engineering bias.

6.3.5 Outcome

The outcome of the manufacturing strategy process was partially a function of the original scope described earlier. In Case 1, the major outcome was the development of focus in manufacturing. The plant was focused by product line; market-based priorities were agreed between marketing and manufacturing; current and desired performance was documented. A set of actions and investment programmes was identified for each plant. In Case 2, a vision of the factory of the future was developed, and a series of programmes and investments needed to reach it were identified and costings were prepared. In Case 3, the outcome varied from site to site, but in most the following outcomes were expected: make versus buy, manufacturing systems redesign, implementation plans and financial appraisal of these plans. The outcome of Case 4 was similar to that of Case 2. A vision of the future factory was developed, and a series of investments and programmes necessary to get there were identified and costed. In all four cases the strategies and accompanying investment proposals were presented to the main board of the company for approval. In Cases 1, 2 and 4, the process was repeated at regular intervals rather than being a one off.

The case data is summarized in Table 6.1.

6.4 DISCUSSION

A review of the above data shows a great deal of commonality between the processes at the various companies, together with some areas of great diversity. We can review these sequentially. All were triggered by a challenge or issue, either internally or externally, and were initiated by at a senior level. It was noted that when it was initiated away from manufacturing, the managing director had previously been director of manufacturing. This leads to a tentative proposition that initiation of a process of developing a manufacturing strategy is more likely when there is strong manufacturing representation at a senior level. As many UK companies are dominated by other functions, this could lead to accounting- and marketing-led companies failing to initiate the dialogue between marketing and manufacturing.

Having initiated the process, a key decision would seem to concern the scope. In particular in the two companies with very large and complex manufacturing sites, the scope was narrower than in the smaller companies. The pattern of the process was very similar across all four cases, with a facilitator, cross-functional teams and the analysis effort located in staff functions. The nature of the cross-functional team was a function of the scope. A broad scope had teams from many functions; a narrow scope had teams from a limited range of functions. The need for facilitators has been identified elsewhere. In the process described by Schroeder and Lehr (1990), an internal facilitator was used. Platts (1988) had to simplify and modify his process considerably to reflect the fact that it would be done without a facilitator. The pattern of meeting and analysis found in all companies is very similar to that

Table 6.1 Summary of case data

	Case 1	*Case 2*	*Case 3*	*Case 4*
Trigger				
Issue	Internal issue	External pressure	External pressure	Internal issue
Person	Managing Director	Manufacturing Diector	Managing Director	Manufacturing Director
Scope	Broad	Narrow	Broad	Narrow
Process				
Leader	Managing Director	Manufacturing Systems	Central Manufacturing Systems	Manufacturing Engineering
Facilitator	Outsider	Outsider	Corporate group	Outsider
Team	Yes	Yes	Yes	Yes
Functions	All	Manufacturing engineering and systems	All	Manufacturing engineering and systems
Interaction	Seminar–analysis–seminar cycle	Analyse, develop, present to board	Workshop–analysis–workshop cycle	Meeting-analysis-meeting cycle
Analysis				
Marketing	Yes	Yes	Yes	Already available
Manufacturing capability	Yes	No	Yes	Yes
Competition	Yes	Yes	Yes	Yes
Paradigm	Hill	CIM	Systems engineering	In-house (Wheelwright)
Outcome	Focus agreed priorities	Vision of future action plans	Make *v.* buy manufacturing system redesign	Vision of future action plans

proposed by Miller (1988). Although Miller proposed two-day workshops, the UK companies used one-day meetings.

The first stage of the analysis in each company seemed to follow a similar pattern of marketing analysis, manufacturing analysis and competitor analysis. Whereas Hill (1989) suggests a sequential analysis, the analysis tended to be done in parallel. Hill proposes two levels of marketing analysis: the marketing strategy, and the order-winning criteria (the market priorities). In all companies the equivalent of both were collected, but the market strategies tended to dominate. In one company the analysis started in manufacturing because marketing refused to become involved. It did become involved later when the manufacturing strategy was presented to the board with an assumed marketing strategy. The second stage of the analysis was to develop a vision of the 'ideal' factory, not the mythical 'factory of the future', but one that responded fully to market needs in the future. The degree to which this was done thoroughly depended on the available skills and knowledge of the company. Case 1 did not possess all the necessary skills to do this. In three of the companies this involved considerable make versus buy analysis.

Despite the common structure of the analysis stage there was a wide variety of paradigms and methodologies used to do the analysis. All four companies were very satisfied with their methods. This may indicate that having a method is more important than the method used. Alternatively, it may indicate that the key element in the process is the analysis of strategic needs and if the steps are followed, the method does not matter.

At the end of the process there is a common pattern of justification and, if necessary, reworking the analysis. It was found that there were a number of important feedback loops, in particular between manufacturing and marketing analysis.

Overall, the analysis indicates a common process of manufacturing strategy with areas of choice in a number of key areas. These findings are listed in Table 6.2. Given the small sample size it is difficult to argue that these are generalizable conclusions. They do however do show considerable consistency with the processes put forward by authors in the area. In particular it is consistent with the five-stage model of Hill, and the pattern of analysis and interaction suggested by Miller. In addition, the process described by Schroeder and Lehr fits this pattern. It is possible to propose that there is a process of formulation of manufacturing strategy that works in practice and is consistent with some of the major writers in the area. This pattern, and the main actions in it, are shown in Fig. 6.1.

Table 6.2 The process of manufacturing strategy formulation: common factors and differences

	Common across sample	*Different between companies*
Set up	Trigger	Scope – narrow or broad
	Leadership – senior manager	
Process	Facilitator	Functions involved (function of scope)
	Locus of process-not line manufacturing	
	Pattern of meetings	
Analysis	Outside-in analysis	May be done prior to process
	Inside-out analysis	Very different paradigms or methodologies used
	Vision of future factory	Scope of vision
		Different methdologies used to develop and define future factory
	Financial justification	
	Presentation to board	
	Feedback loops	Varying degrees of feedback
		Repeated at regular intervals

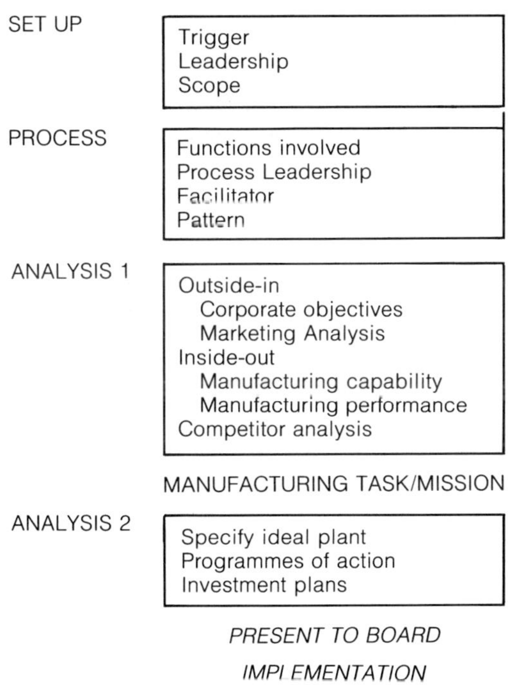

Fig. 6.1 Formulating manufacturing strategy.

6.5 SUMMARY

Empirical data collected from four cases has shown a consistent process of manufacturing strategy development. This is summarized in Fig. 6.1. It has a number of common elements and others that allow considerable scope for choice by the company. Whilst there are obvious limitations in a model derived from four cases, it may not be the only possible model; it does provide a framework that can be used by companies and researchers in the area of manufacturing strategy.

6.6 REFERENCES

Adam, E. and Swamidass, P. (1989) Assessing operations management from a strategic perspective. *Journal of Management*, **15**(2), 181–203.

Hayes, R. and Wheelwright, S. (1984) *Restoring Our Competitive Edge*, Wiley, New York.

Hill, T.J. (1989) *Manufacturing Strategy, Text and Cases*, Irwin, Homewood, IL.

Miller, S.S. (1989) *Competitive Manufacturing, Using Production as a Management Tool*, Van Norstrand Rheinhold, New York.

Platts, K. (1988) *Competitive Manufacturing: A Practical Approach for the Development of a Manufacturing Strategy*, IFS, Bedford.

Schroeder, R.G. and Lahr, T.N. (1990) Development of a manufacturing strategy, a proven process, in *Manufacturing Strategy The Research Agenda for the Next Decade* (ed J. Ettlie *et al.*), Kluwer, Norwell.

Voss, C.A. (1989) Strategic approaches to manufacturing, in *The International Management Development Review*, Management Centre Europe, Brussels, 332.

CHAPTER 7

Operating strategies for the 1990s: elements comprising world-class manufacturing

Aleda V. Roth, Craig A. Giffi and Gregory M. Seal

7.1 ABSTRACT

Over the past decade, world-wide interest in manufacturing has escalated. This sudden interest in manufacturing is due to a single factor – increased competitiveness. Competition levels are rising, significant new competitors are appearing in parts of the world not previously known for manufacturing eminence and, increasingly, manufacturers around the world are thinking and operating with global perspectives. This paper is predicated upon a research project involving a series of interviews with top manufacturing executives in companies perceived by many to be 'world class' today, discussions with key technology vendors, and an extensive international literature search. The research study establishes a framework that underlies the operations and strategies of world-class manufacturers today and outlines a 'state-of-art' assessment of the elements of world-class manufacturing for competing in the 1990s.

7.2 OVERVIEW AND INTRODUCTION

Manufacturing is vital for the health of the United States. It is discussed in the highest levels of government, studied by some of the most respected scholars in academia, and written about in most of the pro-

Source: Roth, Aleda V., Giffi, Craig A. and Seal, Gregory M. (1990)

minent journals published today. Books dealing with manufacturing have become best-sellers, and in corporate circles manufacturing has emerged as a major concern of those charged with guiding the course of their companies into the future. This sudden increase in the visibility of manufacturing is due to a single factor – increased international competitiveness.

In 1985, the President's Commission on Industrial Competitiveness, which was chaired by John Young, President and CEO of Hewlett-Packard, defined competitiveness this way (Boyd, 1986):

> Competitiveness is the degree to which a nation can, under free and fair market conditions, produce goods and services that meet the tests of international markets, while simultaneously maintaining or expanding the real income of its citizens.

This means that any nation can sell anything it wants to sell, to anyone who wants to buy it, anywhere in the world with no restrictions. But it also means that in order to be considered a real competitor, costs must be competitive, but not at the expense of the standard of living of a nation's citizens.

The definition of competitiveness offered by the President's Commission in 1985 is especially relevant today, in that it represents not only a model of theoretical competition, but also a true indicator of the current challenges confronting US manufacturers. Contrary to popular belief, the basic rules of competition have not changed, but the basis for competition and the game plans have. In the game today, the playing field has become global. The players are better athletes, and they are more aggressive. They emerge from every corner of the world.

The combination of advances in transportation which have resulted in a rapid expansion of trade and travel, coupled with the speed of satellite-linked electronic communication has accelerated the movement of manufacturing from a game played nationally to a game played by global competitors. The manufacturing game plan today has changed from one in which the field is defined by national boundaries to one in which there are no boundaries. Manufacturers from every industrialized nation in the world are expanding their markets to territories which heretofore had been considered safe and sheltered. The market today is the world. Companies who want to compete successfully are striving to become world-class manufacturers.

Furthermore, the level of competition in this expanded game is significantly higher than ever before. Players can no longer assume they will participate merely because they have been playing longer than other, newer players. The expanded game plan shows no favourites, no quarter is given, none is asked by newer participants. To compete in this new global manufacturing game is to perform at a 'personal best' level; to win requires a response to the competitive challenges from the best players in the world, to in effect perform at 'world-class' levels.

Today, 'world class' is no longer a meaningless phrase; it is a symbol of the level of manufacturing performance which is being exhibited by the top manufacturers in the world. It is a precursor of the standard of performance which will be expected of all who are to continue as manufacturers in the future.

This paper was based upon the culmination of a research study commissioned by the National Center for Manufacturing Sciences in 1988. The study focused on the elements of world-class manufacturing from the manufacturers' perspectives. An extensive world-wide literature search was undertaken and detailed interviews were conducted with top manufacturing executives in companies perceived by many to be world class today. Field research was conducted to validate the critical components of manufacturing excellence and the impact of these components on business performance. Additionally, technology vendors and system integrators were interviewed to provide perspective on state-of-the-art technology application. The full results of this undertaking are presented in Giffi, Roth and Seal (1990a).

Our research efforts were focused primarily on the infrastructural components of manufacturing strategy, i.e. quality programmes, organizational structures, human resource policies and programmes, production planning and control approaches and techniques, performance measurements, and reward systems. Our review of the structural ('brick and mortar') elements of strategy focused on the trends and implementation issues surrounding advanced manufacturing technology. In addition, the linkages to performance of many of the elements of both structure and infrastructure were identified.

In this paper, we present some of the definitions of world-class manufacturing that were uncovered during our literature search. These definitions provided the theoretical foundation upon which we developed world-class manufacturing framework contained herein. Following the definitions of world-class manufacturing, we summarize a series of operating principles for the 1990s derived from the findings and conclusions reached from the conduct of the research. The operating strategies are not a cookbook for manufacturers, but rather a set of guidelines within an operating framework that we believe differentiates world-class manufacturers from the ordinary. They also serve as a basis for future research in manufacturing strategy.

7.3 DEFINITIONS OF WORLD-CLASS MANUFACTURING

The definition of world-class is a topic under much discussion today, as is the path to effectively becoming a real world-class manufacturer. However, the phrase 'world-class' has become popular in the literature only recently; we found little reference to 'world-class' prior to 1986. In its place we typically found 'manufacturing excellence' as the most common way of referring to the goal of achieving superior manufacturing capabilities.

Each reference on manufacturing excellence we reviewed had its own, slightly different definition of world-class manufacturing. There were no standard definitions of the term 'world-class'. The definitions offered in the literature for both 'world-class' and 'manufacturing excellence' provided a beginning point for our research and provided a foundation upon which to build. In this section, we summarize a select number of the definitions offered that we found most helpful for benchmarking attributes of world-class manufacturing.

Hayes, Wheelwright and Clark (1988) identify these key attributes of world-class manufacturers:

1. *Becoming the best competitor.* 'Being better than almost every other company in your industry in at least one aspect of manufacturing.'
2. *Growing more rapidly and being more profitable than competitors.* World-class companies can measure their superior performance by 'observing how their products do in the marketplace and by observing their cashbox'.
3. *Hiring and retaining the best people.* 'Having workers and managers who are so skilled and effective that other companies are continually seeking to attract them away from your organization.'
4. *Developing a top-notch engineering staff.* 'Being so expert in the design and manufacture of production equipment that equipment suppliers are continually seeking one's advice about possible modifications to their equipment, one's suggestions for new equipment, and one's agreement to be a test site for one of their pilot models.'
5. *Being able to respond quickly and decisively to changing market conditions.* 'Being more nimble than one's competitors in responding to market shifts or pricing changes, and in getting new products out into the market faster than they can.'
6. *Adopting a product and process engineering approach which maximizes the performance of both.* 'Intertwining the design of a new product so closely with the design of its manufacturing process that when competitors "reverse engineer" the product they find that they cannot produce a comparable one in their own factories without major retooling and redesign expenses.'
7. *Continually improving.* 'Continually improving facilities, support systems and skills that were considered to be "optimal" or "state of the art" when first introduced, so that they increasingly surpass their initial capabilities.' Hayes, Wheelwright, and Clark also go on to say that 'the emphasis on continual improvement is the ultimate test of a world class organization'.

Continuous improvement and the **elimination of waste** are the two principles underlying manufacturing excellence as identified by Huge and Anderson (1988). They also identify 'lower cost, higher quality, better service, and more flexibility than competitors' as key factors behind the success of leading Japanese companies both in Japan and in the US.

Richard Schonberger (1986) offers a third definition of the concept. 'Today there is wide agreement . . . that continual improvement in quality, cost, lead time, and customer service is possible, realistic, and necessary', and that 'one more primary goal, improved flexibility, is also part of the package'. In addition, Schonberger states that 'with agreement on the goals, the management challenge is reduced to speeding up the pace of improvement'.

The concepts of **total quality, just-in-time (JIT) manufacturing,** and **people involvement** constitute the substance of manufacturing excellence, according to Robert Hall (1987). Hall believes these concepts represent a fundamentally different way of operating a company and are not techniques which can be 'grafted' on to a current organization. Hall stresses the importance of continuous improvement and concludes his book by stating that 'one interesting aspect of manufacturing excellence is that if such a thing is attainable at all, it will be by those who realize that no such condition exists'.

There are, of course, many other perspectives concerning the definition of world class manufacturing. In our research, manufacturing excellence is broadly defined as follows:

> A dynamic process that provides unique value, competitive advantage and delight to customers and suppliers through the development of internal operations capabilities that foster continuous improvements in human assets, technology, materials and information flows, that are synergistic with the total business, and that provide a sustainable competitive position in the firm's target markets.

Regardless of the definition chosen, the world-class manufacturer today is applying manufacturing concepts designed to demolish the obsolete methods, systems, and cultures of the past that have impeded its competitive progress. World-class manufacturers have a common attribute, namely that they are able to create high-value products and services, and earn a superior return over the long run through the application of competitive manufacturing strategies that win orders in the marketplace.

Based upon our review of the 'experts' opinions', our discussions with senior executives and our observations of world-class manufacturing organizations, we have assessed the elements of manufacturing excellence and have developed what we believe are the operating principles

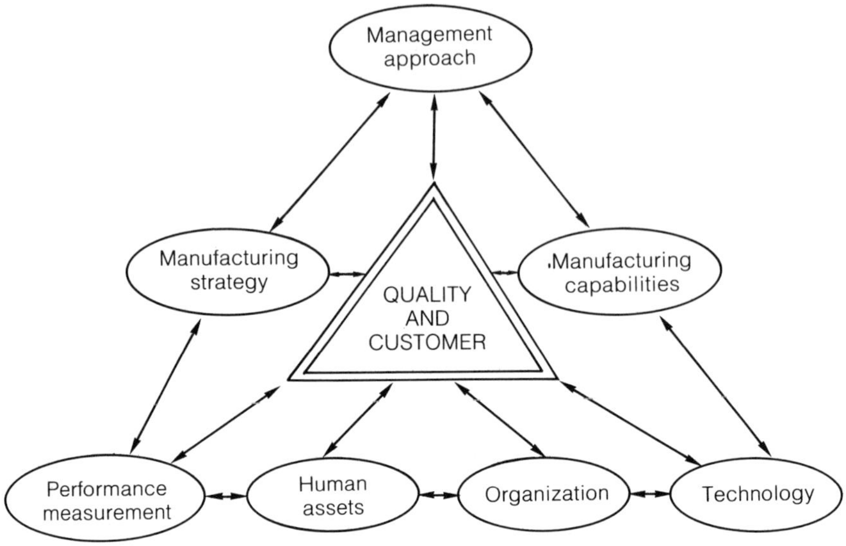

Fig. 7.1 World-class manufacturing framework. Reproduced with permission from Giffi, Roth and Seal (1990a).

for world-class manufacturers for the 1990s. These operating principles are summarized in Section 7.4.

7.4 OPERATING PRINCIPLES FOR THE 1990S

One of the key conclusions reached as a result of our research is that world-class manufacturers, regardless of industry or size, operate within a common framework. This framework, shown in Fig. 7.1, has at its centre the concepts of total commitment and dedication to both quality and customer. We found these concepts to be consistent, strong and inseparable at world-class organizations. Revolving around the heart of this operating framework, we found consistent themes and concepts associated with the management approach, manufacturing strategy, manufacturing capabilities, performance measurement, organization, human assets and technology.

Exhibit 1 operating principles for the 1990s

Quality and customer

1. Define quality in terms of the customers' needs.
2. Integrate the concept of customer closeness into the organization so

that everyone in the organization has a customer, and everyone's goal is to provide quality products and services to 'their' customers.
3. View quality from a global perspective – products, processes and services.

Management approach

1. Develop firm yet open management direction, strategic in thought, and effective in the implementation of innovation.
2. Constantly 'stretch goals' for the organization.
3. Foster an environment where sensible risk taking toward innovation is rewarded.
4. Develop a thorough understanding of the products which are being produced and the critical manufacturing process capabilities required.
5. Develop a systems perspective which treats manufacturing as an extension of a process from establishing customer requirements to satisfactory delivery and preparation of products fit for use.
6. Manage the organization across functional boundaries.

Manufacturing strategy

1. Establish a clearly defined strategic intent; define success in terms of winning orders over the long run.
2. Establish a strategy consistent with the potential for developing needed manufacturing capabilities.
3. Develop a global perspective to competition.
4. Make the strategy a blueprint for action, a pattern of decisions to be executed over time.
5. Put the strategy in writing.
6. Develop the strategy through a participative approach and share it freely with all employees in the organization.
7. Review the strategy on a periodic basis to ensure congruence with current and future goals and capabilities.
8. Allow your strategic intentions to drive the size of your steps.

Organization

1. Focus factories into organizationally flatter structures built around strategic business units.
2. Dissolve the boundaries between management and worker and between functionally segregated staff units; create dynamic cross-functional teams charged with resolving both strategic and operational issues.
3. Embrace the advantages of developing relationships with suppliers, customers and even competitors.

Manufacturing capabilities

1. Make dependability and consistency in quality, delivery and service to your customers the goals of all operations.
2. Develop manufacturing operations that are flexible and able to respond rapidly to changes in products and markets.
3. Restructure the engineering operations to reduce waste and inefficiency, and to improve quality.
4. Consider the environmental impact of all products and processes, making environmental considerations an integral component of the design and production processes.

Performance measurement

1. Focus on competitive variables that the customer sees.
2. Develop measurement systems which encourage continual learning.
3. Increase the vitality of the entire business by focusing attention on integrated business management.
4. Tailor the performance measurement system to the company's strategic action programmes.

Human assets

1. Invest in people.
2. Empower teams of workers to carry out the mission of the organization.
3. Eliminate the terms 'supervisors' and 'supervision'.
4. Evaluate the success of your people by their ability to achieve competitive capabilities, to learn and to adapt to change.
5. Develop accelerated and integrative learning programmes.

Technology

1. Develop an investment strategy for the continual enhancement of technology throughout the organization, based on a clearly defined vision of future competitive requirements.
2. Identify the competitive advantage of the knowledge base that advanced technology can create; simultaneously implement new technology and develop the new knowledge base.
3. Carefully plan technological upgrades to be consistent with infrastructural upgrades.

Adapted from Giffi, Roth and Seal (1990).

Within each of these areas, world-class manufacturers appeared to adhere to a fundamental set of operating principles, Exhibit 1, that differentiates them from their competitors. We found that world-class organizations adopted these principles in total, while their non-world-class counterparts often adopted select principles as if they were derived from a menu of options. The conclusions that lead us to the formulation of the world-class manufacturing framework and operating principles for the 1990s are the topic of the remainder of this paper.

In our view, the operating principles presented here represent the critical success factors for manufacturers in the 1990s. World-class manufacturers in the 1990s will be those manufacturers who not only understand the significance of these principles, but who make them an integral part of their everyday operations.

7.5 CONCLUSIONS

7.5.1 Management approach

> The management leadership of world-class manufacturing organizations is itself 'world class'. The development of manufacturing excellence requires management direction that is firm, innovative, and aggressive. Excellent manufacturing executives have developed a vision of the future that provides consistent guidance to their leadership. From top to bottom, world-class manufacturing organizations share the values and vision of the top manufacturing executive. Top management leadership through shared vision and shared values will continue to differentiate the world class competitors in the 1990s and beyond.

Over and over again as we tried to determine what differentiated the world-class companies from their competitors we returned to the single word 'management'. Whether we were conducting executive interviews or analysing case histories of world-class companies, management appeared to be the discriminating factor. It was management that developed and promoted the superior manufacturing strategies, defined the customer as the most important element to success, had the vision to adopt the right technologies at the right time, reorganized to promote learning, growing, team-oriented work environments, continually monitored the most important indicators of performance, and developed the rewards that promoted the never ending cycle of continuous improvement so critical to success.

This does not mean that the other elements of manufacturing excellence described in this document are not important, or that the right manager can make all your troubles go away. But, after extensive research into what constitutes the elements of manufacturing excellence, we have concluded that management is far and away the most important of those elements.

Management's ability to develop a corporate culture that promotes manufacturing excellence, tolerates, encourages and appropriately rewards risk-taking toward innovation is the catalyst that transforms good manufacturing organizations into world-class competitors.

In the final analysis, it is clear that the operating principles for the 1990s defined in this research will only lead to success when executed by 'world-class' executives who can create and share their own vision of manufacturing excellence, and who are capable of leading their organizations through the continual improvements required to become and remain 'world-class'.

> A hallmark of world-class manufacturers is a closeness both to the product and to the critical process technology. Competitive winners will continue to exhibit this common characteristic in the 1990s. They will enhance their capability to capture and retain a **knowledge base** about products and processes. World-class manufacturers realize that an understanding of product drives manufacturing and service requirements, and an understanding of manufacturing process capabilities can be used to define competitive advantages.

Marketplace 'winners' appear to have a keen understanding of the products they manufacture as well as the manufacturing processes that are critical to their success, now and in the future. This trait was clearly articulated by every senior manager we interviewed. Their understanding of product and process enabled them to differentiate between good and bad strategic moves regarding the development of process technology. Those that could articulate the competitive elements of process technology carefully guarded their unique technological advantage.

The International Manufacturing Futures Survey found that the Japanese develop significantly more of their process technology internally than do their competitors around the world. The research team conducting this survey concluded that the Japanese emphasis on internal development provides them with a unique knowledge of process which is an obvious advantage in the efforts to create unique process characteristics (Miller *et al.*, 1989). The executives to whom we spoke confirmed that the strategic elements of both product and process technology must be defined and segregated. In this way, R&D emphasis for new manufacturing processes can be focused in areas that will create and/or sustain competitive advantage.

World-class manufacturing executives have obtained a thorough knowledge of their products and processes due to a willingness to become involved with the essential details and to 'get their hands dirty' when necessary. Our research found that successful manufacturing managers are not strangers to the shop floor. They have an ability to communicate intelligently about the strategic elements of their products and processes; they can motivate their employees to meet the competitive challenges.

7.5.2 Manufacturing strategy

> A manufacturing strategy is a critical component of world-class manufacturing. World-class competitors have clearly thought out and defined manufacturing strategies and plans. Their strategies are congruent with the overall business goals and objectives and are flexible enough to adapt to change. However, the real winners are those best in strategy implementation.

Ironically, the new tool that appears to be in vogue with executives around the world is **planning**. Not just any type of planning, but strategic manufacturing planning. Long held in contempt by manufacturing executives, planning appears to be returning to its rightful place in world-class companies. One of the most useful tools identified by executives is a well-articulated manufacturing strategy. In our case history documentation and in our interviews with key executives, we found that developing the plan was the most important step along the journey. The planning process was essential whether the case history was Deere and Co., Caterpillar or other industry leaders.

While much debate continues over the best method for developing a manufacturing strategy, several elements appear to be critical:

1. The strategy must be more than a formal statement of policy. It must be a blueprint for action; a pattern of decisions to be executed over time.
2. The strategy must be based upon a realistic assessment of capabilities and priorities.
3. The strategy must deal with structural (brick and mortar) elements, infrastructural ('soft' issues) and integration (coordination) elements.
4. The strategy must be flexible and adaptive over time as the competitive environment changes.
5. The strategy should be a written document. Those manufacturers with documented strategies have been found to be more forceful in their manufacturing efforts than those with no strategy in writing (Roth and Miller, 1988).
6. The strategy should be developed through a participative approach, and freely shared with all employees in the organization.
7. The strategy should be reviewed on a scheduled periodic basis to ensure congruence with current goals and current capabilities; modifications and enhancements should be made as required.
8. Strategy should be implemented to affect change in a gradual process

of small continuous steps over time; 'bet the farm' actions or quantum leaps should generally be avoided.

It is unlikely that manufacturers in the 1990s will simply 'fall into' being world-class competitors. The importance of developing a realistic manufacturing strategy cannot be overstated. The vision of where you are going and how you are going to get there is fundamental to success.

> There will be no safe harbor for those companies unwilling to work toward, or resistant to becoming world class in the 1990s. All manufacturers must face new thresholds of performance. Directly or indirectly the standards for competition will be redefined by world-class companies in each industry. In order to remain viable competitors, aspiring organizations must meet or exceed the newly established thresholds.

Trends to widen gateways into untraditional markets, to move towards competitively favourable trade agreements and partnerships, and to diminish national boundaries are forcing an increasing number of industries into the global arena. The Japanese have capitalized upon these trends to a greater extent than either the Europeans or Americans. Evidence exists that the Japanese will continue to exploit these trends even more so in the future. The International Manufacturing Futures Survey found that The Japanese place a significantly higher degree of emphasis on the development of new products for new markets than other world regions, including the US and Europe (Miller *et al.*, 1989; Roth and Giffi, 1991).

The Europeans appear to be ahead of the US in breaking into Japanese markets. From 1986 to 1987 Japanese imports of European Community products increased nearly 26% measured in dollars or 9% measured in yen, while US imports increased only 8% in dollars and actually declined when measured in yen (Borrus, 1988). The Europeans, while still having a long way to go, are moving rapidly into being truly global competitors. And the South Koreans are now nudging their way into the global arena.

Unfortunately, the recognition of a global economy is not universal in the US. In a June 1988 report to the US Secretary of Defense regarding approaches to bolstering the Defense industrial competitiveness, researchers found that 'many senior managers continue to view the nature of markets as national, not international, and the nature of product requirements as good enough, not world class' (US Department of Defense, 1988).

The need to meet or exceed the evolving standards set by world-class manufacturers is most apparent in highly visible markets such as with electronics or automobiles. Here the products of various competitors are in head-to-head competition, and the final customer chooses the victor.

Many manufacturers which are suppliers of intermediary parts and components are less likely to perceive the need to change; thus, they may be more vulnerable to global attack. Eventually, intermediary parts manufacturers will be forced to comply with increasingly stringent standards of performance.

Today, these standards have begun to solidify in several industries. Just-in-time deliveries and the need to 'qualify' through attaining and sustaining increasingly higher levels of product quality have trickled down to even the lowest levels of manufacturing. In the 1990s, a full definition of world-class manufacturing will become a requirement of manufacturers at every level of product structure.

7.5.3 Quality and customer

> Quality has become a clear prerequisite to competing in the 1990s. Quality will be treated as a commodity. Those without it will not play in any markets. Superior quality validates a company's 'worthiness' to compete.

The number one priority of US manufacturing executives continues to be improved quality. The Japanese competitors continue to define quality as a top priority. However, the Japanese have begun to turn their attention elsewhere, toward cost and flexibility, as they already understand the importance of quality to success, and in general dominate the quality arena (Roth and Giffi, 1991). The Europeans also have quality as their top priority as have many other nations.

In every corner of the world, manufacturers have focused their attention and efforts toward quality as a top priority. With the significant attention and effort being directed at quality world wide, quality expectations are on the rise as are manufacturers' abilities to deliver quality products. The Japanese have demonstrated that superior quality can be attained with current technologies and known techniques. It is reasonable to expect that the 1990s will bring unprecendented levels of quality, and that quality will not just be a characteristic of world-class companies, but of all companies competing in open markets.

> Quality must be defined from the customers' perspective. Closeness to customer is critical to developing and understanding quality requirements. 'Quality is what your customer says it is.' Conformance quality drives perceived quality and perceived quality drives market success.

A hallmark of world-class manufacturers is their never-ending quest to achieve superior quality. They recognize that their customers' requirements, and more importantly their customers' perception, are the drivers of market place success. The Profit Impact of Market Strategy (PIMS) research has clearly demonstrated the importance of perceived quality to market share (Buzzell and Gale, 1987).

In a 1989 survey of 759 manufacturing executives, perceived quality was the top-ranked critical success factor for competing in the 1990s (Siffi, Roth and Seal, 1990b). Our discussions with executives support the notion of perceived quality as a fundamentally important concept. While conformance quality can and does drive perceived quality, the customers' interpretation of relative value is the final determinant.

In one way or another, the quality theme was repeated and extended to a full customer orientation by each executive we interviewed. Each executive had formulated a broader definition of quality than conformance to specifications. Matching marketing and operations capabilities are what differentiates world-class competitors. Determining what the customer needs and wants, which attributes add value to the product–service bundle, and creating operations capabilities to satisfy those requirements are characteristic of world-class companies.

Customer-driven manufacturing has been taken one step further by several excellent manufacturing organizations. Both internal as well as external customers have been defined. Everyone within the organization has a customer. And everyone's goal is to provide quality products and services to their own customers. The concept, while simple in nature, appears to be an indicator of the orientation of world-class competitors.

7.5.4 Manufacturing capabilities

> There is a necessary progression from quality to delivery to flexibility and/or cost. While timing and staging of priorities may vary from company to company, manufacturing excellence in the 1990s requires that a firm achieve multiple capabilities that set it apart from the competition (Roth and Miller, 1990).

Competition appears unlikely to stop at dependable deliveries or delivery speed. After delivery comes flexibility and cost. Flexibility, in terms of both product design and volume, is a major competitive capability in the time-based race.

The Japanese already appear to be focusing on flexibility as a key ingredient of their manufacturing strategies (Miller *et al.*, 1989; Roth and Giffi, 1991). Those manufacturers that can rapidly design new products and introduce them to the marketplace ahead of their competition will be winners. Likewise, those manufacturers that can shift quickly between product lines and product volumes will come out on top.

But flexibility does not appear to be achievable without first developing a foundation of superior product quality and superior delivery performance. Quality is the key to delivery, and flexibility and cost reduction are natural progressions. As more and more manufacturers achieve quality and delivery goals, the importance of both cost and flexibility capabilities will increase dramatically in the competitive battlegrounds of the 1990s. Being good at only one element will not lead to success.

> World-class manufacturers understand the value of designing product and process together, improving both the product and the process, and slashing the time required to push product and process from concept to market. They are breaking down the traditional organizational barriers between product design engineers, production engineers, and manufacturing engineers; removing functional goals and substituting with shared objectives. **Simultaneous engineering** is a differentiating characteristic of world-class manufacturers in the 1990s.

The drive toward time-based competition has rightfully spilled over the boundaries of the production floor into traditional staff areas, as companies who would be world-class try to squeeze time and expense out of all parts of their organization. To this point, manufacturing operations have received the brunt of the scrutiny, and undergone much of the restructuring, while other organizational components have continued relatively unscathed.

Today, however, world-class manufacturers understand that the customer cares little about the length of the manufacturing process and much about how long it will take to receive his product. They also understand that continually shrinking product life cycles can no longer support the lengthy design processes typical of most engineering functions today. Further, they realize that many of the pressures and frus-

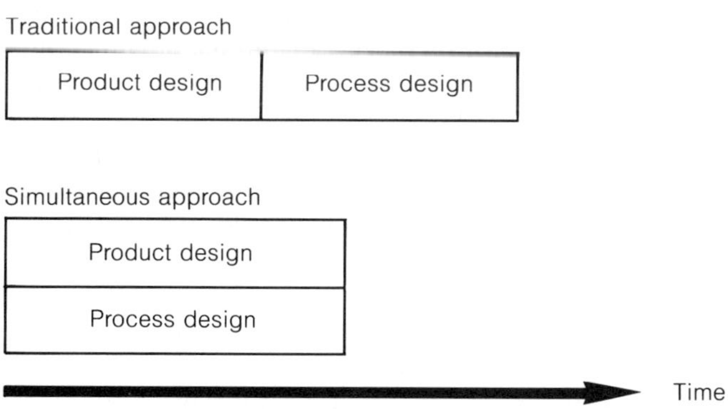

Fig. 7.2 Simultaneous engineering is key.

trations experienced in manufacturing on the shop floor are heavily influenced by the work of process and product designers. While competitive advantage is usually viewed on the shop floor, it is first conceived by cooperative engineering groups.

There are striking resemblances between the historic trend to organize shop floor equipment by process over product, and the sequential, phased, approach to engineering which was formalized by NASA in the 1960s and has been in general use in the US and Europe ever since. Manufacturers have discovered the folly of trying to achieve efficiency and quality by segregating equipment types into different areas of a facility, and have since undertaken a massive desegregation effort to recast production around products rather than functional departments. Competitive pressures have caused a similar scrambling of the engineering organizations of world-class manufacturers to what is currently called a 'simultaneous engineering' structure (Vasilash, 1987). The concept is simple in design, although difficult and non-standard in application, and has been in regular everyday use in Japan for years (see Fig. 7.2).

7.5.5 Performance measurement

> Time will be the measurement by which world-class competitors will differentiate themselves in the 1990s. Time-based competitive capabilities, such as dependable delivery time, delivery speed, and flexibility will be to the 1990s what quality was to the 1980s.

While the degree of emphasis on time-based competitive factors varied on an industry-by-industry basis, all executives we interviewed identified the next competitive battlefield to be that of time and service. This conclusion was supported by our literature search as well as research studies being conducted by many academicians in the field. Both the Manufacturing Futures Study and the North American Survey of Manufacturing Technology found that while quality may be the number one priority currently, time-based competitive capabilities have become increasingly more important in the last several years (Miller and Roth, 1988; Giffi, Roth and Seal, 1990b). Furthermore, the emerging time-based capabilities exhibit significant gaps between current abilities and future requirements. The steps that North American manufacturers must climb in the 1990s (see Fig.7.3) are steep indeed, with delivery-time reliability viewed as both the most important future requirement and among the largest of the steps to climb.

Many executives we have spoken to are focusing on what we would define as a first-level time-based capability; i.e. dependable on-time delivery. This focus is apparent in the significant attention being given to shipping schedules and shop-floor schedule performance. While dependable delivery times are important, they are only a starting point.

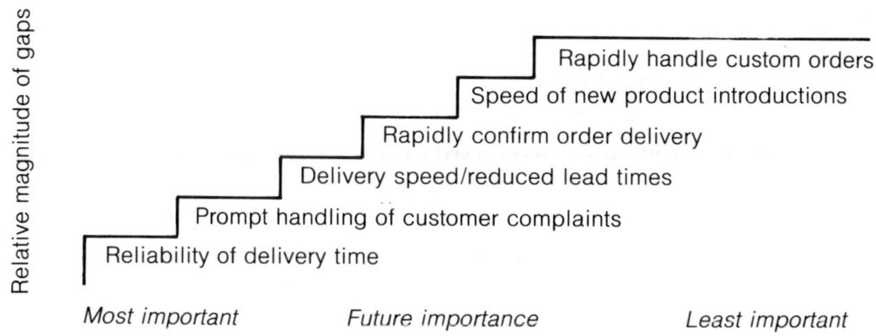

Fig. 7.3 Time-based capabilities indicate significant gaps. Reproduced with permission from Giffi, Roth and Seal (1990b).

After dependable delivery comes delivery speed or reduced cycle time. The capability to produce a high-quality product, on time, and faster than the competition represents a formidable competitive advantage.

In addition to throughput improvement programmes like JIT, aimed at reducing manufacturing cycle time and set-up times, many manufacturers are turning their attention to non-manufacturing areas to improve their delivery dependability and speed. In discussion after discussion, executives described the frustration of having reduced manufacturing cycle time only to have customer orders delivered late because of delays in the order entry or engineering design stages. While many manufacturers had very long manufacturing cycle times, they generally expressed confidence in their ability to reduce those times to competitive levels. There was much less confidence expressed with time reductions in order entry or engineering. Clearly, these are two areas which may lead to a competitive advantage if viable solutions are implemented.

> World-class manufacturers are changing their performance measurement systems to encourage manufacturing excellence. Performance measurement systems in the 1990s will be tailored to a company's strategic action programmes. Cost-accounting systems at leading manufacturers will be dismantled and replaced with **performance management** systems.

World-class performance measurement will be much broader than 'cost' in the 1990s. Each executive with whom we spoke indicated two things: first, the current cost-accounting systems in place are not effective for making management decisions; and, second, efforts are underway, or being considered, which redefine both the financial and non-financial measures of performance. The leading academic minds in this country have turned their attention to redefining cost-accounting systems as have the National Association of Accountants, Computer Aided Manufacturing – International (CAM-I), the 'Big Eight' accounting firms, and a host of others. With such unanimous recognition of a problem and strong support for efforts to change, future performance measurement systems cannot help but look different.

All of our research indicates that accounting systems will be reshaped to look at activities that drive costs rather than departments that collect costs. Cost drivers that have a direct influence on operational performance will be identified and managed. Overhead will not be spread like peanut butter over all manufacturing work centres based upon changes in direct labour input, but will be directly traceable and accountable. Costs will be segregated into value-added and non-value-added categories. Non-value-added costs are those that can be eliminated with no deterioration of product or service. Target costs will be developed that are not based upon internally generated cost standards, but rather on externally defined competitive costs.

Performance systems will be expanded to include the critical non-financial or operational measures of performance. Time-based measures of performance such as manufacturing cycle time, set-up time, on-time delivery, engineering cycle time, and order processing cycle time will become important barometers and regular measures of performance. Quality measures will be expanded to include not only manufacturing quality, but non-manufacturing service quality as well. Customer feedback and performance ratings, currently an integral part of the systems of world-class competitors, will become a common part of the performance measurement systems of all contenders.

Performance measurement systems of world-class competitors in the 1990s will:

1. Track attributes related to customer-driven critical success factors (e.g. quality, on-time delivery and speed).
2. Promote both inter- and intrafunctional integration and communication.
3. Provide feedback on the gaps between 'best-in-class' and the manufacturing unit's own performance over time.
4. Accelerate organizational learning and continuous improvement.

7.5.6 Organization

World-class manufacturers have streamlined the operations of their companies. They now operate from organizational structures which have been made lean and flexible. They have adopted a philosophy of waste reduction. They constantly monitor and evaluate ongoing operations from a value-added perspective. These companies have elected to regain and sharpen their competitive position by removing the historic buffers which dulled management sensitivity to factors critical to success.

There are few in the US who are unfamiliar with the highly publicized Japanese successes. The Japanese have employed methodologies which catapult their performance beyond those of most other industrialized nations. These methodologies have been rigorously studied by researchers and manufacturing companies alike. The central themes embodied in every analysis pertain to frugal operations, high attention to details, the elimination of inefficiency wherever it occurs, and the acceptance of simple solutions wherever possible.

World-class companies understand this central theme. It is not a Japanese creation – it is a common characteristic of all world-class companies. It occurs regardless of the operating philosophy or the products marketed. It is a way of life for the entire operation and not an 'improvement programme of the month'. This way of life endures long after the faddish popularity of the latest buzzword has dissipated.

Of the executives we interviewed, several were experimenting with some type of waste reduction programme. Some called it just-in-time (JIT), and others called it good business. Few had embarked on full scale just-in-time implementation programmes. JIT was most commonly being practised by the electronics manufacturers and auto suppliers, and to a lesser extent in the other industries we observed.

Unfortunately, as viewed by many of the executives to whom we spoke, leanness of operations appeared to be more associated with cost reduction than as a way to 'rightsize' for a lasting competitiveness. Our research found that leanness of operations was the most critical linkage in improving productivity, yet many manufacturers are still struggling with the first steps in that direction. World-class manufacturers in the 1990s will continue to improve their operations on an ongoing basis. They will strive to eliminate waste and create lean organizations that bolster their competitive posture.

> Complexity is a key contributor to poor manufacturing performance. Focus of organization and factory reduces complexity and improves both productivity and quality. World-class manufacturing companies are reducing complexity by developing focused manufacturing organizations.

We found that the concept of the focused factory, first popularized by Wickham Skinner (1974), is as true today as it was 15 years ago. Skinner cautioned manufacturers to focus their competitive resources to obtain productivity improvements. We have found that the increased complexity of most manufacturing organizations has made this axiom of manufacturing even more essential today.

The majority of the executives we interviewed reported a strong trend toward focusing their manufacturing organization. Conventional wisdom derived from our literature search indicated there is no single right or wrong way to organize. In our executive interviews, however, we found a strong bias toward focusing upon products and product lines. This finding held in both traditional job shop and batch manufacturing as well as in process-oriented industries.

We hypothesize that the importance of customer closeness to leading manufacturers is the motivating force behind product-oriented manufacturing organizations. Manufacturing cells were typically defined with a product line as the focus. In the larger, more diverse manufacturing organizations we visited, factories within factories, focused around specific product lines, were being implemented. Process manufacturers had entire plants devoted to the manufacture of a small number of products that shared not only similar manufacturing process technology, but also common customer and product relationships. At the larger manufacturers, efforts to focus the organizations were often accompanied by the implementation of work teams and the reduction of multiple job classifications to a manageable few.

Driving out complexity was always a clear objective at these focused manufacturers. Many had come from a history of very large, very complex factories which produced a multitude of products and required a wide array of product and process knowledge under one roof. When questioned about the loss of economies of scale that may occur in a focused factory, or the reductions of equipment utilization, most executives discounted the notion of any economies, indicating that the size and complexity of those unfocused factories eroded any economies of scale that were supposed to have been present.

Smaller manufacturers appeared much less inclined to reorganize facilities or people around a product focus. 'We're too small' is often the excuse. However, utilizing work teams and reducing multiple job classifications are as applicable for the small manufacturer as for the large. Likewise, those small manufacturers that carefully reviewed their product and product line volumes often found that a particular product or product line focus could be identified, and used as the structure for a focused factory approach. The key is to reduce complexity and focus the critical organization resources around competitive requirements.

World-class manufacturers of the 1990s will overcome the complexities associated with diverse manufacturing by focusing their organizations. Perhaps all manufacturers of the 1990s will heed Skinner's warning of the 1970s. We believe those manufacturers that are deemed world class in the 1990s will demonstrate the importance of focus to creating and sustaining competitive advantage.

> World-class manufacturers have developed organizations with revamped corporate cultures. There is an attitude of teamwork. All employees feel as though they have a stake in the business. The management style is participatory. Successful competitors in the 1990s will be those which are able effectively to dissolve the line

> between management and worker, and between functionally segregated staff units. They will merge individual factions into significant multifaceted, multitalented teams and still reward innovation.

Organizational effectiveness in world-class companies is built on the strengths of individuals which have been directed toward common goals. A characteristic of these organizations is that they not only hire top talent, but also build an organization structure around it which fits the goals and objectives the organization wants to reach.

There is not a single correct organization structure today, although many companies are trying to find one, in an attempted response to increasing competitive pressures. World-class organizations recognize that rather than concentrate on a specific form of organization, or the symmetrical appearance of positions on an organization chart, the single most critical factor is simply the meeting or exceeding of objectives through a team approach.

The skills and knowledge inherent in the human workforce are recognized by many companies as a tremendous competitive resource. World-class companies have not only discovered the existence of this resource, but have found ways to capture it effectively as well. Many companies have introduced non-traditional reward systems to spur employee involvement and share the rewards which solid business health can provide. In all cases, management and labour benefit as concerted effort is expended toward mutual goals.

Manufacturers aiming for world class status in the 1990s are flattening their organization structures, and reducing the number of supervisory personnel they employ. Those supervisors who remain are being re-trained to work in a role where consultative and persuasive skills replace directive, controlling styles. The ability to bolster the flow of information throughout the organization and to communicate is critical.

Excellent manufacturing organizations today are discarding the neat, logical organizational charts, sequential approaches to work, and functional and hierarchical divisions of duties and responsibilities. These have often hindered the ability of the organization to respond to competitive challenges. The organization of a world-class manufacturer, both today and in the 1990s, is one which capitalizes on all employees, and segregates none. Clearly, the most important factor is the fostering of shared motivations and common goals.

> Exercising effective vertical control over supplier networks and the channels of distribution will be a critical element of world-class manufacturing in the 1990s. World-class manufacturers have found that it is not enough to achieve excellence in only their in-house operations. The full effective length of their operations,

including suppliers and distributors of component parts and services, is of equal importance.

In the past, many manufacturers felt that the only way to manage effectively more of the full effective length of their production process was to buy it as part of a strategy of vertical integration. Today, world-class companies still acquire additional parts of their full effective production process when it makes good sense. But they also understand that it may be more advantageous to establish instead strong bonds and a good working relationship with component and service suppliers. In this way, they obtain the advantages of vertical integration without the drain on management's attention required by ownership.

The Japanese are demonstrating an increasing tendency towards very strong vertical control over their supplier networks. All manufacturers that are suppliers to the auto industry and that have felt the impact of GM's 'Targets for Excellence' programme understand the movement toward vertical control.

Increasing vertical control is an element of the strategies of many of the executives we interviewed. This control, exercised by extending back into their supplier base, is taking on many forms. Quality and delivery standards for suppliers, enforced with a vengeance, are the most common approaches being used. But many of the large and small world-class manufacturers are working with their preferred suppliers in the areas of education, training and technology transfer. They are including their suppliers in the up-front product and process design decisions, creating a concurrent engineering capability extending beyond their own organizations.

All manufacturers in the 1990s will exert ever-increasing pressure and control over their suppliers. World-class manufacturers will exercise that control in the form of partnerships toward improving the competitive advantage of both themselves and their suppliers.

> World-class manufacturers have integrated their operations not only with those of their vendors, but also with those of their customers, creating a competitive advantage out of their ability to manage the entire supply chain dynamically. **Supply chain management** is becoming a common technique employed by world-class manufacturers and will become a requirement for world-class competitors of the 1990s. Closeness to customer and to supplier are fundamental to developing manufacturing excellence, and drive the requirement for supply chain management.

Closeness to customers can take on many forms for the world-class manufacturer. However, creating uniqueness in meeting customers' needs through close coordination of the firm's operations with those of its customers is among the most effective competitive weaponry available. The ability to develop intimate linkages with customers provides a critical opportunity to differentiate the firm's capabilities from its competitors. World-class manufacturers have found that being linked to their customers, through such mechanisms as EDI (electronic data interchange) for production and purchasing information, frequent and routine personal communications, joint engineering development teams for both product and process, or missionary field engineering support can be as effective a competitive weapon as low price, fast deliveries or superior quality. World-class competitors are adopting tools like, quality function deployment (QFD), to link valid customer requirement to their design and manufacturing functions.

Several of the executives participating in our research described their ability to manage the entire supply chain by first developing a unique linkage to their customers. In one case this took on the form of developing and building the specialized equipment necessary for the customer to use the company's products in its own manufacturing process. While this executive indicated that his company was not in business to develop and build machinery for its customers, doing so has resulted in an almost irreversible relationship with them. For a customer to change over to a competitor's product, would require a complete replacement of entire segments of the customer's own production process.

Such linkages can not only produce a unique supplier relationship with customers, but also provide invaluable knowledge of customers' manufacturing processes. This knowledge can be used in the development of new products designed to work more effectively within the customers' facilities, further enhancing competitive positioning.

While the debate over the correct amount of vertical integration will continue for years to come, excellent manufacturers will continue to improve their management of the entire supple chain, from vendor through customer. World-class manufacturing organizations in the 1990s will develop unique ways to strengthen their customer and vendor relationships resulting in improved manufacturing operations and competitive advantage.

7.5.7 Human assets

The manufacturing function in the US does not compete effectively for high-quality personnel. There continues to be low regard for manufacturing both internal and external to manufacturing organizations. Attitudes toward manufacturing careers often prevent the best people from beginning or sustaining careers in manufacturing. The future of manufacturing in the United States is tied to the ability of manufacturing to attract the best and brightest people, and to retain them.

For many years the top college graduates in the United States have mistakenly viewed manufacturing as a dirty, smokey, dark place with few challenges and questionable career paths. Too often, the recognition that US manufacturing is in the midst of a major cultural and technological transformation has been overlooked. Consequently, few of the best people have historically embraced a career in manufacturing with fervour. Compounding the problem, the Department of Defense (1988) has recently concluded that US universities have little to offer students interested in careers in manufacturing.

Today the misperception has not changed much. Few of the top young people choose manufacturing over the perceived glamour of consulting or investment banking. Those who are attracted often become disenchanted with organizations they perceive to be hesitant to risk the outcome of significant challenges on the ability of employees who have not had years of seasoning. Through either direct action or an unwillingness to take action, organizations communicate to employees that they are not willing to make the same commitment to the individual as they would ask the individual to make in return.

To be world-class in the 1990s and beyond is not about technology or techniques as much as it is about the strength of the people who will implement and utilize the best of the techniques and technologies which are available today, and who will be charged with responding to the global challenges in the future that many of our industries are now beginning to experience. The need for attracting the best and brightest is clear, as is the need to understand that filling this need is not an option; it will be a requirement for those who wish to be world-class.

> World-class manufacturers are making significant investments in the planning and execution of employee development. They have found that reliance on external institutions alone is an inadequate solution to providing the level of education required. The combination of rapidly increasing technological complexity, a rapidly decreasing pool of young workers, and questionable public schooling effectiveness will prove to be a major challenge for the world-class manufacturers in the 1990s.

Improvements in technology during the past decade, including robots, vision systems, artificial intelligence, CAD, CAM, and a host of other computer related advancements has created a challenge for both white- and blue-collar workers. The trend toward technological advancements will only intensify in the 1990s, increasing the challenge for manufacturers. Some of the ramifications of this challenge will include:

1. The scarcity of trained craftsmen, coupled with increasingly demanding blue-collar job requirements will necessitate the internal generation of skilled workers.

2. Adoption of a team approach to shop floor operations will require training of team members in problem solving and group techniques.
3. Supervisors will need to be developed in order to function in an influencing/consultative role rather than an authoritative/controlling role.
4. Engineers, who today face a daily challenge of obsolescence due to the pace of technological change, will need to be trained continually in both the tools available to them as well as the technological evolutions of product and manufacturing process.
5. Employees of organizations who are part of the vertically controlled manufacturing process of many companies will continually require training in quality improvement techniques, manufacturing process improvement techniques, and product design techniques. Manufacturers realize that their performance, like that of a chain, can only be as strong as the weakest link.

Many believe that US industry lacks adequate programmes to provide continuing professional education and training to engineers and production workers. Continuing education and training programmes in US industrial firms are often weak, ineffective or non-existent according to a recent study by the Department of Defense (1988).

At many of the companies that we visited, extensive in-house training and education programmes are under way. These companies are improving their screening processes for new hires and investing heavily in their employees. Some were experimenting with innovative teaching programmes.

Training programmes for world-class manufacturers of the 1990s will be tailor made to the specific requirements of its products, manufacturing processes, and its customers. Training and education will be a significant component of every manufacturer's operating budget. Leading manufacturers will continuously seek out educational innovations to accelerate the pace of learning. World-class manufacturers will recognize the significant challenge presented by training and education requirements and respond without waiting for government relief or incentives.

7.5.8 Technology

Technology is not the problem nor the entire answer. Infrastructural components of manufacturing are the critical foundation upon which to build a lasting competitive architecture. 'Leapfrogging' the competition by employing technology, without first having developed a sound infrastructure, is a notion only for those unwilling to accept the realities and challenges of the real competitive issues.

While the focus of our study was not on the 'hard' technologies, during our literature search we did review current advanced manufacturing technology trends, we interviewed a number of technology vendors, and discussed the importance of technology with each of the executives who participated in this study. During our interviews, we did not speak to one executive who believed that the reason for the general lack of competitiveness of the US was significantly related to technology.

Numerous benefits can be cited for those companies, both domestic and foreign, that have successfully implemented technology into their manufacturing operations. However, we could find no support for a conclusion that the US was, in general, technologically disadvantaged. While particular industry segments at different times find themselves being set back by their competitor's latest application of technology, this appears to be more the natural ebb and flow of competitive markets than a widespread disease.

This is not to say that keeping up with technology is not important. On the contrary, keeping pace with technology was a real concern voiced from all corners. Ignoring the rapid technological changes occurring was defined by many of our sources as a sure way to erode competitiveness. Continuing to invest in modernizing plant and equipment and looking for advancements in technology to improve competitiveness appear to be an important element of long-term success. But as one executive pointed out, 'the pace of technological change is affecting all competitors equally'. None of the executives was willing to say that 'they are beating us because their technology is better than ours'. In fact, most executives retorted that the global nature of the technology marketplace makes technology equally available to all competitors. Technology transfer, rather than technology itself, is the real obstacle facing manufacturers.

There is overwhelming evidence that managerial issues surrounding the manufacturing infrastructure and its technological readiness are more urgent than those surrounding technology itself. This was the resounding message we received from manufacturing executives and from our literature search. Infrastructure, such as human resource policies and skill levels, organizational structure issues, quality-improvement approaches and cultures, performance measures and measurement systems were all defined as the real challenges facing manufacturers in regaining their competitiveness. Further, a sound infrastructural base was considered paramount to the successful application of technology. The Japanese, typically the cause for US concerns regarding competitive pressure, have long focused first on improving their infrastructures to prepare for eventual technology application.

The problems General Motors created in the early to mid 1980s through their significant application of technology without first adequately establishing a sound infrastructure has received tremendous attention, however, many other companies find themselves making the same mistake. Hayes, Wheelwright and Clark (1988), after years of research into manufacturing competitiveness, report: 'We have seen a number of companies that have been able to build a powerful competitive advantage around their internal capabilities and teamwork, even though their plants and equipment were not exceptional; but we have never seen one that was able to build a sustainable competitive advantage around superior hardware (technology) alone. . . . It is almost impossible for a company to "spend" its way out of a competitive difficulty.'

Attempting to leap-frog the competition via technology to create long-term competitive advantage is not likely to succeed without an adequate infrastructure. Continual improvement addressing both structural and infrastructural elements of manufacturing is the only proven approach to overcoming manufacturing barriers. It need not take the 30 years it took Toyota to become a world-class manufacturer, as Deere, Caterpillar and others have shown, but there is no short cut regarding the basic issues that must be addressed.

> World-class companies understand that the major impediments to the adoption of technology in manufacturing are not technological, but are instead human related. They understand that the task of technology implementation is not complete simply because crates have been unpacked, equipment assembled and installed, and electrical wires connected. Until manufacturing becomes a wholly automated environment where humans visit only periodically, the manufacturing environment is a human one, and technology is the visitor.

There is a growing recognition in the US that people are the common denominator in all manufacturing equations – whether in terms of human performance, humans and technology coexisting, or humans tending to an automated environment. In every case, the fate of the operation is in human hands, and rightfully so.

World-class companies understand that technology can help them become or remain competitive. However, they also know that even when automating an entire production process makes strategic, tactical and competitive sense, humans will continue to play an integral role. They will add needed flexibility to react effectively to shifts in customer wants, provide critical process linkages and add judgment as an ingredient no machine can yet match.

There is no dominance by the US in terms of technology implementation. The Japanese are clearly leading in many aspects of technology implementation, leveraging upon intricacies of the Japanese society. This results in an apparently higher degree of willingness by Japanese workers to accept technology in manufacturing. On the other hand, the

Europeans and Americans continue to struggle with the two major obstacles to implementing automation: the gaining of management approval and employee resistance (Farley *et al.*, 1987).

Other findings regarding technology implementation include (Farley *et al.*, 1987):

1. Perceived problems are greatest in those organizations with relatively weak communications and relatively entrenched senior management.
2. The greatest benefits from automation are anticipated by firms able to justify automation because of quality gains as well as cost savings.
3. Companies that have already automated tend to be large, have a high technology orientation, and show interest in longer term considerations by their involvement in planning and decentralized responsibility.
4. Firms tending to automate are likely currently to be experiencing profit problems.

World-class companies in the 1990s will continue to apply technology strategically to manufacturing environments. They will work energetically to automate. Their implementation strategy will consider the necessary human interfaces. They will work in meticulous detail to develop the appropriate infrastructure that must support the technology. Clearly, as companies aspire to automate they will face hurdles. They must first develop a futuristic perspective about the strategic advantages of technology as they compete in the global arena. Second, they must recognize that the major hurdle to implementing technology may be their own management team.

> Superior communications is a key element of a world-class manufacturing operation. Customer communications as well as interfunctional communications between manufacturing, engineering, marketing, sales, accounting and top management are significantly better at world-class manufacturers than at their competitors. World-class manufacturers will focus on improvements in their information management and communications technologies in the 1990s.

When we asked executives about what they believe should be the key application of technology in the 1990s the typical response was 'communications'. Similarly, we found communications to be a key area executives identified for improving their time-based competitive capabilities in the 1990s.

Interestingly, executives cited that communications between people within the organization, and not between machines, were seen as a critical element of improving competitiveness in the future. Most time delays and most non-manufacturing quality problems were seen as related to communications between people. It was not that machine-to-machine communications via MAP or some other standard was seen as unimportant. Executives are placing a higher emphasis upon 'communications basics' first. One company in our study has defined the communications link between their customers, their sales people and their manufacturing plants to be one of the largest opportunities for improvement in the future. The initial focus of their CIM efforts is bolstering the customer/order entry/manufacturing communications link.

Despite the strong trends in applying technology to upgrade communications on the shop floor, we believe that the greatest payoffs reside in upgrading the communications between people. The requirement for timely management information is projected to increase dramatically in the 1990s as time becomes the basis for competition. Unfortunately, many companies are not prepared for the challenge. Our research indicates that today's world-class manufacturers are indeed better communicators and managers of information. They apply technology judiciously and aggressively to enhance the communications in their organizations. To be a world-class manufacturer in the 1990s will certainly require superior communication and information management capabilities.

7.5.9 External environment

> Some of the most significant obstacles which US manufacturers currently face in rebuilding their competitiveness are those that have been put in place by the Federal Government. Antitrust laws are myopic. Their nationalistic perspective does not consider US manufacturing in the light of international competition. Tax incentives for manufacturers investing in technology and education are limited and appear inadequate for the challenges that lie ahead. International competition requires government to act globally and become a partner in the effort to compete.

The lack of national policies designed to support US companies in their drive to become world-class is a critical issue today. Those policies which restrict manufacturing are a national outrage. That some manufacturers have achieved world-class status despite current government policies is a very positive statement about the resolve of those organizations to succeed. But resources are scarce. It is unrealistic to expect the bulk of US manufacturers to attain world-class status, or even survive, when they are required to match individual resources against those of international consortiums. These consortiums are serious contenders and are concentrating an investment of billions of dollars to ensure their success.

The Department of Defense (1988) recently concluded that:

> restrictions imposed by the Sherman Antitrust Act, the Robinson-Patman Act, and the Clayton Act and other antitrust laws have become increasingly irrelevant and harmful to the international competitiveness of American industry. In particular, antitrust laws and regulations that impede cooperative research and the development by American firms in both process and product technology are harmful.

Restrictions were eased by the provisions of the National Cooperative Research Act of 1984, which has allowed manufacturers some latitude jointly to research, develop and transfer technology through organizations like the National Center for Manufacturing Sciences. But many manufacturers still feel constrained by continuing uncertainty with respect to applications of this body of law, which continues to create substantial risks for companies in joint ventures.

The Federal Government has created a relationship with industry that is characterized by extensive restraints, and requirements to document compliance to the 'rules and regulations'. Significant effort is invested on the part of manufacturers in satisfying Federal paperwork and responding to what many have referred to as 'continual meddling in the manufacturing process'.

The same DOD study referred to above also concluded that:

> other countries, Japan in particular, are much more effective than the United States in achieving industry/government/labor cooperation of process and product development, and, through cooperation, are more effective in implementing new ideas to make manufacturing more efficient, responsive, and technologically advanced.

Tax laws and trade policies as currently structured do not provide an incentive for manufacturers to train their workforces or invest in new equipment and process technology (Department of Defense, 1988). Relative to other industrialized nations, the US tax system provides little or no incentive for long-term investments. The US tax system taxes capital gains and earnings on savings and investments as ordinary income, at relatively high rates, but either does not tax consumption or does so at very low rates. This results in reduced capital for industrial modernization and technological development. The US tax system, through income, property, and labour taxes places a tremendous burden on products made in the US, but not on equivalent products manufactured elsewhere and sold in the United States.

Establishing national programmes to foster US competitiveness is a critical component of an overall national policy agends. Congressional passage of the National Cooperative Research Act of 1984 was a step in the right direction; establishment of the Malcolm Baldridge National

Quality Award in 1987 is further evidence of growing concern for US manufacturing capabilities. However, additional steps which address the significant challenges raised by Japan's Ministry of International Trade and Industry (MITI) and the passage of the Single European Act are necessary, as is a response to the increasing challenges of other industrializing nations who are also organizing groups of companies to compete globally.

The gauntlet has been thrown down by our foreign competitors. When or if the US Congress and the President assist US manufacturers in picking it up, and how they respond will have a tremendous impact on the future of world manufacturing as we now know it, and US aspirations for contending in world-class competition in the future.

7.6 CHALLENGE FOR THE FUTURE

The challenges facing manufacturers are clear. The ability to respond to the challenges both now and in the future will determine the continuing viability of US manufacturing. Lasting, sustained competitiveness cannot be created overnight, and will never be reached if manufacturers maintain a myopic focus on only some of the elements of the manufacturing equation. Manufacturers today must have clear vision and the strongest of commitments to fact up to the hard work and tough decisions required to become and remain world-class.

Our research has found that the elements necessary for becoming a world-class manufacturer are within the grasp of all competitors. Giants like GM, GE and Toyota have no more or less opportunity to achieve world-class capabilities that does the small ten million dollar a year parts manufacturer that serves as a link in the giant's supply chain.

Being world class is more about an approach to competition and knowledge creation than about company size or level of resources available. World-class companies realize they must approach manufacturing from a balanced perspective, placing equal emphasis on both the structural, infrastructural, and integration elements of strategy. They regard customers as the core of their existence. They invest judiciously in technology and in their employees, conscious not to become complacent about their capability advantages relative to their competitors. World-class companies are nimble, responsive and innovative. They closely monitor performance indicators in a continual search for incremental improvement, careful to reward all contributors to continued success.

The final and perhaps most important component of developing world-class capabilities is managerial leadership. World-class companies are managed by leaders who have a clear vision of the future, and who share this vision with all employees. These leaders, through their own convictions regarding manufacturing excellence, are the catalysts that created and now sustain world-class environments. Our research indicates that the drive of many US companies to become competitive at a world-class level has already begun, and that the 1990s are bright with promise for manufacturing in the US.

7.7 ACKNOWLEDGEMENT

Published with the permission of the National Center for Manufacturing Sciences and Business One Irwin. This paper was developed upon research sponsored by the National Center for Manufacturing Sciences, Ann Arbor, Michigan, USA under contract #NCMS-99-MO-1SOA, and summarizes key findings contained in Giffi, Roth and Seal (1990a).

7.8 REFERENCES

Borrus, Amy (1988) Beating America to the punch. *International Management*, **43**, pp. 31–3.

Boyd, Joseph A. (1986) International trade competition: The declining position of the United States and its effect on the American way of life, Presented at Competing in a Global Economy Seminar, Melbourne, Florida, December 11.

Buzzell, Robert D. and Gale, Bradley T. (1987) *The PIMS Principles*, The Free Press, New York.

Farley, John, U., Kahn, Barbara, Lehmann, Donald R. and Moore, William L. (1987) Modeling the choice to automate. *Sloan Management Review*, Winter, pp. 5–15.

Giffi, Craig A., Roth, Aleda V. and Seal, Gregory M. (1990a) *Competing in World Class Manufacturing: America's 21st Century Challenge*, Business One Irwin, Homewood, IL.

Giffi, Craig A., Roth, Aleda and Seal, Gregory M. (1990b) Survey shows priorities are quality, workers, and mature technologies. *CIM Review*, Fall, pp. 5–15.

Hall, Robert W. (1987) *Attaining Manufacturing Excellence*. Dow Jones-Irwin, Homewood, IL.

Hayes, Robert H., Wheelwright, Steven C. and Clark, Kim B. (1988) *Dynamic Manufacturing*, The Free Press, New York.

Huge, Ernest C. and Anderson, Alan D. *The Spirit of Manufacturing Excellence: An Executive's Guide to the New Mind Set*. Dow Jones-Irwin, Homewood IL.

Miller, Jeffery, G. and Roth, Aleda V. (1988) Manufacturing strategies: Executive summary of the 1988 North American Manufacturing Futures Survery. Research Report of the Boston University School of Management Manufacturing Roundtable.

Miller, Jeffery, G., Amano, Akio, DeMeyer, Arnoud, Ferdows, Kasra, Nakane, Jinichiro, and Roth, Aleda V. (1989) Closing the competitive gaps: International report of the Manufacturing Futures Project, in *Managing International Manufacturing* (ed. K. Ferdows), Elsevier, Amsterdam, pp. 153–68.

Roth, Aleda V. and Giffi, Craig A. (1991) *Manufacturing's secret arsenal for the 1990's: Deciphering world-class Japanese and North American manufacturing strategies*, Duke University Working Paper.

Roth, Aleda V. and Miller, Jeffery G. (1988) *Manufacturing Futures Factbook, The 1988 North American Manufacturing Futures Survey*, Boston University School of Management Manufacturing Roundtable.

Roth, Aleda V. and Miller, Jeffrey G. (1990) Manufacturing strategy, manufacturing strength, managerial success, and economic qutcomes, in *Manufacturing Strategy* (ed. J.E. Ettlie, M.C. Burnstein and A. Feigenbaum) Kluwer, Boston.

Schonberger, Richard, J. (1986) *World Class Manufacturing: The Lessons of Simplicity Applied*, The Free Press, New York.

Skinner, Wickham (1974) The focused factory. *Harvard Business Review*, May–June, pp. 113–21.

US Department of Defense, *Bolstering Defense Industrial Competitiveness*, Report to the Secretary of Defense by the Under-Secretary of Defense, July.

Vasilash, Gary S. (1987) Simultaneous Engineering. *Production*, July, pp. 36–41.

CHAPTER 8

A methodological framework for the formulation of an industrial strategy

Francis Jouffroy and Jean-Claude Tarondeau

8.1 INTRODUCTION

Corporate industrial strategy differs from strategy alone to the extent that it puts more emphasis on major technological choices, which deal with products or the industrial processes, be they internal or external, whose mission is to put those products at the customers' disposal. It complements a product–market analysis by taking into account not only design and manufacturing technologies, but also organizations and information systems which drive them.

Industrial strategy objectives can be expressed in terms of product design or quality (degree of adaption to the customers' needs, reliability, style, time to market, etc.), cost, quality of service, quick and on-time deliveries, and finally, adapting product supply to a varying and fluctuating demand and adjusting output to demand.

The main areas of industrial strategy concern value-adding activities (Porter, 1980) characteristic of the two main industry cycles: design cycle and manufacturing cycle. The first one regroups activities which contribute to the emergence of products, their improvement and transformation over a growth and maturity period, their revamping when decline looms ahead, to their disappearance as the product life cycle ends. The second one encompasses all the activities concerned with product manufacturing which aim at satisfying market demand. Both cycles are presented in Fig. 8.1, characterizing a French PLC manufacturer.

Specially commissioned for *Manufacturing Strategy – Process and content*.
Edited by Christopher A. Voss.
Published in 1992 by Chapman & Hall, London ISBN 0 412 43660 4.

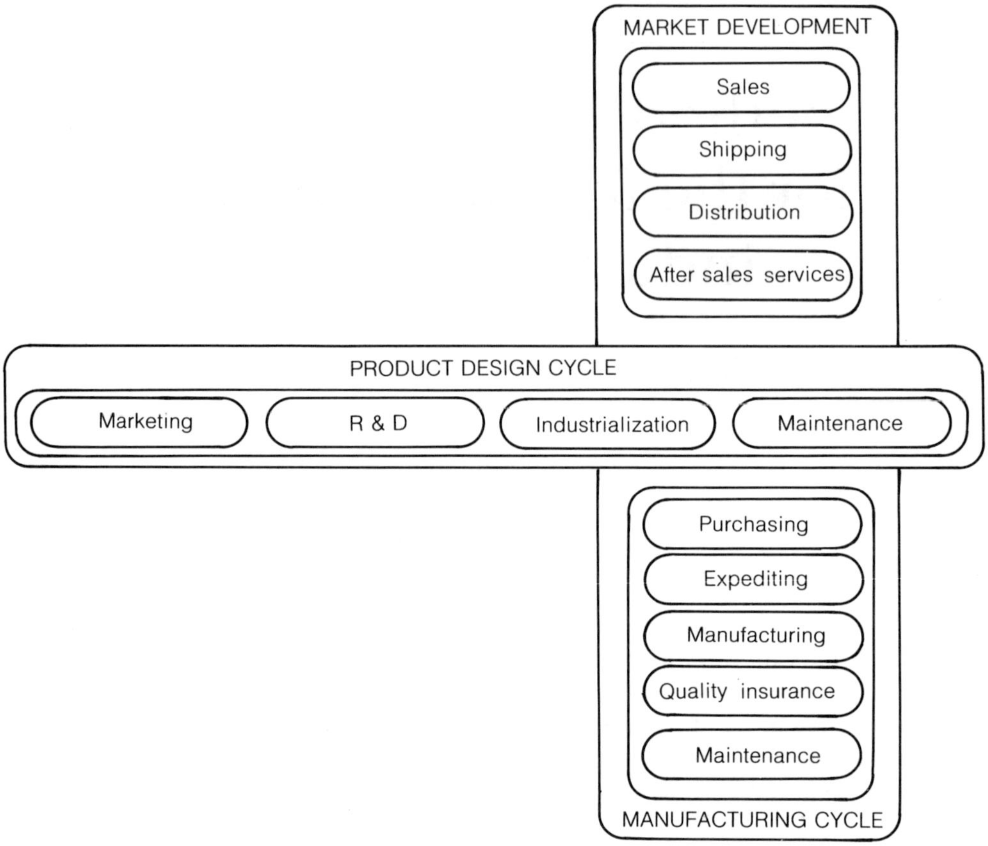

Fig. 8.1 Two industrial cycles, the April case.

The scope of industrial strategy having been defined, we will show briefly the evolution of thoughts and doctrines in this field. Explicit references to research works of specialists in strategy, manufacturing policy; innovation and organization will be made. It is well known that doctrines largely evolved under the influence of information technologies, and through the importing to the West, of managerial and industrial practices successfully implemented in Japan.

Industrial strategy appears today as an orderly and unified theoretical field. However, strategic formulation techniques, which contributed so widely to the popularity of strategic analysis, are still badly missing. The goal of this paper is to present a methodology which helps formulate an industrial strategy, and to illustrate its use in a French company manufacturing programmable controllers.

8.2 INDUSTRIAL STRATEGY: A CONSISTENT BODY OF KNOWLEDGE

8.2.1 A theoretical field getting progressively ordered

As a scientific-domain, organized body of knowledge and a specialized domain of action, industrial strategy first appears in the United States at the end of the 1960s, under the impulse of production management specialists who saw in the deficiencies of strategic thinking devoted to manufacturing policy an essential cause for the relative decline of the US manufacturing sector (Skinner, 1969).

Up until the 1960s, unidimensional manufacturing policy concepts prevailed. The manufacturer would see to it that production costs were minimized, that the best products were realized with the shortest lead times, and that the product range was adapted to meet the diversity of demand imposed by the market. Wickham Skinner (1969) was the first in pointing out that confusion of objectives, wrong economies of scale and the inefficiencies of local optimization. At a time when flexible technologies, just-in-time and total quality principles were still a long way off, he advocated defining a threefold industrial strategy:

1. Choose a set of objectives which are not only consistent between themselves but also with the company goals in terms of service to customers.
2. Design and choose industrial processes and technologies which match the aforementioned objectives.
3. Implement an organizational framework coupled to management systems which permits those processes to be mastered and the set objectives to be reached.

Skinner and his followers (Hayes and Wheelwright, 1979a and 1979b; Banks and Wheelwright, 1979; Hayes and Schmenner, 1978; Schmenner, 1976, 1978, 1979), while they recommended 'focusing' industrial systems, introduced new forms of rationality in the production world. A factory cannot be good at everything. It can, for instance, be excellent in costs terms but this will be detrimental to a capacity to adapt quickly to market fluctuations; or it could be outstanding as far as delivery times are concerned but suffer from over-capacity or costly stocks. Those were the arbitrage or trade-off doctrines (Skinner, 1974). A consistent set of objectives has to be chosen.

Those production objectives options should be in keeping with corporate global strategy (Jelinek and Burstein, 1982; Fine and Hax, 1985). The plant can become a competitive weapon provided it can contribute to establishing a lasting competitive edge.

Choices bearing on production processes are heavily correlated with product characteristics (Hayes and Wheelwright, 1979b). Rigid processes leading to small unit costs will be preferred when products are standardized and produced in mass quantities. Flexible processes will be given preference when products are very diversified and manufactured in small quantities. Thus, the necessary coherence between products, on the one hand, and industrial and logistic processes on the other hand, is clearly evidenced.

It belonged to Abernathy to show that products and processes evolve jointly all along the strategic segment life cycle. The search for an always lower cost, while rigidifying industrial processes, puts stumbling blocks on the product innovation path (Abernathy and Wayne, 1974). There lies the dilemma between productivity and innovation (Abernathy, 1978). This also gives evidence of strong relationships between production processes and product design processes. Not only should products and processes be consistent, but this consistency should be maintained over time.

Researchers in social sciences have shown that product, technology and organization are entangled in a network of strong relationships. Stable or little diversified products realized with fully mastered technologies will imply rigid organizations of a routine type (Woodward, 1965, 1970; Perrow, 1967; Burns and Stalker, 1971). On the contrary, diversified or even custom designed products with frequent update releases, which are realized using experimental technologies, will demand organizations endowed with large adaptation capacities. It stems from this that the corporation which manages a diversified product and technologies portfolio should segment its production system into homogenous or focused productive units. Hence, the search for coherence between products and technologies should spread across all layers of the organization, and to all industrial activities management modes. Therefore, industrial strategy is defined as a search for consistency between products, technologies (or industrial and logistic processes) and organization and management modes (or information systems), as Fig. 8.2 shows. Such a three-pole analysis is used to present generic industrial strategic options (Kotha and Orne, 1989).

Numerous analyses have demonstrated that the ability to market quickly new or revamped products constitutes a substantial competitive edge, if the developments, which we summarized briefly above, originated in the production domain. It is the quality of research, development and industrialization works, accomplished in the product design cycle, which, to a very large extent, determines costs, quality and lead times in the transformation cycle. It is in the design phase when, for instance, product structure choices are made. On the greater or lower degree of commonality between components, assemblies or sub-assemblies depend stocks levels and scale economies realized in purchasing, production and distribution domains. Similarly, a fast rhythm of product renewal will imply choosing flexible technologies

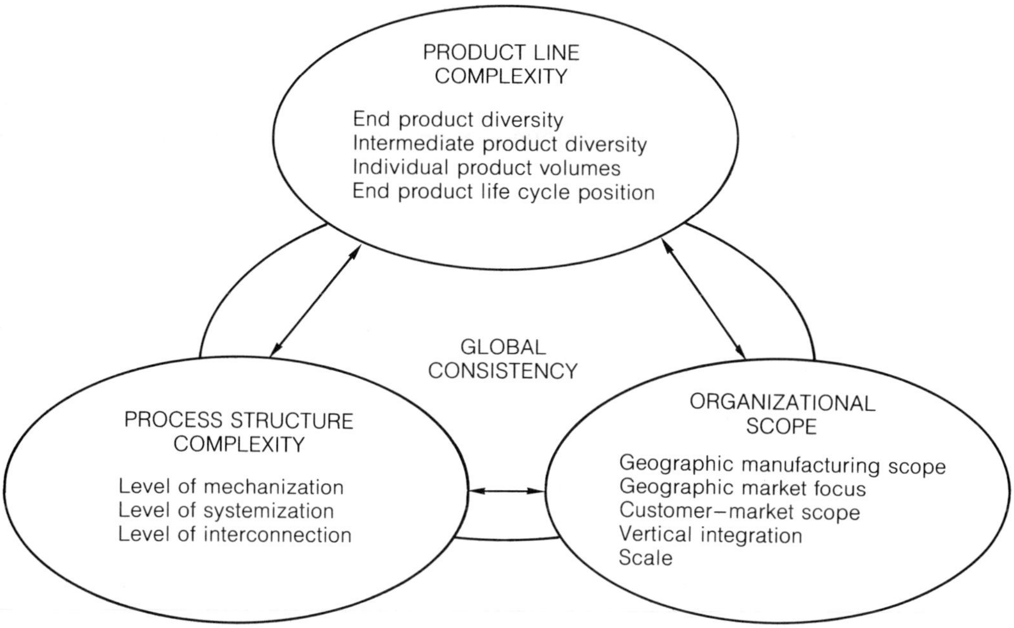

Fig. 8.2 Global consistency in an industrial system. Source: Adapted from Kotha and Orne (1989).

and will limit experience curve effects and the manufacturer's ability to bring costs down. Hence, analysing the manufacturing cycle independently from the design cycle, leads to partial conclusions as well as sub-optimizations. One should analyse them jointly.

Both cycles, which delimit the scope of industrial strategy, are made of activities which generate products by putting to work operating processes distributed across organizations and supervised by information and monitoring systems. Thus, they lend themselves to similar analytical methods.

8.2.2 An environment in mutation

The technological evolution and new, emerging methods (for Western people), introduced successfully by Japanese industrialists, lead us to question the main industrial strategy paradigms. Three major domains are concerned: emerging technologies, quality and material flow management.

Technologies based on electronics and computers (technologies regrouped under the concept of Computer Integrated Manufacturing) substitute the rigid automata which can only repeat endlessly the same program by the intelligent automata which can perform complex, varied and often modified operations. From now on, one can be both flexible and productive at the same time.

The power of data processing systems and their ability to be networked forces us to reconsider old rationales of work breakdown and work specialization. The fractionation of industrial systems and the search for local optima gives way to integration and globalization.

The status and role of quality are changing. So is its definition. Quality is no longer defined as product compliance to technical standards. Quality is the response to customers' expectations, it no longer only concerns products but also associated services, delivery times, administrative processes, and the like. We are now talking of total quality. This new concept makes necessary some doctrinal modifications. Whereas arbitrage theoreticians advocated a necessary trade-off between quality and cost, the new qualiticians demonstrate that total quality is a source of productivity by eliminating the numerous disruptions and costs which originate in non-quality.

The third domain is concerned with materials flow management, that is to say, with all operations (transformations, transports and storages) in the product manufacturing cycle as a whole, from the supplier to the customer. Stocks play a decisive role in industrial development. They disconnect the production system from the stochastic nature of market demand. They authorize scale economies through demand aggregation in economic quantities. On the other hand, they tie up financial means, increase lead times, diminish responsibility by providing operators with a false sense of security. Material handling was largely increased as industrial processes were split. Japanese industrialists will pave the way in this domain. By focusing on interoperations, they reduce process configuration costs at time of changeover between products, they aggregate operations, strive to decrease the specialization of equipment and operating personnel, and invent new material flow management methods where products are pulled by market demand rather than pushed by planning forecasts.

Economic quantities do not disappear, they thin down, sometimes to a single unit. Distribution and work-in-process inventories follow the same track. The duration of industrial cycles is shorter. Forecasts are no longer necessary. The quick adaptation capacity thus achieved limits the need for anticipation. The application of postponed differentiation principles, which consists of specifying products as late as possible, induces designers to find modular solutions far from the old value analysis principles.

These innovations helped new strategies as well as new industrial systems to emerge. Focus principles, which aim at stratifying industrial systems, are progressively replaced by unifying principles such as CIM, TQC and JIT (Tarondeau, 1982). At a superior level of excellence, an industrial system can be excellent in every respect.

8.3 A METHODOLOGY FOR THE FORMULATION OF AN INDUSTRIAL STRATEGY

Based on considerations explained in the first part of this paper we have designed a methodology to analyse a given industrial strategy, reach a diagnosis on projections of this strategy in a given time horizon and formulate new orientations.

8.3.1 A methodology in three steps

The methodology for an industrial strategy formulation involves three phases:

1. Analysis of the existing industrial system,
2. Diagnosis of the projected industrial system, and
3. Formulation of strategic orientations necessary to achieve the desirable future industrial system.

This methodology is presented in Fig. 8.3.

These three phases will be presented in an abstract manner to begin with. Then, we will complement and illustrate them through the case analysis of a French programmable controller manufacturer.

The first step consists of 'taking a photograph' of the existing industrial strategy. In order to do that, one will identify the objectives set on a long-term horizon corresponding to the irreversibility duration of main industrial strategic decisions. The company strengths and weaknesses will be listed so as to verify that the strategies contemplated are well justified, and to identify the potentials for change which may be mobilized to transform industrial strategies.

Most of the analysis will then bear, for each and every product-market segment, on industrial systems, namely on the three main areas of any industrial strategy: products offered on the market; industrial and logistic processes making it possible to put those products or services at the customer's disposal; and organizations and information systems used to create and manage those products, and animate industrial processes.

One will pay particular attention to what we call **structuring constraints**, namely those elements from the present situation, strategy and industrial system, which the company intends not to modify for the long-term horizon selected. In addition, one will identify the decisions already made which might, over various time horizons, modify the picture of the present situation. During that phase, the goal is to build a faithful and rigorous picture which will gain wide support across the company.

In the second phase, one will be concerned with building the so called **projected industrial system**, or the product–process–organization system as it should appear eventually. This projected system results from

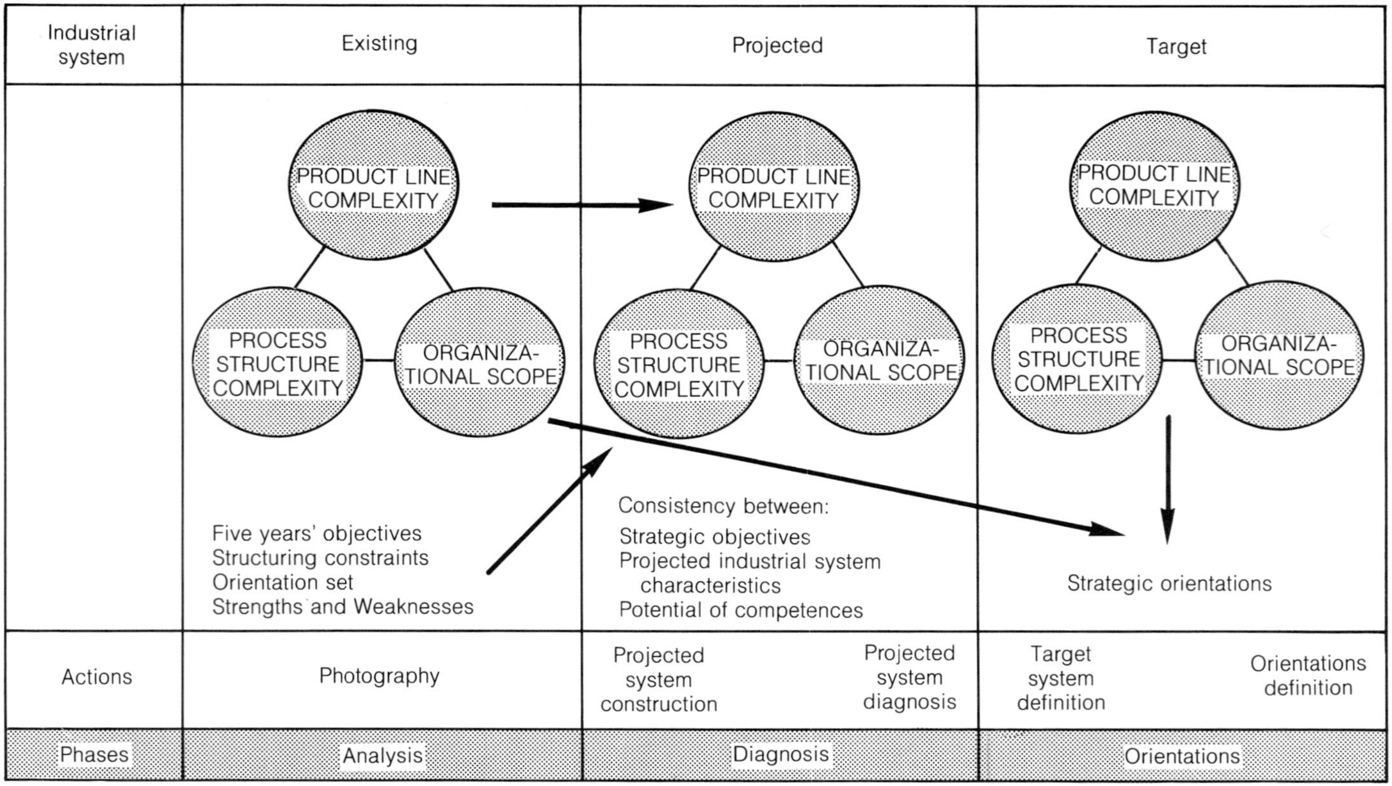

Fig. 8.3 A methodology for industrial strategy formulation.

the existing system distortions caused by the evolution of product–market segments, structuring constraints and already set orientations. The projected system is an abstract model of the future corporate industrial system provided no industrial strategy decision would come to modify the course of observable evolutions. This construct will serve as a base to make a diagnosis of corporate industrial strategy. Such a diagnosis will be obtained by searching possible inconsistencies between the company strategic objectives, the industrial system which should support them and the conditions of use of the available potential. A consensus is easily reached on this diagnosis since it does not imply any criticism of the existing industrial strategy and management.

The identification of inconsistencies between strategy and projected system, and their reduction, help put strong emphasis on the characteristics of the target industrial system which should be the result of the third step. By comparison between present and future situations, one will be able to define the main corporate industrial strategy orientations.

This approach was developed in a medium-sized French company to formulate new industrial strategies bearing on the product design cycle and the logistic cycle at the same time. We will now introduce the main elements of analysis for this latter cycle only.

8.4 AN APPLICATION OF THE CASE OF A PROGRAMMABLE LOGIC CONTROLLER MANUFACTURER

8.4.1 April: An emerging industrial identity

April was incorporated from the merger in June 1984, of SMC, a company which was a subsidiary of Renault, and the Merlin Gerin programmable controller department. At that time, the Renault company which owned SMC through its subsidiary Renault Automation, held 55% in capital while Merlin Gerin held the remaining 45%. In 1987, Renault sold its share to Merlin Gerin which became sole owner.

The merger, which a leading consulting firm had advised to achieve volume effects, made April, in 1987, the leading French PLC manufacturer with approximately 40% of the market share, ahead of La Télémécanique (35%), Siemens (10 to 12%), CGE (8 to 10%) and Allen Bradley (4 to 5%). Siemens is the leader in Europe where it is approximately three times as big as April. Allen Bradley is the world leader and is five times as big as April.

However, April remains a company split on two industrial sites: Castres, a former APSMC location, where a new factory was built in 1983; and Grenoble, a former Merlin Gerin site, where the company settled in its own building in 1986. Each site develops, maintains and manufactures two largely competing product lines. They also have to manage the smooth integration of two strong corporate cultures.

Since 1985, April's market share has tended to level down and new actions will have to be undertaken to remedy that situation. Those actions consist of:

1. Introducing a new generation of controllers which would be substituted for the two old product ranges of SMC and Merlin Gerin which have been marketed simultaneously since the merger;
2. Preparing the release of low- and very low-priced ranges of controllers to ensure a more comprehensive coverage of the PLC market;
3. Looking at, and possibly striking alliances with, other manufacturers, be they in Europe or in Asia;
4. Rethinking April's industrial strategy by taking those developments into account on the one hand, and by the construction of a new industrial site in Grenoble on the other; this new site should offer the opportunity to redefine the breakdown of industrial activities between the Castres and Grenoble sites; and
5. Implementing new industrial strategies permitting the securing of a lasting competitive edge in the French and European markets.

This was the context when we started an action to reformulate the group industrial strategy.

8.4.2 A picture of the initial situation, the first step of industrial strategy analysis

(a) April's products–markets strategy and five-year objectives

The two current product lines were to be replaced by a unique one in 1989. To this unique medium-to-high priced range of PLCs, a low-priced PLC will be added later on. For the coming five years, April remains a designer, manufacturer and seller/distributor, whose products range grows wider to cover the market fully. New products, which April can only distribute, should foster growth.

The PLC market being in the maturity stage, April expects a slow corporate growth for the five coming years.

(b) Structuring constraints

Those constraints are orientations that the company does not intend to question on the five-year horizon which initially was selected for this analysis. They bear, in this case, on products, industrial processes and corporate culture. Future products will have structures similar to the current products. A PLC is made of a rack, a power supply, a central unit, input-output and functional cards. The demand structure under this product composition should not change significantly. Industrial processes can evolve along the lines of the following constraints: there will be no vertical integration; and both industrial sites (Castres and Grenoble) will be kept, with no significant personnel transferring between them. Every recommendation should take April's system of values into account:

1. Very strong emphasis should be put on the search for a high participation level, autonomy, responsibility and motivation for the entire personnel.
2. There should be great concern for rigour in procedures implementation. Those procedures should be clear, easily understandable, regularly evaluated and reassessed.
3. Search for global optimization instead of trying to solve narrow problems whose side effects are not evaluated.

(c) The existing industrial system

Products are built from electronic components, integrated circuits and packaging elements bought on the market. There are approximately 3000 such elementary components in the two existing product lines. Some of them are common to several electronic cards and programmable controllers. Prices are negotiated with suppliers on the basis of one-year provisional consumption. Delivery times on supplied goods vary anywhere from three to seven months. Those elementary products are purchased on a one-year ahead forecast basis and stocked at a components storage location. According to short-term anticipations, the plant launches sub-assemblies (electronic cards for the most part) to production. Those cards are subcontracted to specialized firms. Cards are tested and debugged and burned in a drying oven when they come back before they are sent to a finished products stock. When the customer's order comes in, the PLC, an assembly of cards and packaging elements, is configured, tested, taken to pieces and shipped. There are approximately 200 such finished products. They serve in two separate PLC lines whose configurations are specific for each customer. This industrial process is shown in Fig. 8.4 whereas the product demand structure is presented in Fig. 8.5.

Each plant is fully responsible for the entirety of the industrial functions dealing with its own product line. Functions such as marketing, research and development, distribution–sales and administration are centralized.

The industrial management system is of an MRP type. It determines the net requirements for components and finished products according to

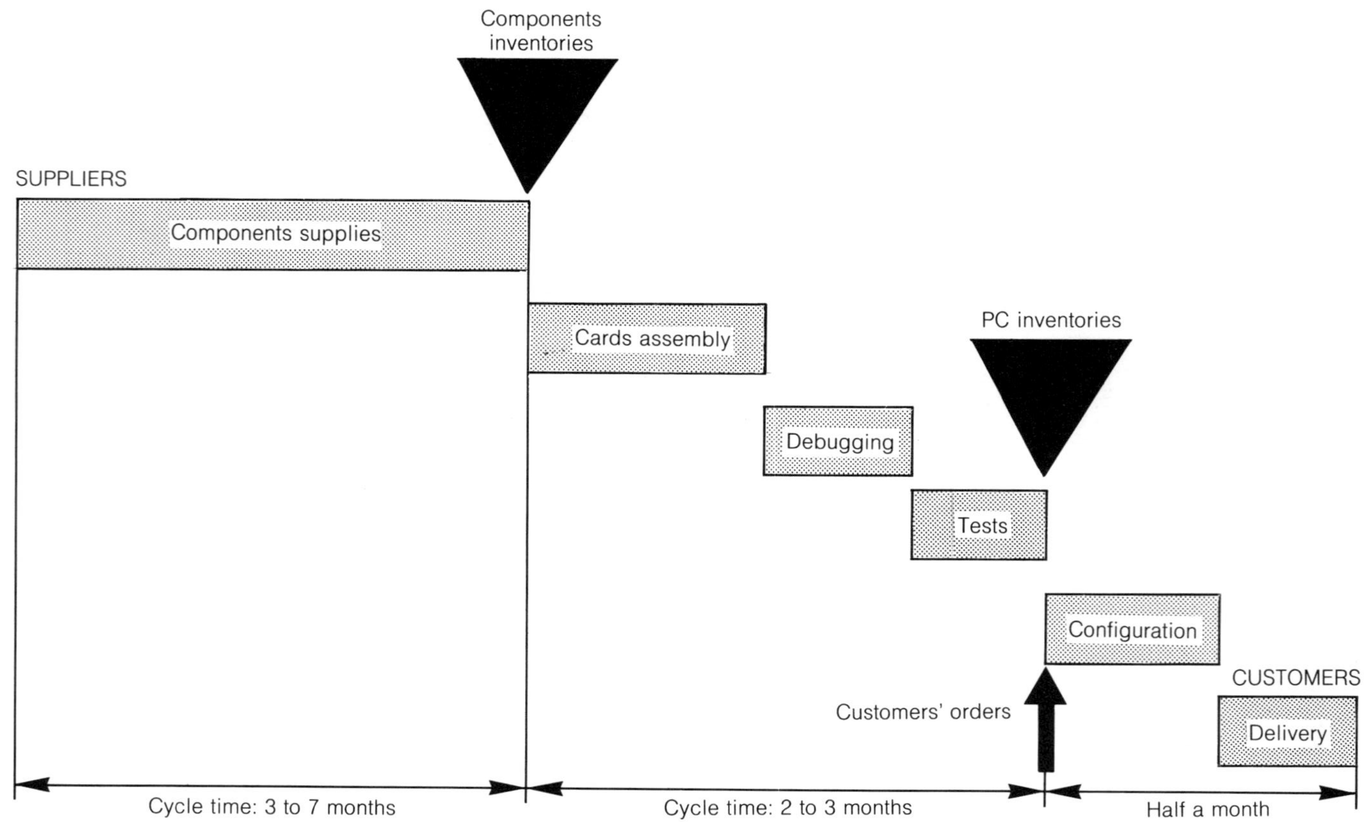

Fig. 8.4 The existing industrial and logistic process.

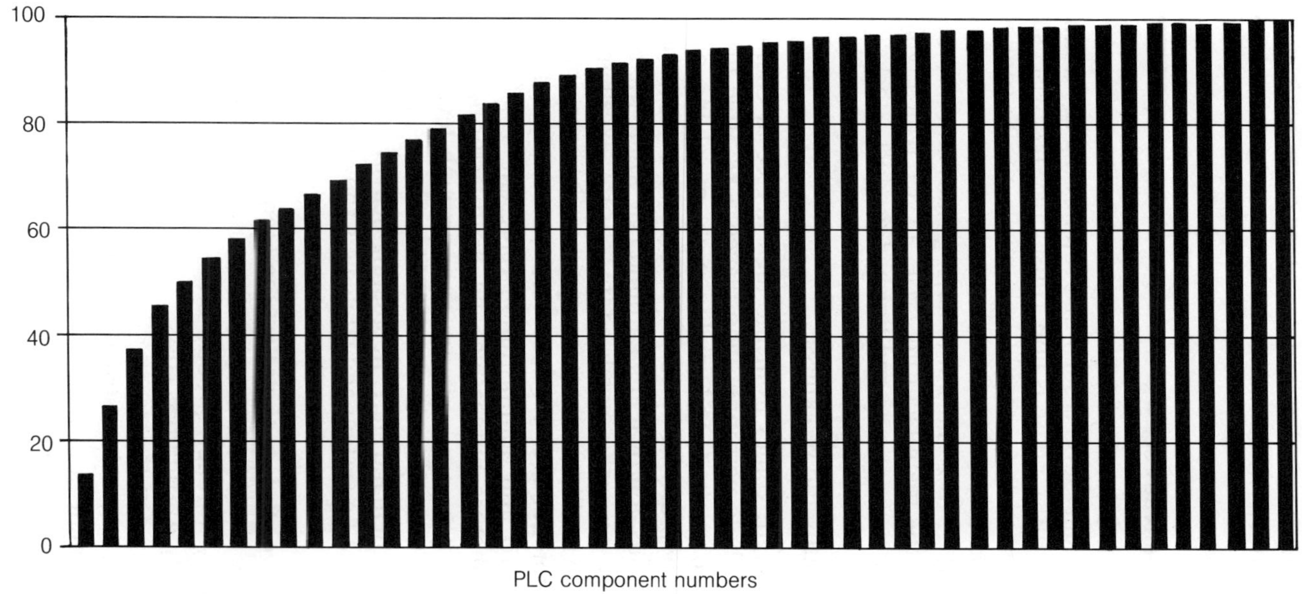

Fig. 8.5 Cumulative demand (%).

sales forecasts computed in terms of generic PLCs. It manages stocks and spots missing products when launching card production or configuring PLCs.

(d) Orientations already set

April is engaged in a total quality programme which largely has born fruit. The cards' quality level is such that suppressing the testing of cards which are in a maturity phase is now considered. Likewise, suppression of tests on configured PLCs should help avoid on-site configuring and expedite PLCs in kit form.

A project called 'A to Z' aims at enlarging skills and jobs of operators who intervene in the manufacturing cycle, a cycle which typically takes place between the components and finished products stocks.

(e) April's strengths and weaknesses

April's main strength comes from the high degree of commitment on the part of the whole personnel who played a key role in the quality plan success. The analysis and quality problem-solving processes are known and widely practised by all employees. Decisions based on group work and consensus approval are a ground rule inside the organization. The company is correctly mobilized to ensure excellent service to customers. However, little attention is paid to delivery times problems, whereas large potentialities for lead time and stock reduction clearly appear. Sales, R&D and administration represent 65% of the cost of products sold, whereas purchasing and manufacturing variable costs account for 25%. The cost structure shows a high level of fixed costs compared to tiny variable costs. This cost structure makes April very vulnerable in case demand starts fluctuating heavily. A summary of this analysis of the existing system and strategy is shown below.

1. April strategy:
   - (i) A defensive strategy to protect the French market share,
   - (ii) A strategy based on service quality and readiness,
   - (iii) A necessity to renew product lines,
   - (iv) A necessity to compress costs.
2. April structuring constraints:
   - (i) No change in product structure in demand behaviour,
   - (ii) No investment in in-house production processes,
   - (iii) Two industrial sites to be maintained without significant transfer between them.
3. April orientations already set:
   - (i) A total quality control programme bearing impressive results,
   - (ii) A programme aimed at developing enlarged skills for operators in the industrial cycle.

4. April strengths and weaknesses:
   (i) High degree of commitment from all personnel,
   (ii) Team work and consensual decision making,
   (iii) Low attention paid to costs and deliveries.

8.4.3 Projected system and diagnosis, the second step of industrial strategic analysis

The construction of the projected industrial system is done as a function of the existing system, and its structuring constraints as well as the orientations already set.

The projected industrial system does not change scale. Weaknesses associated with a high level of fixed costs are not eliminated. More generally, high product costs which limit April's potential for growth are not brought down, since beside fixed costs, purchasing expenses could be cut down only from a better bargaining position.

The target industrial system features a single product line replacing the two lines which served as a foundation for the specialization of industrial activities on each site. This specialization being questioned and both sites being maintained, it is necessary to define a new logic for industrial specialization. This logic should take into account the elimination of some industrial tests and configuration operations on products which have reached maturity. This logic should integrate activities linked to new products, which will have to be stabilized, and to obsolescent products which have to be maintained.

A new distribution of industrial activities will involve organizational changes and will challenge the rationale for double purchasing, supply, inventory and shipping systems. The information system will be impacted on by those organizational changes.

It is fairly clear that already set orientations will not meet the requirements for consistency between corporate objectives and industrial system characteristics. New industrial strategy orientations have to be defined. That definition will be made with reference to a **future system** which is yet to be invented.

Projected industrial system diagnosis:

1. A single product line (made of three PLC types) instead of two:
   (i) Necessity to define a new specialization logic,
   (ii) Necessity to define new roles for the two sites.
2. A very high level of product quality:
   (i) Opportunity to suppress product testing,
   (ii) Opportunity to define new partnership rules with suppliers and subcontractors.
3. A reduction of the manufacturing cycle time:
   (i) Opportunity to improve customer services,

   (ii) Opportunity to reduce lead times and inventories.
4. No change in scale of operations:
   (i) No economies of scale.

8.4.4 April's future industrial system, the third step of industrial strategy analysis

The diagnosis made on the target system pointed out some deficiencies or inconsistencies which the future system should try to remedy. In the April case, the future system should be designed so that fixed costs and product costs are cut down. The system should offer a coherent solution in terms of distribution of industrial activities between sites and within the organization.

(a) Orientations for costs reduction

Fixed costs reduction deals essentially with the commercial and R&D functions. The second function seems to have the critical size required to master successfully the numerous technologies in the PLC business. As far as sales is concerned, leverage means should be considered by resorting to specialized distributors. Low-priced products are frequently distributed this way. For more complex products, an increased distributor involvement implies: designing products accordingly; defining support procedures associated with delivered products; and breaking down tasks between manufacturer and distributor.

Product cost reduction must focus on the elimination of no-value-added operations. This action has already started for test and control operations, PLC configuration tests, and assembly and de-assembly operations that it implies. Potential improvement areas are at the purchasing and inventory levels.

An improvement in purchasing conditions, in a situation where global volume is remaining approximately constant, can be achieved by reducing purchased products diversity. To do this, one will look for ways to increase component commonality thanks to better standardization. The new range of PLCs should be made of 500 to 600 components instead of 3000 in the old range. An increase in purchased quantities together with a reduction in the number of articles managed by the purchasing department, will help bring down purchased components costs. In addition, this policy authorizes diminishing the size of component stocks. Other things being equal, a simulation was done which shows that stocks should go down by 55% when the number of components uses is multiplied by five.

Actions undertaken to increase the versatility and responsibility of operating personnel (manufacturing cycles taking place between components and finished products stocks) led to significant reduction in lead times. When trying to generalize those cycle time reduction actions (lot size diminution, operations grouping, queue waiting reduction, diminution of the duration of administrative operations, and development of intense business relationships with subcontractors) it is conceivable to cut down the manufacturing cycle, from two to three months, (today's duration) to two weeks in the future industrial process whose configuration is presented in Fig. 8.6. As lead times become compatible with lead times of customers' needs, it is possible to assemble the PLC directly from the components in stock, according to the customers' orders. It results from there:

1. Elimination of finished products' inventories,
2. Forecast horizon reduction and forecast quality improvement, and
3. Use of forecasts to purchase components whose commonality reduces forecasts' uncertainty.

Compared to the existing system, the future system will permit a 75% stock cut while maintaining an equivalent service level.

(b) Industrial activities' breakdown

An analysis of demand volumes by product reveals substantial differences between PLC ranges on the one hand, and between elements included in those PLCs on the other hand. Old PLCs continue to be

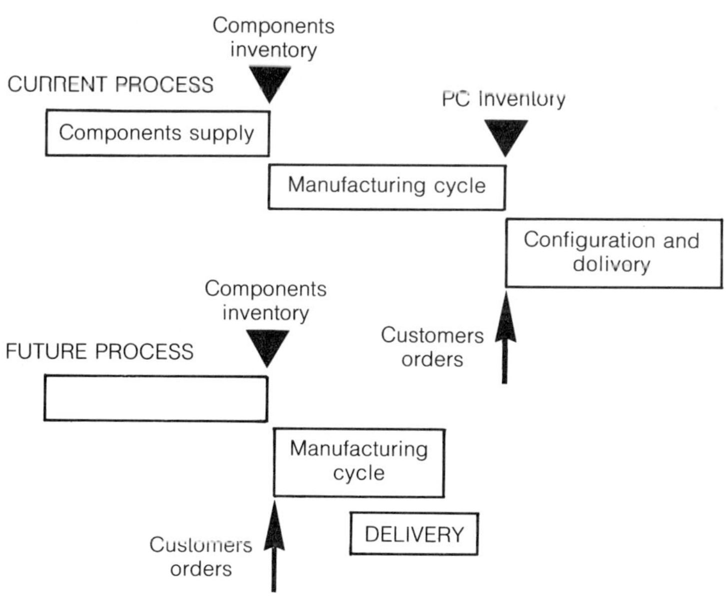

Fig. 8.6 April product–process orientations.

asked for for a long time after they have been withdrawn from the catalogue, during all their life cycle at the customer's. Some PLC elements in the catalogue are sold in large quantities; this is also the case of input–output cards, whose number, per PLC, can reach several dozens. Other components are only for specific and exceptional applications. In the future system, industrial activities will be focused according to this volume criterion.

Focus orientations based on variety and volume criteria:

1. *Grenoble*:
   (i) Design cycle:
      (a) Specific applications product design,
      (b) Maintenance of products nearing the end of their life cycle.
   (ii) Manufacturing cycle:
      (a) Low-volume components manufacturing,
      (b) Specific cards,
      (c) PLC components reaching the end of their life cycle.

2. *Castres*:
   (i) Design cycle:
      (a) Standard applications product design,
   (ii) Manufacturing cycle:
      (a) Large-volume components manufacturing,
      (b) Regrouping and distribution of all products.

Thus, the future system is adapted to April's strategy to which it contributes to better performance. It is capitalizing on corporate strengths mobilized around industrial cycles' reduction. It obeys a logic of focus on volume and product life cycle. This new industrial strategy implementation plan is logically derived from a set of planned actions necessary to the conversion of the current industrial system into the future system.

8.5 CONCLUSION

This industrial strategy formulation method contains numerous advantages. It integrates the company's strategic objectives, takes its capacities and particular competences into account, offers a conceptual framework for analysis, ensures consistency between products–industrial processes–organization and information systems, integrates the time dimension and helps build methodically an action plan.

8.6 REFERENCES

Abernathy, W.J. and Wayne, K. (1974) Limits of the learning curve. *Harvard Business Review*, September–October, 109–19.

Abernathy, W.J. (1978) *The Productivity Dilemma, Roadblock to Innovation in the Automobile Industry,* Johns Hopkins University Press, Baltimore.

Banks, R.L. and Wheelwright, S.C. (1979) Operations versus strategy – trading tomorrow for today. *Harvard Business Review,* May–June, 112–20.

Burns, T. and Stalker, G.M. (1971) *The Management of Innovation,* Tavistock, London.

Fine, C.H. and Hax, A.C. (1985) Manufacturing strategy: a methodology and an illustration. *Interfaces,* **15** (6).

Hayes, R.H. and Schmenner, R.W. (1978) How you should organize manufacturing. *Harvard Business Review,* January–February, 105–18.

Hayes, R.H. and Wheelwright, S.C. (1979a) Link manufacturing process and product life cycle. *Harvard Business Review,* January–February.

Hayes, R.H. and Wheelwright, S.C. (1979b) The dynamic of process product life cycles. *Harvard Business Review,* March–April, 127–36.

Jelinek, M. and Burstein, M.C. (1982) The production administrative structure: A paradigm for strategic fit. *Academic of Management Review,* **7** (2), 242–52.

Kotha, S. and Orne, D. (1989) Generic manufacturing strategies: A conceptual synthesis. *Strategic Management Journal,* **10**, 211–31.

Perrow, C. (1967) A framework for the comparative analysis of organizations. *American Sociological Review,* **32**, 194–98.

Porter, M.E. (1984) *Competitive Strategy: Techniques for Analysing Industries and Competitors,* Free Press, New York.

Schmenner, R.W. (1976) Before you build a big factory. *Harvard Business Review,* July–August, 100–4.

Schmenner, R.W. (1978) *Revisiting the focused factory,* Harvard Business School, Working paper No. 78–25.

Schmenner, R.W. (1979) Look beyond the obvious plant location. *Harvard Business Review,* January–February, 126–32.

Skinner, W. (1969) Manufacturing – missing link in corporate strategy. *Harvard Business Review,* May–June.

Skinner, W. (1974) The focused factory. *Harvard Business Review,* May–June.

Tarondeau, J.C. (1982) *Produits et Technologies, Choix Politiques de l'Entreprise Industrielle,* Dalloz, Paris.

Woodward, J. (1965) *Industrial Organization – Theory and Practice,* Oxford University Press, Oxford.

Woodward, J. (1970) *Industrial Organization – Behaviour and Control,* Oxford University Press, Oxford.

PART THREE

The Industry and International Context

Those familiar with the development of the field of Manufacturing Strategy will know that the early study and teaching took place in the context of industry-based studies. Any manufacturing firm operates in the context its own industry, the technologies available, the economics, and the different product and business policies of the competitors in that industry. These approaches have now been taken forward considerably by business school economists, and have been widely popularized through the work of people such as Porter. The central role that the automotive industry has played in the development of new approaches to manufacture and to the world economy, has led to a revival in the use of industry-based approaches in manufacturing strategy. In this book this approach is illustrated by the work of Dan Jones and his colleagues. They have focused on the world automotive industry. Their results, described in this part of the book, have been based on the work of a large multi-functional, multinational team.

The emergence of the Japanese as the major force in manufacturing during the 1980s has led to a rapid realization that there are major differences between the manufacturing strategies pursued by companies in different countries. In addition, the context in which a manufacturing strategy will be set can differ widely from country to country. The paper by Dan Jones, as well as being an industry study, illustrates the different approaches pursued by the Japanese. On the same theme, but at the other extreme, Attila Chikan and Krisztina Demeter discuss the changing political and economic environment in Eastern Europe and its impact on manufacturing strategies. Arnoud de Meyer and his colleagues in France have conducted regular surveys of the particular manufacturing strategies being pursued by European firms. The results of his most recent survey are presented and discussed.

Finally, an increasing number of companies do not operate in their home country but have to develop manufacturing strategies that cover the additional issues of manufacturing focus and configuration on a global scale. Frank DuBois and Mike Oliff have a major project looking at this area and their paper reports their latest work.

CHAPTER 9

Beyond the Toyota production system: the era of lean production

Daniel T. Jones

9.1 THE INTERNATIONAL MOTOR VEHICLE PROGRAMME

At the end of the MIT Future of the Automobile Programme in 1984 a number of us who had written up the final results were convinced that the competitive advantage of the leading Japanese auto producers derived from their superior organization of the production process (Altshuler *et al.*, 1984). Like others such as Schonberger (1982) we were convinced that the Toyota Production System, and derivatives of it, lay at the heart of this competitive advantage and moreover that it would be possible to transfer it abroad. We were also convinced that this heralded a more widespread change in production organization that would be felt throughout the economies of the west, in the same way that mass production caused such a stir in Europe earlier this century.

However, it proved very difficult to convince the sceptics, with only scattered anecdotal evidence and at that time only a limited Japanese auto manufacturing presence abroad. Even macro-level comparisons of the $2000 cost differential between a car imported from Japan versus one emerging from a US factory were not enough. For some time this led to arguments about levelling the playing field, unfair advantages deriving from unique cultural circumstances in Japan and an exchange rate unduly favouring the Japanese. As long as the competitive threat was from across the ocean people were preoccupied with finding excuses for the gap rather than facing up to realities.

Specially commissioned for *Manufacturing Strategy – Process and content*.
Edited by Christopher A. Voss.
Published in 1992 by Chapman & Hall, London ISBN 0 412 43660 4.

We therefore determined that the only way to demonstrate the full significance of such an important change in manufacturing organization was to develop techniques for systematically measuring the performance gap between Western and Japanese firms and analysing the factors responsible. This led to the MIT International Motor Vehicle Programme – possibly the largest global industry study ever undertaken – a five-year, fourteen-country research programme sponsored to the tune of $5m by almost all the auto producers and many governments around the world.

During the programme a number of studies were carried out into manufacturing performance, product development, component supply structures, automotive distribution and product strategy. The book summarizing the final results of this work, combined with an evaluation of the strategic choices facing firms and governments in a rapidly changing multi-regional world, is now in press (Womack, Jones and Roos, 1990). However, the detailed results of each study are also available from MIT in nearly 100 IMVP research papers.[1] This paper develops some of the arguments made in the book.

9.2 A DECADE OF TECHNIQUES

The 1980s might in retrospect be called the decade of techniques – dozens of new techniques and buzz words were enthusiastically advocated, many of Japanese origin, each offering the key to step changes in manufacturing performance. Understandably many grew tired of hearing about and trying to implement a succession of these new techniques, particularly when they did not work as well as they had been led to believe. They often seemed to fail when implemented in isolation.

In many cases there was nothing wrong with the techniques, but they turned out to be only parts of a much larger, more fundamental change in the organization of manufacturing. When one looked for the origins of the ideas behind most of these techniques the authors had almost invariably drawn on the work of Taiichi Ohno, who as the production chief at Toyota was the architect of the Toyota Production System. In fact Taiichi Ohno, who died very recently, may turn out to be one of the most significant figures of the post-war world.

Perhaps it was inevitable that we in the West had first to take this system to pieces in order to understand how it worked and to see whether it would work away from its Japanese context. However now we understand the pieces and have seen many of them successfully implemented in the West perhaps it is time to see how they fit together

[1] A full list of IMVP Papers is available from the IMVP Office, Center for Technology, Policy and Industrial Development, Building E-40, MIT, Cambridge MA02139, USA, tel. (1) 617 253 8540, Fax (1) 617 253 7140.

again – in order to appreciate the full significance of this completely different production system.

Like many others our research began in manufacturing itself, comparing the performance of a large sample of assembly plants around the world. However as we began to look at product development, product strategy, supply systems and distribution it became clear that these pieces are intimately linked and organized on similar principles. Indeed it is their interaction and integration that lies behind superior performance. Integration may well become one of the themes of this decade.

It was precisely the opposite that lay behind the organization of mass production – differentiation of tasks on the shop floor and differentiation of functions in management. By achieving complete and consistent interchangeability of parts, making them simple to put together while eliminating fitting, minutely subdividing tasks and introducing the moving assembly line Henry Ford ushered in a new era of mass production in 1913, that was orders of magnitude more productive than the previous era of craft production (Hounshell, 1984). However, it was left to Alfred Sloan to add the management superstructure, functional specialization and marketing system necessary to run such a huge undertaking (Sloan, 1964; Chandler, 1962). Only then was the system of mass production that we are all familiar with complete. That system has now come to the end of the road – the era of mass production is now over.

We have come to call the system that is now sweeping the world **lean production**. Why? Why not just call it the Toyota Production System; and why lean? First, in the same way as Ford's system was not complete without Sloan's contribution, we have a hunch that Ohno's system will not be complete until it has been implanted across the globe. It will take another decade before Honda, Toyota and others have full top-to-bottom design and production systems operating in both North America and Europe.

Second, the system is best described as lean because as we shall see it uses less of everything – half the time and effort to design the product and half the human effort and tooling to make it, with half the defects and less than half the inventories. As a result the lean production system can offer the customer twice the number of products that only need to be built in half the normal volume per model.

In fact, we found this two-to-one difference in performance between mass production and lean production in every activity – from R & D through design, manufacturing, component supply and distribution – and in every type of activity from the organization of project teams, through the job shop production of body dies to volume production. We also found this two-to-one gap in activities such as body welding which is highly automated and assembly which at present is not.

So what exactly is lean production? Let's first look at the evidence before trying to summarize its characteristics.

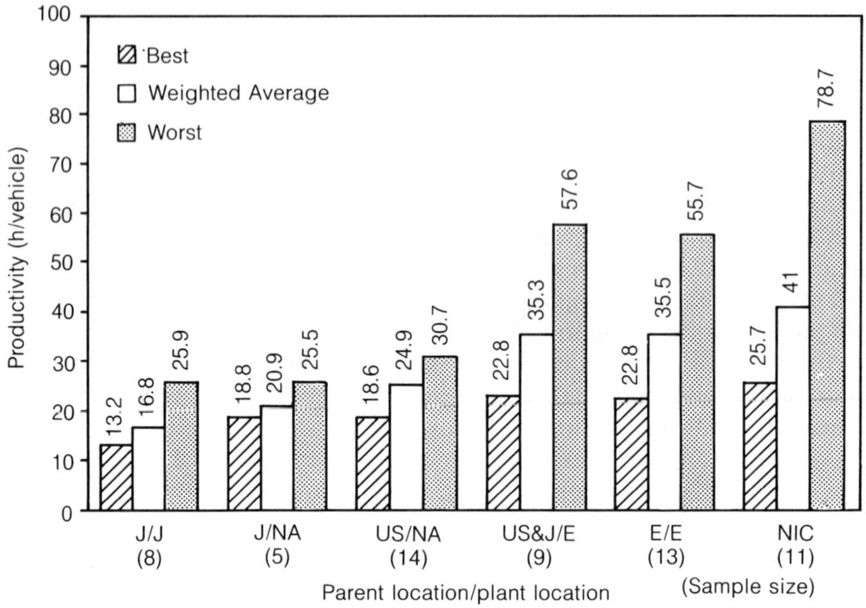

Fig. 9.1 Assembly plant productivity, 1989. This does not include the European specialist producers. J = Japan, NA = North America, E = Europe, NIC = Mexico, Brazil, Taiwan and South Korea. Sources: IMVP (1989) and the J.D. Powers Initial Quality Survey, 1989.

9.3 MANUFACTURING

Our assembly plant survey began with a detailed study of the General Motors/Toyota NUMMI plant in California by a former NUMMI engineer, John Krafcik (1986), who compared it with a traditionally run GM plant and its Toyota sister plant in Japan. This revealed that NUMMI had nearly matched the productivity and was equalling the quality of the Toyota Takaoka plant, while the GM plant required one third more hours and delivered three times the number of defects per car.

Gradually this survey was broadened to include over 80 plants in 15 different countries (Krafcik and McDuffie, 1989). Plant visits were followed up by detailed questionnaires covering all aspects of the operation of the plant. The data were reviewed with plant management, who were also told how they performed against regional averages. To ensure they were 'comparing apples with apples' the data were adjusted for a standard set of activities, a constant level of vertical integration, a standard-sized product with a standard level of options, and made allowances for differences in relief systems and absenteeism.

The results in terms of productivity and quality are shown in Figs. 9.1 and 9.2. A number of points stand out from these comparisons:

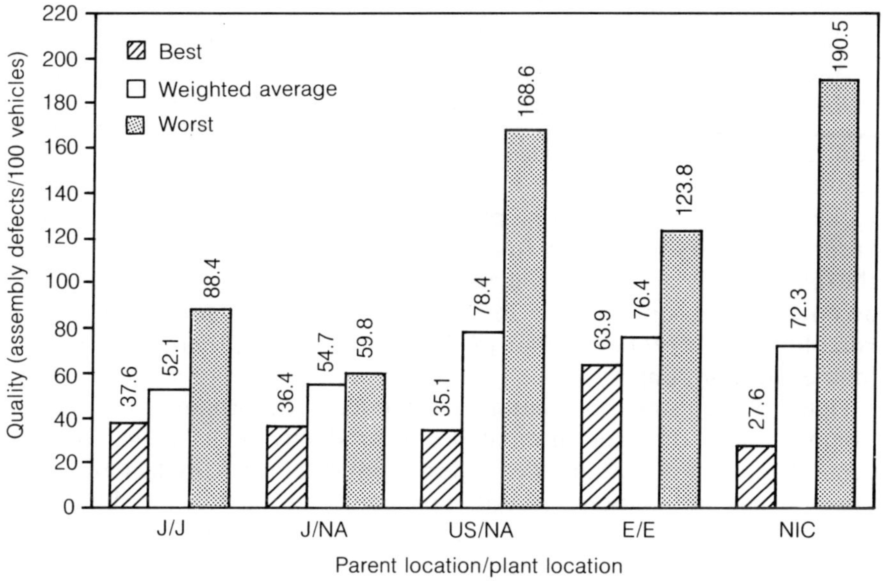

Fig. 9.2 Assembly plant quality, 1989. This does not include the European specialist producers and refers only to assembly defects. J = Japan, NA = North America, E = Europe, NIC = Mexico, Brazil, Taiwan and South Korea. Sources: IMVP (1989) and the J.D. Powers Initial Quality Survey, 1989.

1. There is a two-to-one gap between the best and worst plant in Japan – not all Japanese producers are lean.
2. The Japanese have almost matched the productivity and have equalled the quality of their domestic plants in their transplant facilities in North America – it can be done abroad with a local workforce.
3. Our research indicates that the US average was probably over 30 hours per car at the beginning of the 1980s and that under pressure from the transplants some US plants have made impressive improvements in productivity – now equalling the performance of the transplants – it can be done in older plants as well as on green field sites.
4. On average, the Europeans still take over twice the number of hours as the Japanese to assemble a car, with only a handful of plants within striking distance of the Japanese – without the demonstration effect of local transplants progress towards closing the gap is more difficult.
5. A few plants in Mexico and South Korea have also made impressive progress, particularly where they have a close association with a Japanese lean producer – it is possible to transfer lean production to any location.

6. There appears to be no correlation between productivity and quality, however, almost all the Japanese plants achieve both high productivity and high quality – indicating that there is no tradeoff between them for lean producers, indeed they go hand-in-hand.

Krafcik and McDuffie carried out detailed statistical analysis of the results to find the reasons for these performance differentials. They looked at five possible factors, plant scale, build complexity, the level of automation, product design (manufacturability) and the organization of the plant. They found that plant scale and build complexity did not account for the gap, and that in fact many Japanese plants build a more complex mix of products than in the west – making their performance look even more impressive.

Figure 9.3 shows the relationship between the proportion of direct assembly steps that are automated and productivity. It is clear that in general more automation means less human effort – in fact they estimated that automation accounts for about one third of the difference in productivity. However the level of automation is not a guarantee of less effort being required, as shown by the huge difference between the best and worst plant for any given level of automation.

The level of automation is of course only part of the story. A lot depends on the type of automation and how it is used. We found some

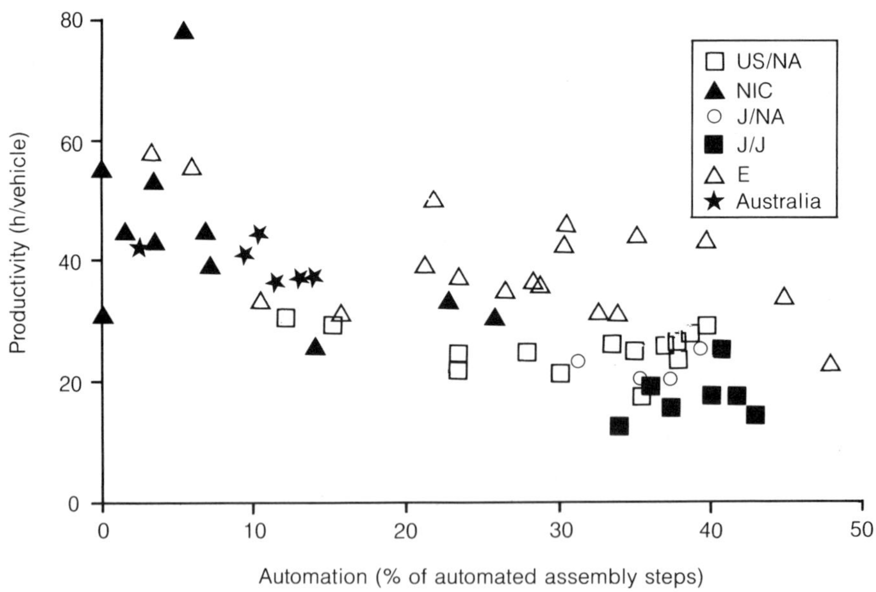

Fig. 9.3 Productivity versus automation, 1989. The percentage of automated assembly steps excludes materials handling. Reproduced with permission from IMVP (1989).

unimpressive results from highly automated showpiece plants in the West, using both dedicated and flexible (AGV-robotic) equipment. It is well known that GM's dream of closing the gap with the Japanese through high technology in its GM-10 plants has been a dismal failure (Keller, 1989). In fact those US plants that have closed the gap with the transplants did so without a high level of automation, whereas the high-technology plants still perform poorly.

Krafcik found the same contrast as other studies – by Jaikumar and Tidd of FMS and Robotic assembly systems – between the Japanese and Western approaches to automation Krafcik, 1989; Jaikumar, 1986; Tidd, 1989). The Japanese start by automating the simple tasks, not the most complex, and use what they call 'featherlight' equipment, built in-house specifically for that task, rather than expensive general-purpose equipment purchased from outside vendors. The equipment is introduced with a high level of shop-floor involvement and once debugged is rapidly adopted for all similar functions within the firm's plants. Western firms tend to wait until the introduction of a new model to install new tooling – compounding the difficulties in doing so. The Japanese manage to achieve a much higher machine utilization and build a more complex mix of products, making full use of the rapid tool change and flexibility built into the equipment.

Product manufacturability is difficult to measure precisely. Krafcik (1990) conducted a survey of the tear-down departments of auto producers around the globe, asking how they ranked the products of their competitors in terms of manufacturability. This resulted in a surprisingly consistent ranking of firms. When combined with the age of the product being built in each plant it also explained a third of the productivity differences between plants.

Finally, McDuffie and Krafcik collected information on different aspects of plant management, including such things as the size of the repair area, the size of buffers, the amount of inventory and frequency of delivery, the degree of employee involvement, team working and job rotation, worker responsibility for quality, machine programming and maintenance, pay systems, status and the amount of training (McDuffie, 1989). Some of these are reported in Table 9.1. These were then combined into a management index indicating the extent to which plants were organized along Western, buffered lines or Japanese, lean, principles. Statistical analysis confirmed that lean plant organization accounted for the missing third of the productivity gap.

From all our many plant visits and data analyses we concluded that there were five key principles that characterized lean production organization:

1. The maximum number of tasks and responsibilities are transferred to those actually adding value to the product on the line.

Table 9.1 Assembly plant performance

Averages for plants in each region	*Japanese Japan*	*Japanese US*	*US US*	*All Europe*
Performance				
Productivity (hours/vehicle)	16.8	21.2	25.1	36.2
Quality (assembly defects per 100 Vehicles)	52.1	54.7	78.4	76.4
Layout				
Space utilization (sq ft/vehicle/year)	5.7	9.1	7.8	7.8
Size of repair area (percentage of assembly space)	4.1	4.9	12.9	14.4
Paint to assembly buffer (percentage of one shift output)	16.9	21.1	21.6	57.2
Inventory level (days, for 8 parts)	0.2	1.6	2.9	2.0
Workforce				
Extent of teamworking (percentage of workforce in teams)	69.3	71.3	17.3	0.6
Extent of job rotation (0 = none, 4 = frequent)	3.0	2.7	0.9	1.9
Suggestions per employee	61.6	1.4	0.4	0.4
Number of skilled and production job grades	11.9	8.7	67.1	14.8
Training for new production workers (hours)	380.3	370.0	46.4	173.3
Total absenteeism (%)	5.0	4.8	11.7	12.1
Automation				
Welding (percentage of direct steps)	86.2	85.0	76.2	76.6
Painting (percentage of direct steps)	54.6	40.7	33.6	38.2
Assembly (percentage of direct steps)	1.7	1.1	1.2	3.1
Welding flexibility (percentage of welds applied by man or robot)	72.6	82.4	62.3	58.2

Source: IMVP, World Assembly Plant Survey, 1989, and the 1989 J.D. Power Initial Quality Survey.

2. There is an effective system for immediately detecting defects and problems and tracing them to their root cause to make sure they do not recur. Small lot production, JIT and the zero-defects objective, while avoiding waste, rework and scrap, are an essential discipline within the plant to uncover problems. What Toyota calls the 'Five whys' technique is a key to solving them.
3. There is a comprehensive information system so everyone can respond quickly to any problems and to understand the overall situation of the plant.
4. None of this is possible unless the workforce is organized into work teams, that need to be trained to do all the jobs in their area –

including machine repair, quality checking, materials ordering and housekeeping – and to carry out their own problem solving.

5. Such a high level of involvement in proactively solving problems cannot work without a strong reciprocal sense of obligation between the firm and its employees.

9.4 PRODUCT DEVELOPMENT

Simultaneously with our assembly plant survey Professor Kim Clark and Takahiro Fujimoto at the Harvard Business School were conducting a similar survey of 29 different product-development projects in auto firms in Japan, North America and Europe (Clark, Fujimoto and Chew, 1987; Fujimoto, 1989). The main results from their analysis are summarized in Table 9.2. Again to compare like with like they adjusted each project for the size of the car, the number of body types, the number of carryover or common parts and the proportion of the engineering carried out by suppliers.

They concluded that in the mid 1980s, for a comparable project, the Europeans and Americans took about 60 months from concept to launch compared with 46 in Japan and about 3 million engineering man hours compared with 1.7 million in Japan. It is interesting to note that the Japanese in fact use less common and carryover parts when they design a car every four years than the Americans and Europeans do every 10 years. Also, as we shall discuss later, suppliers play a much greater role in carrying out detailed engineering of individual components. This makes the Japanese performance after adjusting for these even more impressive.

A similar two-to-one performance difference can be observed in both die making and building prototypes, typical job shop activities. Also the Japanese manage to sort out most of the crucial compromises and tradeoffs early in the development process so that there are far fewer engineering changes just prior to or after launch, when of course they are most expensive to make. Attention to detail and to manufacturability, so important in assembly plant performance, also pays off when it comes to recovering past levels of productivity and quality after launch, which for the Japanese is a matter of a few months but on average stretches to a year in Europe, often longer. So not only do the Japanese design a car in a shorter time and with less effort but the resulting design is also of much higher 'quality' in terms of completeness and manufacturability when it enters production.

Our own research and the detailed statistical analysis by Clark and Fujimoto indicate that there are perhaps six factors responsible for this superior lean product development process.

Table 9.2 Product development performance

Averages	*Japanese*	*US*	*European volume*	*European specialists*
Design performance[1]				
Engineering hours (millions)	1.7	3.1	2.9	3.1
Lead time (months)	46.2	60.4	57.3	64.4
Project scope				
Number of body types	2.3	1.7	2.7	1.3
Common and carryover parts (%)	18	38	28	30
Engineering carried out by suppliers (percentage of total hours)	51	14	37	32
Project characteristics				
Size of the project team	485	903		904
Die lead time (months)	13.8	25.0		28.0
Cost of engineering changes in body dies (percentage of total die cost)	10–20	30–50		10–30
Prototype lead time (months)	6.2	12.4		10.9
Product launch delayed (percentage of projects)	17	50		33
Return to normal productivity after launch (months)	4	5		12
Return to normal quality after launch (months)	1.4	11		12

[1] From concept study to the start of sales, after adjusting to take account of differences in the size of the car (its final selling price), the number of body types, the proportion of common and carryover parts and engineering hours carried out by suppliers.
Sources: Clark, Fujimoto and Chew (1987, pp. 741, 755 and 759) and Fujimoto (1989, Tables 7.1, 7.4 and 7.8).

1. The *first* is project leadership. Lean producers organize their development projects around one strong individual, called a Shusa in Toyota, who has total responsibility for the project from start to finish, and most likely for that model throughout its successive redesigns every four years. So there is a powerful product champion who stamps his personality on the car, mobilizes all the resources necessary to design it and who can defend the project against interference from top management. This contrasts with the traditional functional organization of product development in a mass producer, where if there is a project coordinator at all he is in a weak position to get the functions to work together.
2. The *second* is teamworking. The Shusa assembles a small team of specialists from every functional department and from marketing, manufacturing, selected suppliers, etc. These team members, while

retaining their designation to their original departments, are wholly committed to the team for the life of the project. In fact their promotion is determined by their work in the project team and not by their performance as specialists in their original department – clearly implying a very different notion of a career. The team is also a lot smaller; when Clark looked at his data by the type of organization he found that the Shusa had a team of only 333 people, compared with 1400 in a firm organized by traditional function divisions (Clark Fujimoto and Chew, 1989).

3. The *third* is simultaneous engineering. Instead of each step in the chain of activities waiting for the previous step to complete its work an integrated team can work on different steps in parallel. Not only does this save time but a unified team avoids all the wasted time as each department rechecks the work of the previous department to make sure they have not been landed with a problem that will be blamed on them.
4. The *fourth* is good communication – which is obviously much easier in a smaller team of people working towards a common goal without having to look over their shoulders at their department heads. However, it is not just the amount of information that is shared between the actors involved or the frequency of their interaction but also the quality of that information that is important. That depends very much on how the management accounting system is used. Lean producers use the way they measure costs and value as a key tool to improve performance, and not simply as a way of reporting results.[2]

The **target cost method** used by lean producers to design a car works on a market price minus basis, rather than a cost plus basis. The costs of each component and of each production step are analysed in great detail using value engineering techniques and the necessary tradeoffs to meet the target price are made early on. This of course only works if the cost analysis data are as accurate and representative as possible and if there is a willingness by each actor to share this information. Making sure manufacturing is represented in the development team right from the beginning overcomes the problem of production witholding data or trying to build some slack into their costings. The same goes for suppliers – which we shall discuss in a minute (Nishiguchi, 1989).

Of course everyone tries to analyse costs but lean production makes it easier to do so accurately. Where set-up times have been honed to always take only a few minutes, instead of anything from several minutes to several hours, and where production runs are frequent, short and interrupted, cost estimators do not have to wait

[2] I am grateful to Richard Hervey of Sigma Associates for bringing this point to my attention.

around for hours, days of even weeks to average the performance over several production runs. They quickly can collect data that is accurate and representative. Indeed collecting the data.can be left to the machine operators themselves. This makes it possible to do a complete cost analysis several times a year and to provide an accurate basis for analysing possibilities for further cost reductions, both in design and production.

5. The *fifth* is the devolution of much of the responsibility for detailed design and engineering to the suppliers, who have the most knowledge about making the part. Key suppliers send **resident engineers** to work with the assembler as a part of the design team to ensure the successful integration of these parts into the final product.
6. The *sixth* is the way the Shusa system is customer oriented and not driven by the requirements of manufacturing in the way that mass production certainly was. This is achieved through the full integration of the selling process into the design process. As we shall see the dealers, like the suppliers, are regarded as an integral part of the whole process – not only are there representatives from the sales channel in the design team but also the dealers themselves collect a large amount of information on their customers and trends in the market and feed this back to the team. Product planning is not an isolated activity that has to conduct its own market research and customer evaluation of design proposals.

9.5 THE SUPPLY CHAIN

Table 9.3 shows that supplier performance, in terms of lead time, small lot production, teamworking, quality and inventories, etc., is very similar to assembly performance – on all the factors shown Western suppliers are off by at least a two-to-one margin. This is probably not surprising as in most cases Western assemblers have for years held the whip hand over their suppliers. In other words the assemblers get the supply structure they deserve.

One significant difference is the fact that lean producers only deal with a very limited number of suppliers directly – between 300 and 350. Most Western auto producers still have between 1000 and 1500 suppliers, down from over 2000 a few years ago. Of course this difference reflects the fact that the Japanese components industry is vertically organized into three or four tiers. However, it means that lean assemblers can have a very different kind of relationship with their suppliers than can mass producers.

Figure 9.6 also shows the extent to which Japanese suppliers are involved in the product development process and the extent to which design responsibility is devolved to them. The contrast is greatest

Table 9.3 Supplier performance

Averages for each region	*Japanese Japan*	*Japanese US*	*US US*	*All Europe*
Supplier performance[1]				
Die change times (minutes)	7.9	21.4	114.3	123.7
Lead time for new dies (weeks)	11.1	19.3	34.5	40.0
Job classifications	2.9	3.4	9.5	5.1
Machines per worker	7.4	4.1	2.5	2.7
Inventory levels (days)	1.5	4.0	8.1	16.3
No. of daily JIT deliveries	7.9	1.6	1.6	0.7
Parts defects/100 cars[2]	24	NA	33	62
Supplier involvement in design[3]				
Engineering carried out by suppliers (percentage of total hours)	51	NA	14	35
Supplier propriety parts (%)	8	NA	3	7
Black box parts (%)	62	NA	16	39
Assembler designed parts (%)	30	NA	81	54
Supplier/assembler relations[4]				
Number of suppliers per assembly plant	170	238	509	442
Inventory level (days, for 8 parts)	0.2	1.6	2.9	2.0
Proportion of parts delivered just-in-time (%)	45.0	35.4	14.8	7.9
Proportion of parts single sourced (%)	12.1	98.0	69.3	32.9

[1] From a matched sample of 54 supplier plants in Japan (18), US (10 US owned and 8 Japanese owned) and Europe (18) (Nishiguchi, 1989).
[2] Calculated from the 1988 J.D. Power Initial Quality Survey.
[3] From the survey of 29 product development projects by Clark and Fujimoto (Clark, Fujimoto and Chew, 1989).
[4] From the IMVP (1989).

between the Japanese and Americans, who keep a strict hold on design, with the Europeans somewhere in between.

However this very strong reliance on suppliers in Japan could not work without a very different kind of relationship with the assemblers. This is indeed the case, though the nature of the relationship has been misunderstood badly in the West to be based on trust alone. It is in fact based on something far more solid than that. There are two key elements upon which supplier relations are built in Japan (Nishiguchi, 1989).

The first is a rational system of jointly analysing cost, determining prices and sharing profits that supersedes power-based bargaining. The joint objective between assembler and supplier is to meet a market-determined price for the vehicle and all its components. This is achieved and the necessary tradeoffs agreed through detailed joint analysis of costs as described above. This pursuit of cost reductions does not stop after the car has entered production – indeed contracts for parts are based on an agreed falling price over the life of the model. Why does the supplier share all this information with the supplier? Surely this is a recipe for disaster for the supplier?

Not at all, Japanese producers respect the need for the supplier to make a reasonable profit – indeed, they depend on them to a far larger extent as we have seen. What they agree with the supplier is a formula for sharing the gains from cost reductions over the life of the product. Any profit realized through joint cost reduction activities between the assembler and supplier are shared between them. The profits from any additional cost reductions achieved by the supplier alone are kept by the supplier. He knows that he is not going to get ripped off by the assembler, and has every incentive to innovate; this is quite a contrast to a Western supplier facing an assembler who neither knows or cares about the particular circumstances of the supplier as long as he can squeeze him on price.

In addition to agreeing a formula for sharing profits Japanese assemblers treat their suppliers, as they do their employees, as a fixed cost – sharing the bad times with them as well as the good times. The Japanese assemblers depend on the cooperation of their suppliers not just for low-cost parts and commitment to constant improvement. To fully utilize their highly flexible manufacturing system the assemblers require suppliers consistently to deliver parts with zero defects, which with no buffers could quickly stop the line, and to be able to match their flexibility in switching production from one part to another. In this way total system utilization is high and sudden changes in production are avoided. A commitment by the assembler to keeping total production volume as constant as possible – even though the mix may change frequently – has an enormous benefit for suppliers for whom the impact of fluctuations is magnified many times.

The second key to ensuring a constant commitment to pursuing efficiency is the supplier grading systems employed by all Japanese assemblers. These are relatively simple. Suppliers are rated by the number of defective parts delivered to the assembly line, the percentage of on-time deliveries in the proper quantity and sequence and their performance in reducing costs. Suppliers regularly evaluate their scores and highlight those problem areas for attention, often with the help of engineers on loan from the assembler or through the resident engineers they have stationed in the assembler's plant.

Another way in which Japanese assemblers maintain the pressure on their suppliers to improve their performance constantly is through multiple sourcing. This will come as a considerable surprise to Western observers who believed the Japanese used single sourcing to gain economies of scale. In fact, the pattern in Japan varies: Toyota often dual sources even the same part number, whereas Honda and Nissan only go as far as sourcing the same kind of part not the same part number, from competing suppliers. However, these competing suppliers are not played off against each other for short-term price advantage as in the West (price is determined through joint cost analysis) but to make sure none of them lets them down on quality and delivery reliability.

Western assemblers have responded to intensified competition by resorting to such things as across the board price cuts, increased outsourcing, rationalization of their supplier base, single sourcing, the use of SPC and insistence on JIT delivery. However all these reforms involve pushing the existing system of supplier relations to the limit rather than fundamentally changing the way the system works. Progress towards lean supply systems remains blocked by the unwillingness of Western assemblers to give up the power-based bargaining with suppliers that they have relied on for so long. For example, comparing the level of inventories in Figs. 9.4 and 9.6 shows how successful the Western assemblers have been in reducing their parts inventory to two to three days while the suppliers still carry eight to 16 days supply. No wonder suppliers remain sceptical about JIT.

9.6 DISTRIBUTION

Automotive retailing in Japan is little understood in the West and has usually been dismissed as having nothing to teach us. First, it involves door-to-door selling, something the West feels uncomfortable with and second it appears to be inefficient, as sales per salesperson are much lower than in the West. However a closer look with the above lean principles in mind reveals quite the opposite. Door-to-door selling is declining in Japan and anyway is not the key to understanding its efficiency. More research, however, is needed before one can firm up the conclusions presented here.

The first thing that is striking is how few dealer firms the assemblers deal with in Japan – there are only 1700 dealer firms in the whole country, compared with 17 600 in the USA, a market between two and three times as large. Each of these dealer firms has eight or more outlets of comparable size in terms of sales to the average US dealer.

A Japanese assembler selling two million vehicles is dealing with perhaps 300 firms compared with 3600 in an equivalent firm in the West. Moreover each of the Japanese assemblers sells its cars through between three and five dealer channels, each with their own product range, marketing staff, dealer network and customer base. The Toyota Corolla chain for example sold 600 000 vehicles through 78 dealer firms with 1335 outlets.

As with suppliers the assembler can have a very different relationship with such a small number of firms. The relationships are close and long lasting, with the dealers being seen as an extension of the design and manufacturing operation. Western assemblers are wary of losing the arms length bargaining position with hundreds of small dealers, none of whom can exert a great deal of influence over them.

The second difference is in the selling process. This is organized on a team basis, rather than individual salespeople selling on commission. The teams receive a high level of training and maintain personal contact with their customers. Not only do they sell them a vehicle but they also sell a bundle of services – in fact everything to do with running the vehicle until it is traded in. Again, the emphasis is not on the price that is cleverly negotiated by the salesman in a one-off transaction but on giving the customer a good deal on a number of transactions over many years. Dealers aim to keep the customer loyal, to fix problems that arise and thereby maximize their income over a series of transactions. In addition to selling, the salespeople are collecting a vast amount of data on their customer base. This is systematically analysed and fed back to the product development team designing the next model.

The third difference is in the relationship with the factory. Cars are almost all custom ordered in Japan, so the salesperson does not have to persuade the customer to buy the car the factory has already made that is standing on his parking lot, even though it is not quite what the customer wanted. Not only are cars custom ordered but they are delivered in a week to 10 days from the factory. Frequent input of orders and the flexibility of the production process not only means the customer gets what he wants fast but it also avoids the huge hiccups as salespeople try and guess the market and overreact when cars begin to stick. Not surprisingly, the total number of cars in stock in Japan amounts to less than 21 days' sales, compared with 67 days' in the USA.

What we have then is a distribution system that provides the customer with a high level of service and a high level of feedback to the manufacturer. When product planning, marketing and distribution costs, together with the benefits of more accurate matching of production with demand and better production scheduling are added in, the Japanese system is probably delivering a higher level of service for a much lower real cost than traditional arms-length distribution in the West.

9.7 PRODUCT STRATEGY

The full competitive implications of this lean production system are not appreciated until one looks at the product strategy that it opens up. Figures 9.4 and 9.5 show that the Japanese producers have used their more efficient product development process and their more flexible and efficient production facilities constantly to increase the number of

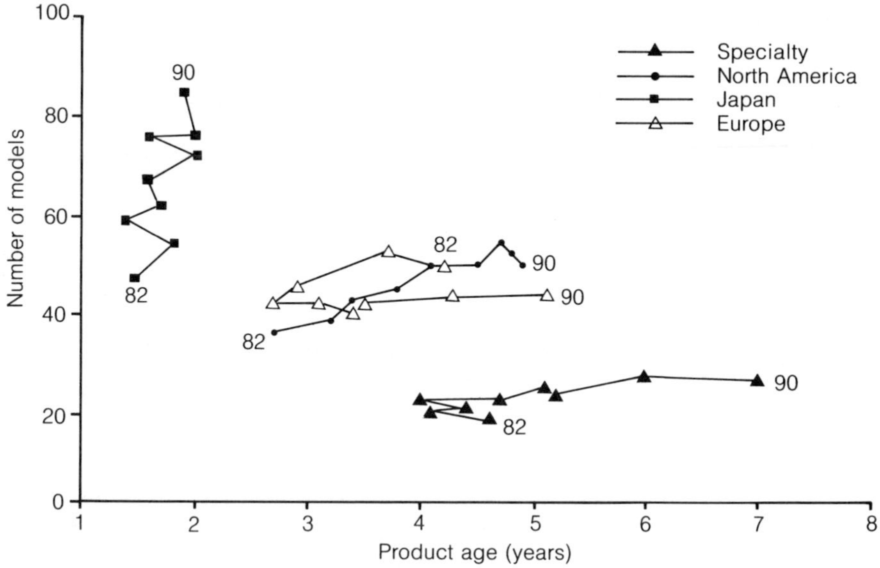

Fig. 9.4 Product strategies, 1982–1990. All modes developed in Japan, Europe and North America are counted by the location of their parent, e.g. Opel and Ford of Europe's models are counted as US. A model is defined as having different sheet metal from others, though variants such as station wagons are not counted separately. The calculation of the average age is weighted by the sales volume per model. Estimation of life of model volume is arrived at by multiplying annual production volumes times twice the average age. Source: Calculated by A. Sheriff from data in Automobile Review, Geneva, 1990 for Womack, Jones and Roos (1990).

models they offer. Between 1982 and 1990 they more or less doubled the number of models they offered while at the same time replacing them regularly every four years. Western manufacturers in contrast have spent the 1980s living off the volume generated by their existing models and their replacements once a decade.

What is even more surprising is that by the end of the decade the Western firms were on average producing each of their models at a volume of about 200 000 units, compared with about 120 000 for each Japanese model. Marry this with the different replacement cycles and it becomes clear that the Western firms produce two million copies of each design over the life of the model, compared with less than half a million for each Japanese model – less than the European specialists.

Clearly, the Japanese are not operating under the same constraints in terms of spreading large development and tooling costs over large volume output as Western firms. They are not chasing 'focussed factories' or 'world cars', not long ago seen as the strategy for survival

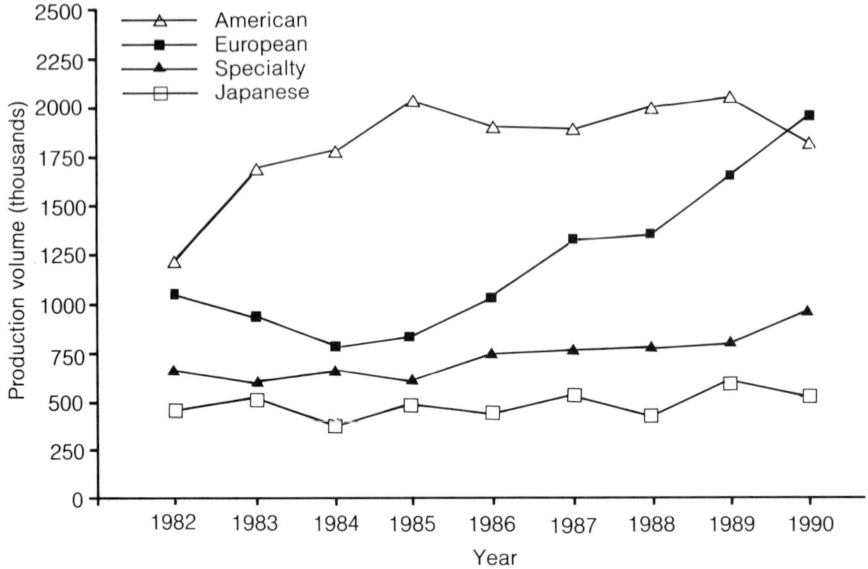

Fig. 9.5 Life-of-the-model production volume per car. All models developed in Japan, Europe and North America are counted by the location of their parent, e.g. Opel and Ford of Europe's models are counted as US. A model is defined as having different sheet metal from others, though variants such as station wagons are not counted separately. The calculation of the average age is weighted by the sales volume per model. Estimation of life of model volume is arrived at by multiplying annual production volumes times twice the average age. Source: Calculated by A. Sheriff from data in Automobile Review, Geneva, 1990 for Womack, Jones and Roos (1990).

for mass producers. Perhaps we should not be surprised however as this strategy of product proliferation is exactly what we saw in other sectors the Japanese came to dominate, such as cameras, motor cycles and electronic goods. In an increasingly affluent market the ability to target specific market segments, to follow trends every four years and to introduce new technologies more often must be a very powerful weapon indeed.

9.8 THE CHARACTERISTICS OF LEAN PRODUCTION

So we have now seen that all of the activities described contribute to making the total lean production system work – indeed it cannot fully operate until all of them are in place and work together. We might summarize the essential characteristics of the whole lean production system as follows:

1. It is customer driven, not driven by the needs of manufacturing.
2. All activities are organized and focussed on a product line basis led by a product champion, with functional departments playing a secondary, servicing role.
3. All activities are team based and the organization is horizontally and not vertically oriented.
4. The whole system involves fewer actors, all of whom are integrated with each other – 330 engineers in the product-development team versus 1400, 340 suppliers versus 1500, about 300 dealer principals versus 3600 (to sell 2 million vehicles) and 2000 assembly employees versus between 3000 and 5500 (for a plant assembling 250 000 units a year).
5. There is a high level of information exchanged between all the actors and a transparent and real cost structure.
6. The activities are coordinated and evaluated by the flow of work through the team or the plant, rather than by each department meeting its plan targets in isolation.
7. The discipline necessary for the system to function and expose problems is provided by JIT and Total Quality in the plant and supplier and dealer performance evaluation.
8. Wherever possible responsibility is devolved to the lowest level possible, in the plant or to suppliers.
9. The system is based on stable production volumes but with a great deal of flexibility.
10. Relations with employees, suppliers and dealers are based on reciprocal obligations that are the result of treating them as fixed costs.

In other words the system is not built on the notion of interchangeable parts, employees, suppliers or dealers who can be dispensed with when times get tough. Mass production was a system for strangers – workers had no reason to communicate with their neighbours, engineers minimized the need for contact with other specialists, suppliers shared as little as possible with the assembler and the salesman certainly had no intention of seeing the customer again soon (if he did it would most likely be over a warranty claim) and he would almost certainly not still be there when you went back to buy your next car. Lean production depends on direct face-to-face contact between every actor in the system for it to work. By giving the employee the responsibility and the skills to improve his own work environment the final division between manual and white collar work disappears. Everyone is valued for their willingness and ability to solve complex problems.

9.9 THE DIFFUSION OF LEAN PRODUCTION

The competitive implications of this leap in manufacturing best practice are now becoming obvious. The Japanese have successfully transferred their manufacturing system not only to North America and Europe but also to countries such as South Korea and Mexico. The Japanese have opened 11 new plants in North America with a total announced capacity of 2.7 million units. They now account for over 20% of North American auto production, up from nothing eight years ago. In the last three years General Motors and Chrysler have announced the closure of an equal number of plants and their combined market share continues to fall.

It is also clear that the leading Japanese firms have no intention of stopping at assembling kits of Japanese parts in North America. Honda has led the way, opening three assembly plants in North America plus an engine plant. It will shortly reach 75% local content by purchasing parts locally. It has also established a 500-man product-development facility and a 200-man process-engineering operation. It will shortly be building the Accord Coupé and Estate, which have been designed and the tooling built in the USA, which will also be exported back to Japan and to Europe. By the end of the 1990s the leading Japanese firms will have established full top-to-bottom design and manufacturing complexes in North America.

The pattern in Europe will develop in the same manner, only a decade later. So far, the Japanese have announced plans to build over a million vehicles a year in Europe from the mid-1990s. Several of them are already establishing design centres and engine plants. However, so far, the impact of these Japanese plants has not yet been felt and the impact of Japanese competition muted in Europe by a series of national restrictions. When they are ready to operate as full European players in the latter half of this decade all bets will be off.

What about the response of the US and European producers – do they have a chance to catch up? The answer is that they do if they move quickly. But how does one catch up? The case of Ford is an interesting example which illustrates what has to be done. In 1981 they were staring bankruptcy in the face and had no cash for investment in high technology solutions favoured by GM. Luckily for them they had a minority shareholding in Mazda, acquired several years earlier. Using this Ford sent joint management and union teams to Mazda to study their operations in detail. What they learnt was that their problems were organizational. They then introduced many of the techniques they learnt from Mazda in their plants back home, even going so far as to video every production step in an assembly plant to show to the workers back home.

Another lesson they learnt was at Hermosillio in Mexico. There they built a new plant based on a copy of a Mazda plant, to build a Mazda designed product on Mazda principles, to be sold through Ford's dealerships in Canada. Once it was running it was turned over to Ford managers and the Mazda staff left. This proved to be another very useful learning situation for Ford, conveniently located away from Detroit and not threatening anyone.

Ford soon learned how to turn around old plants, though they have not adopted all the elements of lean production. They have also rationalized their supplier base to achieve similar cost levels to the Japanese plants in the USA. What they have found more difficult is to close the gap in product development.

The lessons from the Ford case are:

1. That a creative crisis is needed to make the switch to new ways of working possible;
2. That the demonstration effect of a lean plant located down the road is also important to show that it can be done here and to blow away all the remaining excuses; and
3. That it is necessary to have access to learning situations in order to build up a critical mass of operator-level experience of these new techniques. The most obvious way for an assembler to do so is through a joint venture with a Japanese partner. A supplier has the additional possibility of learning by supplying a Japanese assembler in Europe or North America.

There is no doubt that the move from mass production to lean production will bring with it a significant improvement in productivity and living standards. However it is also clear that this transition will not be easy. A very salutary lesson is contained in an interesting study by Wayne Lewchuk (1987). He demonstrates how difficult the transition from craft to mass production was in the UK. Indeed, he shows that the UK motor industry never really adopted full mass production systems until the early 1980s, when craft distinctions and piece-rate pay systems were finally abolished at Rover under Michael Edwards.

Only for a brief moment at Trafford Park in 1915 did productivity equal that in Detroit. However Ford's UK managers, who had less sympathy with and understanding of mass production, were not able to sustain it and by 1922 things had 'gone native'. If the standard bearer could not sustain it, it is not surprising that others were not able to either. By the time post-war trade barriers fell in the late 1960s it became clear that the Europeans had been more successful in implementing mass production and the decline of the UK motor industry began in earnest. It took another decade for things to get bad enough for the necessary action to be taken. By that time it was too late to maintain an independent, locally owned motor industry.

9.10 ACKNOWLEDGEMENT

This paper draws heavily on the insights of all my IMVP colleagues, especially Dr James P. Womack of MIT with whom I have shared so much over the last decade. However, I bear full responsibility for this particular interpretation of our results.

9.11 REFERENCES

Altshuler, A., Anderson, M., Jones, D.T., Roos, D. and Womack, J.P. (1984) *The Future of the Automobile*, MIT Press and Allen and Unwin, Worcester, Mass.

Chandler, A.D. (1962) *Strategy and Structure*, MIT Press, Cambridge, Mass.

Clark, Fujimoto and Chew (1987) Product development in thc world auto industry. *Brooklyn Papers on Economic Activity*, **3**.

Fujimoto, T. (1989) Organization for effective product development: the case of the global motor industry. PhD thesis, Harvard University.

Houndshell, D. (1984) *From the American System to Mass Production, 1800–1932*, Johns Hopkins University Press, Baltimore.

Jaikumar, R. (1986) Post industrial manufacturing. *Harvard Business Review*, November–December, 69–76.

Keller, M. (1989) *Rude Awakening*, Morrow.

Krafcik, J. (1986) Learning from NUMMI. IMVP paper.

Krafcik, J. (1989) A comparative analysis of assembly plant automation. IMVP paper.

Krafcik, J. (1990) The effect of design manufacturability on production and quality. IMVP paper.

Krafcik, J. and McDuffie, J.P. (1989) Explaining high performance manufacturing. IMVP paper.

Lewchuk, W. (1987) *American Technology and the British Motor Vehicle Industry*, Cambridge University Press.

McDuffie, J.P. (1989) Worldwide trends in production system management. IMVP paper.

Nishiguchi, T. (1989) Strategic dualism: an alternative in industrial societies. PhD thesis, Oxford University.

Schonberger, R.J. (1982) *Japanese Manufacturing Techniques*, Free Press, New York.

Sloan, A.P. (1964) *My Years with General Motors*, Doubleday, New York.

Tidd, J. (1989) Technological trajectories and emerging paradigms: robotic assembly as an example of flexible manufacturing. PhD thesis, University of Sussex.

Womack, J.P., Jones, D.T. and Roos, D. (1990) *The Machine that Changed the World*, Rawson Macmillan, New York.

CHAPTER 10

In the attraction of the market economy: manufacturing strategies in Hungary

Attila Chikán and Krisztina Demeter

10.1 INTRODUCTION

Current political and economic processes in Hungary have had a great impact on the strategy and business policy of companies. Instead of the governmental paternalism companies experienced earlier they must turn to the market much more intensively than before in order to survive. The consequences of the rapid changes to manufacturing strategy will be followed in this paper partly on the basis of personal experiences and partly with the help of two recently conducted surveys. One of them concentrates on manufacturing and inventory management practices (this was completed by questionnaires at 78 companies), while the other one deals with production resource slacks (based on a questionnaire filled out at 29 companies). This second survey is extremely important from our point of view, since it was conducted three times, in 1975, 1983 and at the turn of 1989 and 1990, and thus it shows changes in slack management policies over 15 years.

The construction of the paper is as follows: first, overall corporate strategies, and, as a part of corporate strategy, the elements of the manufacturing strategy are discussed in general. Then we turn to planned economies and take a look at the characteristics of manufacturing strat-

Specially commissioned for *Manufacturing Strategy – Process and content*.
Edited by Christopher A. Voss.
Published in 1992 by Chapman & Hall, London ISBN 0 412 43660 4.

egies in Hungary. Before reaching conclusions the features of the transition period of today's Hungary are examined.

10.2 ELEMENTS OF MANUFACTURING STRATEGY

'Strategy is a set of plans and policies by which a company aims to gain advantages over its competitors. Generally a strategy includes plans for products and the marketing of these products to a particular set of customers' (Skinner, 1969). Certainly, manufacturing strategy is influenced highly by overall corporate strategy; in fact, according to our view it is one of the most essential parts of it (especially in planned economies, as we will see later). (This view is not shared by everybody, even though those previously denying the importance of a manufacturing strategy have mostly been convinced by current development of the field). The most important elements of the manufacturing strategy are the following (on the basis of Skinner (1969) and Glueck (1980)):

1. *Selection of plant location and manufacturing hardware.* This element generally leads to an investment type problem and thus it is in very close connection with financial strategy.
2. *Determining critical focuses of managerial control* (quantity, quality, cost, productivity, etc.). This aspect is very important from the point of view of establishing the objectives which should be followed. The main measures of performance are also derived from here.
3. *Selection of control systems and organization,* which determines the means by which management wants to achieve its goals, and also reflects the forecasted response from the organization's side.

Let us take a look now how these strategic elements appear in a planned economy and how the current changes in the political and economic structure of Hungary affect manufacturing strategy.

10.3 CHARACTERISTICS OF MANUFACTURING STRATEGIES IN A PLANNED ECONOMY

In this section we describe the main characteristics of the manufacturing strategy of companies in planned economies. Of course, all economies have a different history and had various specific characteristics, but some general conclusions can certainly be drawn.

We would also like to call attention to that the operation of planned economies went through very important changes, also regarding manufacturing. Here we examine the planned economies after those fundamental reforms which eliminated rigid central planning (under which we could not even talk about manufacturing strategy – companies simply executed decisions made high in the state hierarchy) and gave more room for independent company decisions. (That has been the case in Hungary and in Yugoslavia from the late 1960s, in Poland somewhat later, while in other countries only recently.) Some consequences of the dramatic changes in 1988–1990 will be discussed in Section 10.4.

10.3.1 Selection of plant location and manufacturing hardware

These decisions were only very rarely made entirely freely by companies. Establishment of new plants, capacity expansion, etc., needed resources usually far beyond the financial power of companies. Obtaining credit was rather strictly regulated by government. So, for manufacturing strategy, plants and equipment were usually considered as an outside constraint.

Some basic background of these facts, or rather the process of decisions on investment or production hardware were the following:

1. The company made a decision on investing in fixed assets. If it itself possessed enough money the decision could be executed. If not, it needed the money from somewhere else.
2. Why did companies not have enough money? Based on the rules of the regulation system a very high proportion of company profit was centralized. We have to add, however, that profits can be centralized only from companies which produce some profit. Thus the basic cause of the lack of money is the redistribution policy which reallocates money from the strong to the weak companies. This policy conserved the structure of the industry and made companies poor.
3. Since, in the majority of cases, companies did not have enough money for investment they turned to other sources, namely to the government. Since the bank system was centralized and monopolized by the government, applying for credit meant basically the same as asking for state funds. The right government office collected investment demands of companies and selected those which would get the money.
4. What kind of facts affected governmental decisions? The following are some of them:
   (i) Current governmental policy preferences (which industry has to be developed?);
   (ii) Current financial position of the government. This asks two questions, namely, 'How much money they can redistribute?' and, more important, 'How much hard currency can they provide for investment purposes?' The latter is really very important because only very few companies had the right to import goods directly and thus even 'rich' companies needed governmental support.

It was very rare that these decisions were affected by any economic fact. They did not really take into account either the position of the given company (one exception was the export power to the market economies) or the possible return on the given investment.

Summarizing this problem, companies generally needed financial support from outside to execute investment decisions and this fact led to a queue at these sources, reinforcing the paternalistic relationship between companies and the government, and also to the squandering of investment capital.

10.3.2 Critical focuses of manufacturing management

It is well known that planned economies were typical shortage economies. The general overdemand made companies put the main emphasis on increasing the quantity, the volume of production. Neither the need nor the possibility existed for innovation: the same old product could be sold for a long time, to fulfil extensive demand. Also, cost considerations did not play an important role: producers, who on a shortage market were always stronger than buyers could raise their price without difficulty if their costs increased. Profits of companies depended mostly on the availability of state support, and if profitability of actual production does not play a significant role, productivity is not a major issue either.

Under such conditions company efforts to increase production volume were restricted by non-economic considerations, mostly by the availability of resources. The main strategic issue was to ensure a continous supply, i.e. to obtain the necessary resources. This is well reflected in Table 10.1, which is from the evaluation of the survey about manufacturing practices; after the trivial consumer order due date, material availability plays the second most important role in production scheduling – far ahead of any other aspect.

Companies' behaviour is much different depending on which market they want to sell in. In case of domestic and COMECON market they usually promise long delivery time, while in case of selling to market economies these lead times are much shorter. Our survey has shown the following results (Table 10.2).

This situation is of course not the consequence of something like a subjective preference of the companies, but the different requirements of the market. This is illustrated by Table 10.3, answering the question, 'How are the delivery times determined?'

The efforts put into meeting demand in time are different in the different directions – as a consequence, the results are different as well – as illustrated in Table 10.4.

Further consequences of the situation discussed above for the production system are: 1. companies offer long delivery time for consumers so

Table 10.1 Priorities in production scheduling

Aspects	*Perentage of companies mentioning a given aspect*
Consumer order due dates	91.0
Processing time required	35.9
Similarity of set-ups	3.8
Material availability	61.5
Marketing preferences	16.7
First-come first-served	9.0
Selling price of item	6.4
Management directive	17.9

Table 10.2 Delivery time promised in different market orientations

Delivery time in days	*Market economies*	*COMECON countries*	*Internal*
<60	40.8	20.0	29.8
60–180	44.9	44.4	54.3
>180	14.3	35.6	15.8

Table 10.3 Determinants of delivery lead times

	Percentage of companies		
Lead time is mainly determined mostly by	*Market economies*	*COMECON countries*	*Internal consumers*
Consumer	83.8	50.0	38.2
The company	0.0	12.6	23.6
Negotiation	16.2	37.4	38.2

Table 10.4 Percentage of companies which have various percentages of late orders

Percentage of ordes late	*Market economies*	*COMECON countries*	*Internal consumers*
<10%	78.8	62.8	30.8
10–20%	16.7	23.1	39.7
>20%	5.1	14.1	29.4

as to have time to ensure the availability of resources; and 2. since there is a great difference between the various market relations, those companies which sell to more than one type of market (and these are the majority of the Hungarian companies) make themselves flexible by rescheduling their production when necessary to ensure the timeliness of delivery to the more important relation. This, however, already leads us to the third aspect of manufacturing strategy.

10.3.3 Organization and control

The characteristic feature of control systems and organizations is a rigid, unified and formal operation with 1. lots of just-in-case slacks, and 2. an important informal system, which are also the consequences of shortage.

1. Slacks of production resources on the one hand serve as buffers against shortages, to protect production against disturbances – while on the other hand the shortage in one kind of resource can cause unusability of another resource, this way causing unintended slacks. As for the use of slacks, the second survey shows that it mainly goes on by using overtime: if short-term shortages occur in input materials or parts, companies wait until the necessary input arrives and then they try to catch up by working overtime. What does this kind of operation mean? First, it means large fluctuations in production. Second, such fluctuations automatically lead to high excess capacity and manpower. So concerning the question of slacks we can state the following:

   (i) Uncertainties of input materials are the main problem during the operation of companies, and the large input resource slacks are only the consequences of this situation.
   (ii) The problem with input materials causes the frequent fluctuations in production, which on the one hand lead to excess capacity, manpower and inventories, but on the other hand make companies feel it is still not enough.

   Thus the main task of a manufacturing strategy in shortage economies is to solve the problem of input materials, because handling this question of production becomes more balanced and the feeling of the lack of resources can be eliminated. We will come back to the question of slacks in Section 10.4.
2. The manufacturing strategy of companies in shortage economies leads to two important consequences concerning organizational flexibility. The first is the operation of an informal decision system, built organically into the system (Barancsi, 1986). This informal system makes it possible to handle various market conditions and to react relatively fast even in large companies. Second, also because of the frequent changes in priorities, or on the field of materials, the sales-production-inventory-purchasing processes in the vast majority of cases go on under the leadership of a **production manager**, who is usually at a high level in the hierarchy. In this way production

plays the main role and has an important position in the informal system as well.

10.4 CHANGES DURING THE TRANSITION PERIOD

This period is still full of contradictions. Its main character is that the institutions and rules of the old system do not work any more – but, on the other hand, those of the new, market-oriented system do not work yet. As for manufacturing strategy, we experience changes in all the three fields.

Companies are already entirely independent in investments into plants and equipment. This means two things: 1. the state neither has the aim nor the money to support companies any more, and 2. more and more companies have the right to export and import goods and thus they have to produce the needed hard currency for their own purchasing. However, taxes and interest rates are still very high under the conditions of the current monetary restriction, so companies have no money to invest. In the meantime the capital market is still not strong enough. High inflation does not help the transitory process either.

A lot of news can be heard about company transformations and privatization but entirely optimistic conclusions would be misleading. The majority of transformations is carried out not on the basis of long-term economic considerations but to reach various short-term advantages (e.g. tax reductions). A lot of large companies have begun their decentralization processes but they only give legal form to the illegal one, which has worked until this time. This type of transformation again does not lead to large investments. (Of course there are exceptions in both cases.) The only field where we can see dynamic movements, and motivation to invest is the circle of individual (as opposed to institutional) entrepreneurs. However, most of them needed financial support from somewhere to carry out their aims. In this situation we can hardly observe any change in the current production structure of the economy even though it would be very important to do so.

Export orientation has become stronger, with a changing weight of directions. Soviet and CMEA direction is not so important any more in the form it existed in the past, especially since an increasing proportion of CMEA trading is in hard currency. This fact helps to restructure Hungarian industry, although it is a hard task, with increasing unemployment and uncertainties. The changing orientation has been modifying company behaviour. Companies have to look after buyers if they want to survive, and thus the sellers' market will slowly disappear. This fact leads to new priorities in production planning and to restructuring of inventories. Production management will not have to give preferences to orders from market economies, since markets will not be different.

Environmental changes have consequences on the third examined aspect as well. The changes in resource allocation and utilization, in the whole economic environment, cause changes in slacks. These changes can be seen clearly from our second survey if we look at some causes of slacks during the three examined periods. Management now cares more about internal company operations instead of trying to get subsidies and preferences from governmental sources. The financial position of internal operations has become much more important than it was before, which shows a slow movement towards a market economy. In this new situation, in which management can concentrate on internal processes, more structured/controlled operations can be achieved which give a better chance to use integrated production/inventory systems, and in this way effective methods developed in market economies for market situations can be used.

10.5 CONCLUSION

Manufacturing traditionally has had a more central role in Hungary (and in other countries in Eastern Europe) than is usual in the companies of the developed industrialized countries. We have to emphasize that this dominant role of production is not the consequence of strong production but of the weakness of other functions like finance and marketing, which stems from environmental conditions. The result of this situation in these days is that manufacturing has a great responsibility for the success of the transition process. It seems that we can experience some important changes in attitude and experiment in implementing up-to-date methodology. (However, it is still open whether manufacturing will be an asset or a liability in the process of becoming a market economy. Positive attitude and experiments are still not enough.) Changing both the physical structures (equipments, machinery, inventories) and the corporate culture require more time and further efforts.

10.6 REFERENCES

Barancsi, É. (1986) Formal and informal systems of inventory management in Hungary – a case study, in *Inventory in Theory and Practice* (ed. A. Chikán) Elsevier, Amsterdam, Budapest.

Chikán, A. and Demeter, K. (1989) Production-inventory management in the Hungarian Industry – from an international perspective, in *Proceedings of the 'Konferenz fur Lagerhaltungssysteme und Logistik'*, Mathematische Gesellschaft der Deutschen Demokratischen Republic, Leipzig, 56–68.

Chikán, A. and Whybark, D.C. (1990) Cross-national comparison of production-inventory management practices in Inventories: theories and applications (Proceedings of the Fifth International Symposium on Inventories), Budapest, 1988; published also as a special issue of *Engineering Costs and Production Economics*, **19**, 149–57.

Glueck, W.F. (1980) *Business Policy and Strategic Management*, 3rd edn, McGraw-Hill, New York.

Kornai, J. (1980) *The Economics of Shortage*, North Holland, Amsterdam.

Skinner, W. (1969) *Manufacturing – missing link in corporate strategy, Harvard Business Review*, May–June, 136–45.

CHAPTER 11

An empirical investigation of manufacturing strategies in European industry

Arnoud De Meyer

11.1 ABSTRACT

The need for the development of a manufacturing strategy has been stressed greatly over the last decennium. Few authors have, however, attempted to define generic manufacturing strategies. Roth and Miller have proposed an empirical approach to determine manufacturing strategies on the basis of a North American sample. In this paper we apply their method on a European sample in order to determine the robustness of their results both over time and in different geographical regions. The proposed method proves to be useful, but does not lead to generic strategies for both Europe and North America. This may be explained by bias in the samples or by differences between the strategies of European and North American firms. For the European sample the results prove to be stable over time. We can distinguish three categories: high-performance products manufacturers, manufacturing innovators and a marketing oriented group of manufacturers.

11.2 INTRODUCTION

Over the last decade the strategic role of manufacturing has received more attention in the operations management literature. This increased attention is partly due to the mounting awareness that manufacturing

Specially commissioned for *Manufacturing Strategy – Process and content*.
Edited by Christopher A. Voss.
Published in 1992 by Chapman & Hall, London ISBN 0 412 43660 4.

has an important role to play in improving the sustainable competitive advantage of the company or the industry. A team from MIT came recently to the conclusion, on the basis of eight industry studies (Derouzos, Lester and Solow, 1989), that the revival of the US industry must come from a fresh look at the production system, a new industrial paradigm as they called it. If manufacturing has a strategic role to play, it is conceivable that a company goes beyond using manufacturing as a tool to implement a wider business strategy, and that it develops a manufacturing strategy proper.

The content of a manufacturing strategy was probably first defined by Skinner's (1966) seminal work and further refined by Buffa (1984), Schmenner (1982), Hill (1985), and Wheelwright (1978, 1981). In particular Hayes, Wheelwright and Clark (Hayes and Wheelwright, 1984; Hayes, Wheelwright and Clark, 1988) have made an important contribution in analysing the role of manufacturing in relation to the other functions of the company and the determination of what a manufacturing strategy consists of. Although most of this work is clear in defining the components of a manufacturing strategy, it does not attempt to (and probably does not intend to) explore whether one can define generic manufacturing strategies. Most of the work on manufacturing strategy discusses the need to make trade-offs between different capabilities (Skinner, 1978), the need for focus (Skinner, 1974), lists the categories of decisions and policies (Wheelwright, 1978; Schmenner, 1979; De Meyer and Ferdows, 1987), or emphasizes the change in manufacturing strategy according to the life-cycle (Hill, 1985). Few papers actually define the content of a manufacturing strategy, or take a position in determining which manufacturing strategies are superior.

Hayes and Wheelwright (1984) define four stages in the development of manufacturing's strategic role. In their classification the third and fourth stages are the ones where manufacturing is either derived from a business strategy, or is aggressively used to build corporate competence. These two stages appear to be based on superior manufacturing strategies. But what it precisely takes to be in stage three or four, and how one gets there are treated in a fairly general way in their papers.

Roth and Miller (1989) have attempted to determine whether generic manufacturing strategies exist and what their content is. They performed their analysis on the basis of the North American Manufacturing Futures Survey database. In this paper we intend to use their methodology to explore whether 1. we can derive some generic manufacturing strategies for the European manufacturing companies, 2. to what extent these strategies are similar to the North American ones, and 3. whether there is any indication about the success of these generic strategies.

11.3 GENERIC MANUFACTURING STRATEGIES

From the onset of the study of manufacturing strategy practitioners have asked whether one could derive a set of internally coherent or generic manufacturing strategies, which could lead to successful performance. The existence of generic strategies, as proposed by Porter (1980) for the general business strategy of the firm has been doubted by some, and in particular scholars have developed an alternative view on the existence of across-industry generic strategies (Dierickx and Cool, 1989). On the other hand Van Lommel and De Brabander (1983) or Jacquemin, de Ghellinck and Huveneers (1980) have shown that industry classifications do not matter in order to describe a company's performance, and that only the intrinsic characteristics and strategy of the firm can explain performance.

The manufacturing community has always more or less accepted the existence of differentiated manufacturing strategies. One of the many contributions made by Skinner (1969) is the idea of manufacturing missions. He asserts that companies can pursue different types of manufacturing missions such as quality, cost efficiency, flexibility or dependability. Although these missions are not mutually exclusive, Skinner indicated that a company cannot pursue these different missions with equal vigour in the same factory and that the company will have to choose with which priority and with which emphasis the different missions will be implemented. Miller (1983) proposed a set of seven missions, and argues basically the same case for making trade-offs. Consequently, one could derive from this line of thought that in manufacturing, these different missions could be the nucleus of generic manufacturing strategies: low-cost manufacturing, high-quality manufacturing, highly flexible manufacturing and highly dependable manufacturing.

The need to make trade-offs has come under some attack over the last decade. Practice has shown that some companies have pursued successfully combinations of these missions. It is generally accepted these days that low-cost manufacturing and high-quality manufacturing do not have to exclude each other, and that in contrast pursuing high quality is the basis for the creation of other strengths such as low costs or dependability (Ferdows and De Meyer, 1989). De Meyer *et al.* (1989) have suggested that in addition in most recent years the combination of low cost and highly flexible manufacturing is vigorously developed by some of the world's excellent manufacturing companies.

A different approach to determining generic manufacturing strategies is based on the idea of technological life-cycles as proposed by Utterback and Abernathy (1975) and Abernathy (1978). Simplifying, one can say that this work leads to two types of approaches in manufacturing: one based on creating flexibility (in the earlier stages of the technological life-cycle), and one based on creating efficiency in the later stages of the technological life-cycle. Hill's (1985) life-cycle planning sees a shift from product to process to product focus in the plant design, which is somewhat similar to ideas of a technological life-cycle.

A third contribution to the determination of generic manufacturing strategies was made by Stobaugh and Telesio (1983) who derived, on the basis of a review of 100 case studies, three groups of international manufacturers: cost-, technology- and market-driven strategy groups.

Except for the last study most of the studies of manufacturing strategies were based on small-sample case-based research. Roth and Miller (1989) have used the data collected in 1987 in the context of the American Manufacturing Futures project to determine empirically a taxonomy of manufacturing strategies, and this on a sample of 188 North-American companies. They follow Cool and Schendel's (1987) assertion that to determine strategic groups at the business level one should use two sets of elements: scope of the business and resource commitments.

In order to cluster the respondents' manufacturing strategies Roth and Miller used competitive priorities in manufacturing for the next five years, as expressed by senior manufacturing managers, as an operationalization of scope. The relative emphasis placed on manufacturing action plans were used as an operationalization of resource commitments. In their results they distinguish three groups, which they call caretakers, marketeers and innovators.

Caretakers tend to compete on price and place relatively low emphasis on modern production improvement methods.

Marketeers seek to obtain broader distribution, to offer broader product lines, and to be responsive to changing volume requirements. Marketeers plan on strengthening their manufacturing operations through infrastructural changes, such as worker motivation programmes, job enlargement and enrichment, and intend to attack quality problems.

Innovators place emphasis on their ability to make quick changes in the product design, and focus as well on providing high performance products. To carry out their strategic orientations, the innovators' manufacturing strategy places more emphasis on CAD and developing new processes for new products.

Roth and Miller found that though the taxonomy they propose is influenced by industry, it is not dominated by it, and applies to a broad number of competitive circumstances.

This work has the merit of proposing a method to determine generic manufacturing strategies in the framework suggested by Cool and Schendel. The groups they have derived coincide to a large extent with the groups proposed by Stobaugh and Telesio. But there is no guarantee or verification that the groups which they have determined are not the consequence of a bias in their sample, or of the particular economic situation of the year during which they have collected data. To check for some of these potential biases we propose to use their methodology and apply it to two sets of data collected through a similar survey administered in Europe.

11.4 METHODS AND SAMPLE

The data for this study were obtained from the European Manufacturing Futures Survey (Ferdows and De Meyer, 1988; and De Meyer and Ferdows, 1990).[1] The questionnaire which is used to collect the data has three sections which are relevant to this analysis. The first set relates to the competitive priorities or capabilities which the respondents think will be important to compete in their industry over the next five years. The importance measures indicate the relative importance (on a seven-point scale) attributed to each capability that the manufacturing firm chooses to emphasize in appealing to customers and competing in the market place. The nine capabilities are similar to those suggested by Buffa (1984), Skinner (1969, 1985) and Hayes and Wheelwright (1984) (see Table 11.1).

The second set of data are demographic data about the business unit, such as investments in R&D, growth rates and profitability. The third set of data are a closed list of 38 action plans:

Table 11.1 Manufacturing capabilities

Mnemonic name	*Capability as proposed in questionnaire*
Low prices	Ability to offer low prices
New product development speed	Ability to make rapid design changes and/or introduce new products quickly
Volume flexibility	Ability to change volume rapidly
Conformance	Ability to offer consistent quality
Performance	Ability to provide high performance products
Delivery speed	Ability to deliver products quickly
Delivery dependability	Ability to deliver on time
Broad product line	Ability to deliver a broad product line
Speed of production changes	Ability to change production plans quickly

[1] The European Manufacturing Futures Project is part of a larger project, the Global Manufacturing Futures Project, initiated by J.G. Miller at Boston University, and carried out in Europe in collaboration with INSEAD, Fontainebleau, (K. Ferdows and A. De Meyer) and in Japan at Waseda University, Tokyo (J. Nakane).

1. Giving workers a broad range of tasks (job enlargement),
2. Giving workers more planning responsibility (job enrichment),
3. Changing labour–management relationships,
4. Manufacturing reorganization,
5. Worker safety,
6. Worker training,
7. Management training,
8. Supervisor training,
9. Preventive maintenance,
10. Zero defects,
11. Manufacturing lead-time reduction,
12. Vendor lead-time reduction,
13. Computer-aided manufacturing,
14. Computer-aided design,
15. Reducing setup/changeover time,
16. Value analysis/product redesign,
17. Group technology,
18. Capacity expansion,
19. Reducing size of manufacturing units,
20. Plant relocation,
21. Developing new processes for new products,
22. Developing new processes for old products,
23. Narrowing product lines/standardizing,
24. Defining a manufacturing strategy,
25. Integrating information systems between manufacturing and other functions,
26. Integrating information systems within manufacturing,
27. Reconditioning of physical plants,
28. Just-in-time,
29. Robots,
30. Flexible manufacturing systems,
31. Closing plants,
32. Statistical quality control (process),
33. Statistical quality control (product),
34. Improving new product introduction capability,
35. Quality circles,
36. Automating jobs,
37. Production/inventory control systems,
38. Reducing the size of manufacturing workforce (including hourly and salaried).

In both 1987 and 1988 we asked companies what was the relative emphasis that they had placed in the last year and would place on each of these action plans in the coming five years. The set of action plans is obviously non-exhaustive, but is the result of brainstorming by the researchers involved in the project. As such it is probably a satisfactory closed list of potential action programmes in manufacturing. The participants could indicate their relative priorities of emphasis for past and future action programmes in 1987 and future action programmes in 1988 on a seven-point self-anchoring scale, where 1 indicates no importance and 7 indicates critically important. For the past action programmes for 1988, the respondent was simply asked to indicate whether or not a significant emphasis had been placed on the particular action programme.

Roth and Miller have proposed using the competitive priorities to cluster the respondents in a limited number of groups. To interpret the groups, they use cluster results plus the differences in emphasis between the groups placed on the action programmes. In Cool and Schendel's framework the scope of the business unit is thus used to do the initial clustering, while the scope and resource commitments are used to interpret the cluster results. This method assumes thus a certain hierarchy between scope and resource commitments. Although this is not completely in agreement with the strategic framework Roth and Miller have adopted, it follows the traditional Skinner framework that argues that the relation between manufacturing policies (i.e. the action plans) and the business strategy is a hierarchical one, with the manufacturing missions (i.e. the competitive priorities) as intervening variables (Skinner, 1969).

The European sample consists of 225 observations in 1987, of which 211 were used, and of 184 observations in 1988 of which 176 were used. Some of the questionnaires were excluded because of largely incomplete data. The original panel to whom the questionnaire was sent consisted of the 100 largest companies in each of the four major European industrial countries and the 50 largest manufacturing companies in the other European countries.

The samples are definitely not representative of European industry as a whole. They are not biased either towards industries or to particular countries. As such they provide an interesting cross-section of European industry. The sample is however biased towards large companies, and towards the more profitable ones (De Meyer and Ferdows, 1990). If we assume that manufacturing makes a difference to the performance of a company, this implies that the generic strategies we eventually will derive form perhaps an incomplete set. Our subset is indeed biased towards the more successful manufacturing strategies.

The method used to determine the clusters was the SAS Fastclus procedure applied to a pooled within-cluster covariance matrix determined through SAS AceCLUS. The differences between groups were tested on the competitive priorities with a Scheffe test. The reader can consult Roth and Miller (1989) for more details on the particular part of the statistical analysis. We used, however, more strict significance limits than they did. The significance of the differences in action plans and demographic data between the groups was tested with an Anova procedure, followed by a Tukey range test.

11.5 IDENTIFYING THE GENERIC STRATEGIES

11.5.1 Clustering

Clustering requires the researcher to determine the number of clusters. To keep comparability with the North American results, a similar set of criteria was used to determine the optimal number of clusters. Three types of criteria were used:

1. Lehmann's suggestion to keep the number of clusters between $n/30$ and $n/60$, where n is the sample size,
2. Pronounced increases in the tightness of the clusters as measured by the pseudo F-statistic,
3. Managerial interpretability of the clusters on the basis of the differences between the centroids. In that way we obtain for both 1987 and 1988 three clusters. The resultant clusters are described in Table 11.2.

If we take a conservative attitude and keep the 1% level of significance as a criterion, then neither low prices, nor volume flexibility contribute to an overall separation of the clusters. In 1988 conformance, speed of delivery and speed of production changes have no separating power either.

To make a first effort to interpret the three clusters we will use the differences between the clusters as measured by a Scheffe test, as well as the absolute ranking of the competitive priorities in each group. The differences between the three clusters in 1987 on the basis of a Scheffe test (5% significance level) can be summarized as follows:

1. Companies in Cluster 1 on the average score significantly higher than those in Cluster 3 on conformance, delivery dependability and broad product line. They score significantly lower than those in Cluster 3 on performance.
2. Cluster 2 scores significantly higher than Cluster 1 on speed of new product development and broad product line. It also scores significantly higher than Cluster 3 on conformance and delivery speed.
3. In absolute terms Cluster 1 puts conformance, delivery dependability, delivery speed and speed of production changes on top, while performance is the lowest competitive priority.
4. Cluster 2 puts conformance, performance, delivery dependability and

Table 11.2 Group centroids of the groups (competitive abilities)[1]

Competitive ability	*Group 1*		*Group 2*		*Group 3*		*F value*
1987	$n = 37$		$n = 132$		$n = 42$		
	Mean	Rank	Mean	Rank	Mean	Rank	
Low prices	5.1	5	5.0	8	4.9	5	0.37
New product development speed	4.5	6	5.4	5	4.8	6	4.79[2]
Volume flexibility	4.1	8	4.7	9	4.6	7	1.99
Conformance	6.6	1	6.6	1	5.9	2	15.87[3]
Performance	3.5	9	6.3	2	6.1	1	134.33[3]
Delivery speed	5.5	3	5.7	4	5.0	4	5.07[2]
Delivery dependability	5.7	2	6.1	3	4.6	8	29.31[3]
Broad product line	4.4	7	5.3	6	3.2	9	40.33[2]
Speed of product change	5.3	4	5.1	7	5.3	3	0.44
1988	$n = 35$		$n = 114$		$n = 27$		
	Mean	Rank	Mean	Rank	Mean	Rank	
Low prices	5.1	7	5.0	7	4.7	6	0.45
New product development speed	5.3	6	5.6	5	3.7	9	22.34[3]
Volume flexibility	5.5	5	4.8	9	4.4	7	3.68[4]
Conformance	6.4	1	6.4	1	6.4	1	0.97
Performance	3.8	9	6.2	2	5.7	3	75.98[3]
Delivery speed	6.1	2	6.0	4	4.7	5	26.76[3]
Delivery dependability	5.7	4	6.0	3	6.4	2	3.96[4]
Broad product line	4.4	8	5.0	8	4.1	8	5.22[2]
Speed of product changes	5.8	3	5.1	6	5.1	4	4.19[4]

[1] Means give mean centroid (on a scale from 1 to 7); rank gives ranking for the competitive priorities of groups.
[2] $p < 0.01$.
[3] $p < 0.001$.
[4] $p < 0.05$.

delivery speed high on the list. Low prices comes only at the eighth place in the rank order.
5. Cluster 3 puts conformance and performance on top, as well as speed of production plan changes and speed of delivery.

It appears that the second cluster is relatively more responsive and reliable. Although speed of new product development is not high on its list, it scores significantly higher than for the other clusters. The first cluster emphasizes consistency in quality and delivery, but clearly not performance. The third cluster is performance oriented, and emphasizes, in absolute terms, speed of delivery and adaptation to the customers' requests. Although quality conformance is high on the list, this cluster scores significantly lower on this competitive priority than do the other clusters.

When we compare these findings with the results of the North American survey, then we can detect some similarity between Cluster 2 and what Roth and Miller call the innovators. It is however difficult to assimilate Clusters 1 and 3 with the caretakers, or with the marketeers.

The same analysis on the 1988 data leads to the following conclusions:

1. Companies in Cluster 1 score significantly higher than companies in Cluster 2 on speed of production plan changes, but they score lower than Cluster 2 on performance.
2. Cluster 1 scores higher than Cluster 3 on speed of new product development, volume flexibility and delivery speed. But the companies in Cluster 1 score lower on performance and delivery dependability.
3. Cluster 2 scores higher than Cluster 3 on speed of new product development, performance delivery speed and broad product line.
4. In absolute terms Cluster 1 emerges in the 1988 data as one where performance is not so important, but where conformance, and in particular speed is very important. Speed of delivery, speed of production plan changes and speed of new product development indeed score high. Although for new product development this is only on a comparative basis with other clusters, not on an absolute basis.
5. Cluster 2, which is again the largest group, scores high on conformance, performance, delivery dependability and, in comparison with Cluster 3, on speed of new product development.
6. The third cluster stands out by its similarity with the second cluster, except for the very low emphasis on speed of new product development.

The comparison with the US results does not become easier. While Cluster 2, which forms the biggest group, can with some goodwill be identified with the US group of innovators, it becomes more difficult to see a lot of commonality between the two other European clusters and the groups of caretakers and marketeers. One of the reasons for this is that in the European sample the emphasis on low prices as a competitive priority seems to be low all the time, and does not have any power to distinguish between the groups.

The comparison between the two clusterings in Europe shows a high stability. The items about which the groups differ significantly change somewhat, but the general picture remains unchanged. Speed of new product development, performance, delivery speed and broad product line have in both cases separating power. The most important difference is that conformance has separating power in 1987, but not in 1988. Apart from some insignificant switches, the rank order of the competitive priorities remains unchanged.[2] The two Clusters 1 stand out by a low emphasis on performance, and a high emphasis on conformance and speed. The two Clusters 2 have both the largest 'membership' and have an identical ranking of the competitive priorities, except for broad product line. They seem to emphasize quality both in consistency, performance and dependability. And compared to the others they put relatively more emphasis on the speed with which new products are developed. The third cluster stands out by its emphasis on performance.

Here we see from 1987 to 1988 a strong increase in the emphasis on delivery dependability. Discounting for some noise due to the fact that we have two unidentical samples, one finds a high stability in the clustering. If we had found the opposite result, we would have had some doubts about the instrument used, because one can safely assume that manufacturing strategies do not change dramatically from one year to another. The irreversibility of some of the investments in manufacturing are one of the reasons why one would expect this stability. That does not mean, of course, that an individual company or business unit could not review its manufacturing strategy, but one would not expect drastic changes for a large group of companies given the relative stability of the external environment in 1987–1988.

11.5.2 Further analysis of the clusters

In order to better understand the characteristics of the clusters we will now analyse how the three groups, which were derived for the 1987 and 1988 samples, allocate their management efforts. An Anova was used to find the action variables which were significantly influenced by the categorization into clusters, and a Tukey range test was then used to find the pairwise difference between the individual clusters. For the

[2] A non-parametric test was carried out to test the significance of the ranking. This test is omitted here since it would only further complicate the results, and would not contribute to the interpretation. We have paid attention only to the extremes of the list.

1988 past-action variables a chi-square test was used. The results of these tests are summarized in Table 11.3.

In 1987 Cluster 1 turned out to be a cluster which scores lower on a wide range of items than either one or both of the other clusters, except for the current market share. With respect to action plans they have some common characteristics with Roth and Millers's caretakers. They seem to defend a market share with means other than an aggressive manufacturing strategy. In terms of manufacturing they seem simply to rely on their dependability. They seem to invest less, comparatively speaking, in modern production systems or techniques.

The second and third clusters are more pronounced with respect to their manufacturing efforts. Compared with Cluster 1, companies in Cluster 2 seems to have applied the newest ideas in operations management. They give more planning responsibility to their workers, they pursue zero defects and improved vendor quality, reduction of manufacturing and vendor lead times, have emphasized more strongly the capability of introducing new products, have developed more new processes for new products, and have invested more heavily in CAD, CAM, JIT, FMS and robots.

Comparing Cluster 2 with Cluster 3, the list is less long: there is a higher emphasis on worker safety, manufacturing lead-time reduction and reconditioning of physical plants. The third cluster emphasizes more than the first cluster the investments in CAD and FMS, and invests more in R&D as a percentage of sales and envisions a higher growth in the proportion of new products to be introduced in the market as compared to previous years.

These differences in past-action programmes are confirmed, though somewhat less pronounced, in the future-action programmes. Cluster 2 emphasizes manufacturing and vendor lead-time reduction more than Cluster 1, and adds to the emphasis on worker safety and reconditioning of physical plants a higher emphasis on JIT and vendor quality, in comparison to Cluster 3. Cluster 3 tends to emphasize FMS more than does Cluster 1.

For the 96 respondents that answered both in 1987 and 1988, we checked whether the differences in clusters had some impact on performance in 1988 or on the type of resource commitments, i.e. action plans pursued. The only performance indicator which was influenced significantly by the clustering was growth: the companies in Cluster 3 had a significantly higher growth than the companies in Clusters 1 or 2. Cluster 3 invested in 1988 more in R&D as a percentage of sales than the two other clusters, intended to emphasize robots more heavily in the future, than did Cluster 1, while Cluster 1 had put more emphasis on manufacturing reorganization.

We propose the following summary out of all this. The first cluster of 37 companies emphasizes a manufacturing scope of dependability and

Table 11.3 Comparison and emphasis on action plans and business unit characteristics

Clusters compared	*1987*	*1988*
Past action plans and business unit characteristics		
1 > 2	None	None
1 > 3	Current market share	None
2 > 1	Give workers more planning responsibility	None
	Zero defects	
	Manufacturing lead time reduction	
	Vendor lead time reduction	
	CAM and CAD	
	Developing new processes for new products	
	JIT	
	FMS	
	Introduction of robots	
	Improving new product	
	Introduction capabilities	
	Vendor quality	
2 > 3	Worker safety	None
	Manufacturing lead time reduction	
	Reconditioning physical plants	
3 > 1	CAD	Reconditioning physical plant
	FMS	
	Past proportion of new products as a percentage of sales	
	R&D spending as a percentage of sales	
Future action plans		
1 > 2	None	None
1 > 3	None	None
2 > 1	Manufacturing lead time reduction	Value analysis and product redesign
	Vendor lead time reduction	Worker safety
	FMS	Integration of information systems
		Across functions
	JIT	Reconditioning physical plant
2 > 3	Worker safety	CAD
	Reconditioning physical plant	Plant relocation
	JIT	Integration of information systems across functions
	Vendor quality	
3 > 1	FMS	Worker safety
3 > 2	None	None

serving the market. Starting with a larger market share, they emphasize less creative development in manufacturing, and have emphasized one year later the reorganization of their manufacturing. Cluster 2 is a cluster of innovative manufacturers and are similar to the innovators in the US sample. The size of this group seems to be high, but this can perhaps be explained by the probability that a voluntary questionnaire introduces a bias towards the more innovative manufacturers. Cluster 3 emphasizes product performance as its scope, and is on many action plans somewhere in between Clusters 1 and 2. It clearly emphasizes more than Cluster 1 the use of FMS, CAD, etc., and invests more in R&D. They seem to be more technology driven. Although we found very few differences in performance in general, this group seems to grow faster than the others.

Let us now turn to the 1988 sample. In terms of past-action plans there were few differences: Cluster 3 scores higher than Cluster 1 on reconditioning of physical plants. However, the differences we observed for the action plans in 1987 are confirmed in the future-action plans in 1988. The large Cluster 2 scores higher in emphasis than Cluster 1 on value analysis and product redesign, worker safety, integration of information systems across functions and reconditioning of physical plants. Cluster 3 scores higher in emphasis on worker safety than does Cluster 1. Cluster 2 scores higher than Cluster 3 on CAD, plant relocation and integration of information systems across functions. Again, it seems that the second cluster is more active in manufacturing terms. The first cluster is putting less emphasis on manufacturing programmes than the other clusters. Cluster 3 is again somewhere in between Clusters 1 and 2.

11.6 DISCUSSION

The analysis has allowed us to derive three meaningful clusters. The present analysis supports the approach proposed by Roth and Miller. As with the North American analysis, we were able to determine three clusters. These clusters show a high stability from one year to another, even though the two samples are not completely identical.

Labelling clusters is always somewhat arbitrary, but on the basis of the analysis we would propose the following. The second cluster, which is also the biggest in both 1987 and 1988, can be called **manufacturing innovators**. Both in competitive priorities and action plans they seem to strive for what others have called world-class manufacturing (Schonberger, 1987). Our analysis does not enable us to say whether or not they are world-class manufacturers. It simply shows that relatively speaking they pay more attention to priorities and action plans which are usually identified with world-class manufacturing.

The third cluster is a group of focused manufacturers which emphasize the performance of their products. They seem to be a bit more oriented towards the deployment of technology in their emphasis on CAD and FMS, and strive for a good production process characterized by worker safety. The difference with the second cluster is, however, not so large. Let us label this group the **high-performance products group**.

The first cluster seems to be a cluster where manufacturing is relatively speaking not so much the key to the competitive strategy. In terms of priorities, conformance and speed of reaction to customers' demand are emphasized (in absolute terms, although not more than the other clusters in relative terms). But the resource commitments in manufacturing, measured as the emphasis on action plans, is lower than in the other groups across the board. Given that their market share is bigger than that of other groups, one could hypothesize that these manufacturers are more marketing oriented in their strategy. This is confirmed by their competitive priorities which emphasize a better service to the customer (such as consistent quality and speed of delivery). Let us call them the **marketing-oriented group**.

One of our objectives was to check whether the North American clustering would be robust. To a certain extent the manufacturing innovators coincide with the North American innovators. The marketing-oriented group has quite a number of analogies with the North American caretakers, but has in its priorities something of the marketeers. Consequently, we would not be so negative about them as Roth and Miller are about the caretakers. Our last group is one that does not appear in the North American analysis. It is a group pursuing top performing products. Some of the companies in this group stand for high-performance products, for which some European manufacturers are well known. It is perhaps a more pronounced European strategy.

These differences between the European and North American results raise the issue of whether we can really talk about generic manufacturing strategies, and if they are generic manufacturing strategies, how many of them do exist? With the exception of the European high-performance product group and the North American caretaker group, none of the clusters which are derived can easily be identified with the typical strategies which were proposed earlier. The caretakers are perhaps low-cost producers, and the high-performance product group are perhaps high-quality producers, but manufacturing innovators, marketeers or marketing-oriented groups do not really coincide with any of the strategies proposed. In the life-cycle model of Hill (1985) or Abernathy (1987), marketeers could perhaps be identified with process oriented manufacturers. In that case one would expect them to place high emphasis on process improvements, which we do not observe. And, although one would not expect the largest cluster (of *large* companies) to be at the beginning of a technological life-cycle, innovators could be identified with companies in the beginning of the technological life-cycle. In that case we would expect them to put emphasis on flexibility. We can observe this partially in the data.

The difference between Europe and North America is of course intriguing. If one accepts the convergence hypothesis, i.e. that management concepts, strategies and approaches in different industrial regions would evolve towards each other if the external conditions became more similar, then Europe and North America would be expected to have similar, if not identical, generic strategies. The convergence hypothesis has been under attack, especially since the more careful study of Asian and, in particular, Japanese management approaches. Perhaps our findings contribute some empirical support for rejecting the convergence hypothesis.

We would suggest cautiously that indeed both this study and North American findings suggest the existence of generic manufacturing strategies. However, we suspect that even if we merge the results of the two studies the list of four approaches (innovators, caretakers, high-performance product suppliers, marketeers) is not exhaustive.

Finally, there seems to be little or no relation in our study between the classification into groups and performance. Except for a higher growth for the high-performance products group in comparison with the marketing oriented group, there was no significant difference for a set of more than 15 performance indicators (going from more general performance indicators such as profit and growth, to more manufacturing-specific indicators such as inventory reduction and quality improvement). None of the three strategies we determined on the basis of the European sample seems to be superior either in the short or medium term.

11.7 CONCLUSION

Manufacturing strategy has received increased attention over the last decade. Many papers about the components of manufacturing or the trade-offs to be made in manufacturing have been published, but few empirical studies exist which attempt to determine some typical manufacturing strategies. Following a similar North American study we derived on the basis of two European samples three typical manufacturing strategies:

1. focused on high-performance products,
2. based on manufacturing innovativeness, and
3. marketing oriented.

Some similarity with the North American results exists, but there are also clear distinctions. Further empirical studies seem to be needed to determine whether generic manufacturing strategies exist, and if so what a complete set of generic manufacturing strategies would be.

11.8 REFERENCES

Abernathy, W.J. (1978) *The Productivity Dilemma*, John Hopkins University Press, Baltimore.

Buffa, E.S. (1984) *Meeting the Competitive Challenge*, Dow Jones-Irwin, New York.

Cool, K. and Schendel, D. (1987) Strategic group formation and performance: the case of the U.S. pharmaceutical industry, 1963–1982. *Management Science*, **9**, 1102–24.

De Meyer, A. and Ferdows, K. (1987) Managerial focal points in manufacturing strategy. *International Journal of Production Research*, **25**(11), 1152–62.

De Meyer, A. and Ferdows, K. (1990) Lasting improvements in manufacturing performance. *Journal of Operations Management*, **9**(2), 168–84.

De Meyer, A., Nakane, J., Miller, J.G. and Ferdows, K. (1989) Flexibility: the next competitive battle. *Strategic Management Journal*, **10**(2), 135–44.

Derouzos, M.L., Lester, R.K. and Solow, R. M. (1989) *Made in America, Regaining the Productive Edge*, MIT Press, Cambridge, MA.

Dierickx, I. and Cool, K. (1989) Asset stock accumulation and sustainability of competitive advantage. *Management Science*, **35**(12), 1504–13. (It was formerly an INSEAD Working Paper.)

Ferdows, K. and De Meyer, A. (1988) European manufacturers: the dangers of complacency. *The Columbia Journal of World Business*, **33**(3), 15–24.

Ferdows, K. and De Meyer, A. (1989) *Lasting Improvement in Manufacturing Performance: In Search of a New Theory*, INSEAD Working Paper 89/04.

Hayes, R.H. and Wheelwright, S.C. (1984) *Restoring Our Competitive Edge: Competing Through Manufacturing*, Wiley, New York.

Hayes, R.H., Wheelwright, S.C. and Clark, K.B. (1988) *Dynamic Manufacturing, Creating the Learning Organisation*, The Free Press, New York.

Hill, T.J. (1985) *Manufacturing Strategy: The Strategic Management of the Manufacturing Function*, Macmillan, London.

Jacquemin, A.P., de Ghellinck, E. and Huveneers, C. (1980) Concentration and profitability in a small open economy, *Journal of Industrial Economics*, **29**(2), 131–44.

Miller, S.S. (1983) Make your plant manager's job manageable. *Harvard Business Review*, January–February, 68–74.

Porter, M. (1980) *Competitive Strategy*, The Free Press, New York.

Roth, A.V. and Miller, J.G. (1989) *A Taxonomy of Manufacturing Strategies*, Paper presented at the ninth Conference of the Strategic Management Society, San Francisco, October.

Schmenner, R.W. (1979) Look beyond the obvious in plant location. *Harvard Business Review*, January–February, 126–32.

Schemenner, R.W. (1982) *Making Business Location Decisions*, Prentice Hall, Englewood Cliffs, NJ.

Schonberger, R.J. (1987) *World Class Manufacturing*, The Free Press, New York.

Skinner, W. (1966) Production under pressure. *Harvard Business Review*, November–December, 139–46.

Skinner, W. (1969) Manufacturing – the missing link in manufacturing strategy. *Harvard Business Review*, September–October, 136–45.

Skinner, W. (1974) The focused factory. *Harvard Business Review*, May–June, 113–21.

Skinner, W. (1978) *Manufacturing in the Corporate Strategy*, Wiley, New York.

Skinner, W. (1985) *Manufacturing the Formidable Competitive Weapon*, Wiley, New York.

Stobaugh, R. and Telesio, P. (1983) Match manufacturing policies and product strategy. *Harvard Business Review*, March–April, 133–20.

Utterback, J. and Abernathy, W.J. (1975) A dynamic model of process and product innovation. *Omega*, **3**(6), 639–56.

Van Lommel, E. and De Brabander, B. (1983) Conjunctuurgevoeligheid van de marktstructurele technologische en operationele determinanten van de Financiele Structuur en-resultaten in de Belgische Industrie, in *Competitief Ondernemen*, Kluwer, Norwell.

Wheelwright, S.C. (1978) Reflecting corporate strategy in manufacturing decisions, *Business Horizons*, February, 57–66.

Wheelwright, S.C. (1981) Japan – where operations are really strategic. *Harvard Business Review*, July–August.

CHAPTER 12

International manufacturing configuration and competitive priorities

Frank L. DuBois and Michael D. Oliff

A great deal of attention has been focused on the spread of US and non-US owned manufacturing facilities around the world to serve both domestic and non-domestic markets. The international dispersal of manufacturing assets has progressed to the extent that as much as 60% of the US trade deficit in 1985 could be traced to intracompany shipments from US majority owned manufacturing affiliates abroad (Shepherd and Hutchins, 1988). Recent research has highlighted the growing competitive forces in the global economy and the possibility for firms to combine competitive advantage with comparative advantage on a worldwide basis (Jolly, 1988; Kogut, 1984, 1985a, 1985b; Grunwald and Flann, 1985; Flaherty, 1986). However, in a review of the current literature on manufacturing strategy, Anderson, Cleveland and Schroeder (1989) make no mention of any existing research on the international dimension of manufacturing strategy. Moreover, Adam and Swamidass (1989) identify the international context of manufacturing strategy as a 'missing theme' in the operations management literature.

With this in mind, research was initiated to determine the underlying reasons for the international dispersal of manufacturing and to investigate the types of international manufacturing[1] configurations

[1] The term 'international manufacturing' usually connotes the presence of autonomous manufacturing units dispersed globally to serve separate markets. 'Global manufacturing', on the other hand implies production of a globally standardized product requiring the coordination and interdependency of international manufacturing operations to achieve the synergies of combined resources (Flaherty, 1986).

Specially commissioned for *Manufacturing Strategy – Process and content.*
Edited by Christopher A. Voss.
Published in 1992 by Chapman & Hall, London ISBN 0 412 43660 4.

used by US-based business units to serve international and domestic markets. As a preliminary step a conceptual framework was developed based on a review of the available research in international business and manufacturing strategy.

Research questions were then formulated and a research design developed to address the numerous issues presented in the conceptual framework. Data was collected using on-site focused interviews with manufacturing executives and international personnel in 16 business units representing four technologically diverse industries. Case studies were written based on the results of these interviews and archival data that was available for each of the firms. The first section of this paper reviews the development of the conceptual framework, the research questions and research design. The second section of the paper presents the results of the interview and case study segment of the research.

12.1 A CONCEPTUAL FRAMEWORK

Figure 12.1 presents the conceptual framework used to guide the research. This framework was developed based on a review of the relevant international business and manufacturing strategy literature and is argued to be representative of the critical decision areas confronting

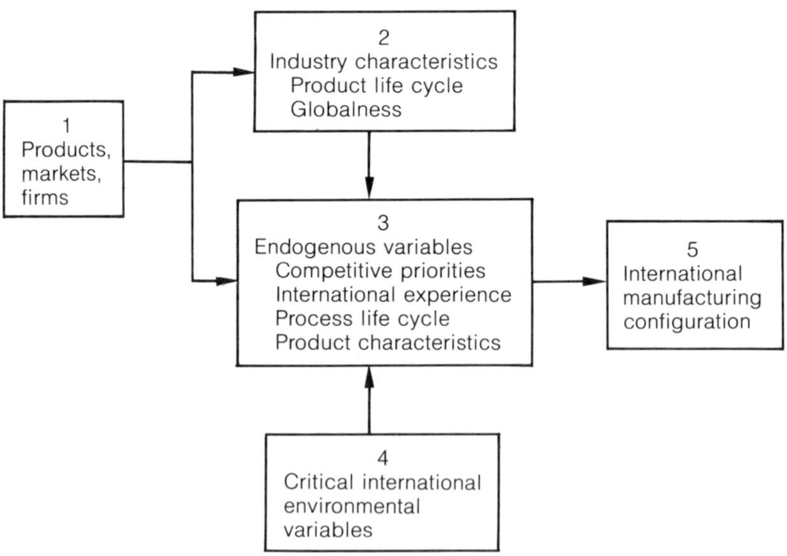

Fig. 12.1 A conceptual framework for international manufacturing configuration.

the organization making international manufacturing configuration decisions.

In Box 1, the industry setting is introduced. At this level, the products, markets and competitors are identified and the strengths and weaknesses of industry members are evaluated. Any industry study must start with an assessment of the current characteristics of the major industry leaders.

Box 2 identifies the characteristics of competition in the industry. According to Porter (1980) the relevant (immediate) environment is the industry. By taking this perspective, the intensity of competition in an industry is determined by five basic forces: 1. the threat of new entrants, 2. the bargaining power of suppliers, 3. the bargaining power of buyers, 4. the threat of substitute products or services, and 5. rivalry among the incumbent firms in the industry. By challenging the business unit to develop a response and articulate a coherent strategy the relative intensity of these forces will directly influence the type of manufacturing strategy and hence the configuration option(s) chosen.

Another aspect of the industry involves the global nature of the competition. For some businesses, competition may be on a nation-specific or if on a larger scale, what Hout, Porter and Rudden (1982) term a **multi-domestic basis**. This implies competitive forces that are nation- or region-specific which have their own distinctive characteristics and competitive imperatives.

Also included in Box 2 is the stage of product evolution as captured in the international product cycle of Vernon (1966 and 1979) and Wells (1968 and 1972). The international product cycle is relevant to the analysis because it concerns the relationship between the stage of the product in its life cycle and the location of production. This argument proposes that as the product matures, its production moves first to other developed countries, and then to less-developed countries as demand increases and manufacturing competitive priorities shift to other dimensions. As product characteristics stabilize and industry competition shifts to price and manufacturing efficiency, manufacturers seek to reduce manufacturing costs by engaging in offshore manufacture and hence, foreign direct investment (FDI) or other forms of international involvement (licensing or subcontracting) start to occur.

In the framework, an arrow links Box 2 with Box 3 because of the direct relationship between the endogenous variables in Box 3 with the exogenous variables of Box 2. Box 1 is linked to both Box 2 and Box 3 because it represents the characteristics of the industry: the products, markets, and firms. In Box 3, the endogenous variables unique to the business unit and relevant to the formulation of business unit strategy are recognized. The effects of these variables on international manufacturing configuration depends on the exogenous **market-based** variables and the exogenous international environmental variables shown in

Box 4. The variables in Box 3 include the competitive strategies and priorities (Wheelwright, 1978 and 1984a) that the business unit devises to respond to the competitive forces in Box 2:

1. *Efficiency/cost.* This encompasses both cost and capital efficiency and is measured by such factors as return on sales, inventory turnover, and return on assets. The major focus of this priority is reduction in manufacturing costs.
2. *Dependability.* The dependability of a company's products and its delivery and price promises. Often measured in terms of the percent of on-time deliveries.
3. *Quality.* Refers to product quality and reliability, service quality, speed of delivery, and maintenance quality.
4. *Flexibility.* Refers to the ability of the process to make different kinds of products and/or to adjust the volume of output as demand fluctuates.

Also included in Box 3 are other endogenous variables relevant to manufacturing configuration strategies: the amount of international experience of the firm (Moxon, 1974; Davidson, 1980), the type of production process used (Hayes and Wheelwright, 1984), and the characteristics of the product(s) produced (Moxon, 1974; Thompson and Harding, 1985).

Decisions which accompany an organization's choice of competitive priorities will have significant implications in terms of the coalignment of the international manufacturing configuration with this strategy. Failure to support a competitive strategy by improper allocation of manufacturing resources, in the absence of some overwhelming competitive advantage, in most cases leads to suboptimal performance and lost market share (Skinner, 1969, 1978, 1985; Stobaugh and Telesio, 1983; Hayes and Wheelwright 1984; Hill 1985; Gunn 1987). The labour content of the production process and the skill level required of labour are also recognized to have significant implications on the location of manufacturing (Drucker, 1971; Moxon, 1975; Davis, 1979; Fitzpatrick, 1983; Ehrenthal and Newman, 1988).

Closely related to the design of the production process and the competitive priorities of the business unit is the stage of the product life cycle (PLC) (Utterback and Abernathy, 1975; Hayes and Wheelwright, 1979 and 1984). This variable is included as an endogenous variable because of the impact it has on a firm's process choice. We argue that PLC considerations enter into the framework at both the exogenous and endogenous levels. It is an exogenous variable to the extent that it can be used as a proxy for the stage of industry evolution (Vernon, 1966; Wells, 1968) and an endogenous variable with regard to the type of production process selected that best suits the competitive needs of the business unit (Abernathy and Townsend, 1975; Hayes and Wheelwright, 1979 and 1984).

The amount of international experience of the firm is included as an endogenous variable because of its influence on the propensity of the organization to engage in international manufacturing activities. Organizations with extensive experience in international activities are much less likely to be resistant to or perceive the risks as being higher in international manufacturing activities (Moxon, 1974). Davidson (1980), for example, found that the amount of international experience of MNEs correlated highly with subsequent investments in international manufacturing activities.

Product characteristics may also influence the location of production. Moxon (1974 and 1975) in his study of the behaviour of electronics firms, observed a positive relationship between the value-to-weight ratio of products and the likelihood that the product would be produced offshore. Products having a high ratio of value to weight were more likely to be produced farther from the major consuming markets because transportation cost penalties were not large enough to offset lower manufacturing costs (ITA, 1983; Haug, 1986).

The environmental variables unique to operations in particular national environments are presented as the major aspect of Box 4. International environmental variables influence international configuration decisions because of their influence on the endogenous variables in Box 3. These variables are classified into four distinct categories using a classification scheme developed by Farmer and Richman (1965). The categories are: 1. political-legal variables; 2. educational variables; 3. socio-cultural variables; and 4. economic variables. The critical environmental variables apply to facilities operating in countries different from that where the home office is located. They may also apply to situations where a domestic facility has interactions with the international environment, e.g. through exporting or international procurement activities.[2]

Finally, Box 5 represents the result of the combined effects of the endogenous business unit specific variables, the forces driving industry competition, and international environmental constraints. Examples of the types of international manufacturing configurations that may evolve as the result of the interaction of these forces are presented below:

1. *Home country manufacturing.* This is a centralized strategy consisting of almost all manufacturing taking place in the company's country of

[2] Because of space limitations readers are referred to the original work by Farmer and Richman for a complete explanation of each of these variables. See also Weigand for a more current treatment of the role of international incentives in influencing the location decision.

origin with a high degree of global sourcing taking place through third parties.

2. *Regional manufacturing.* This strategy takes the view that the world is composed of various regions with essentially common characteristics (similar markets, common distribution requirements, government policies) and develops manufacturing operations in these areas at the product or component level.
3. *Coordinated global manufacturing.* This is the classic strategy for taking advantage of low labour cost areas. This strategy typically leads to the firm having separate manufacturing facilities for lower level components and subassemblies. Outputs are then transferred to other facilities for final assembly.
4. *Combined regional and coordinated global manufacturing.* This involves a mix of the second and third strategies with component and subassembly manufacturing in low cost regions and final assembly in regions close to markets.

12.2 RESEARCH QUESTIONS

To investigate the validity of the conceptual framework, the following research questions were generated. These were used to guide the data collection phase of the research.

1. What are the key internal and external factors that influence a business-unit in configuring its manufacturing assets?

1a. Is there a relationship between business-level competitive strategies as manifested in differing emphasis on low manufacturing cost, quality of the product produced, dependability of the production process to meet due dates, production process flexibility to alter products and/or production volumes, and the international manufacturing configuration that results?

1b. Is there a relationship between the stage of the product life cycle and the international manufacturing configuration used by the business unit?

1c. To what extent does international experience play a role in international manufacturing configuration decisions?

1d. What is the role of characteristics of the product in international manufacturing configuration decision?

2. What variables in the general international environment – as manifested by political-legal, socio-cultural, educational, and economic variables – are most critical to business units making IMC decisions? How do these variables influence the location of manufacturing activity?

12.3 RESEARCH METHODOLOGY

A multiple-industry, multiple-case approach was used in an effort to improve the external validity of the framework (Eisenhardt, 1989). In this study, the sampling frame was broadened to include polar opposites in terms of product characteristics (e.g. hand tools versus electronic components) and process technology levels (e.g. apparel versus personal computers). Cross-industry comparisons, lent strength to the emergent framework. As Eisenhardt (1989) states, '. . . given the limited number of cases which can usually be studied [because of cost and time constraints], it makes sense to choose cases such as extreme situations and polar types in which the process of interest is "transparently observable"' (Pettigrew, 1988).

Data was collected from 16 business units operating in four industries[3]. The four industries selected – microcomputers, electrical/electronic components, apparel and non-power hand tools – provided a distinct contrast in terms of the characteristics of the products manufactured, product life cycles and the role that technology plays in the formulation of competitive strategy (see Table 12.1 for details).

12.4 GENERAL FINDINGS

Separating the business units into two categories according to technological intensity permitted a useful comparison of international manufacturing strategies. Placing concrete boundaries between industry groupings in terms of market focus, however, proved somewhat problematic. Some apparel firms manufactured work clothes for sale to industrial laundries, some hand tool firms targeted a portion of their output to the consumer markets, and in the case of personal computers, it was difficult to separate the consumer market from the industrial (business) market.

The technological intensity of the industry had implications for the complexity of the production process and the degree of rationalization of manufacturing resources that could be achieved. As such, international manufacturing configuration strategies differed considerably across groups.

To say that the international manufacturing configuration chosen by the business units in each industry was dependent on the characteristics and historical precedents of each individual business unit would be an understatement. However, upon close examination, some broad patterns emerged that serve to differentiate configuration patterns

[3] To assure confidentiality, names of each business unit are disguised.

Table 12.1 Characteristics of the industry sample

Industry	*Product characteristics*	*Market and industry characteristics*	*Process characteristics*
Low-tech-firms			
Hand tools Includes business units producing hand powered tools such as pliers, sockets, screwdivers, and open-end wrences.	Various types. Usually low technology manufactured for the purpose of removing and installing fasteners. Many styles of special purpose tooling made for specific applications.	Very fragmented at the professional level. Numerous family owned firms making specialty tools. Very competitive professional and 'do-it-yourself' markets. Little import competition at the professional end of the market. High import competition at the 'do-it-yourself' end of the market mainly from Pacific Rim countries.	Large batch oriented type of process. High level of labour content in premium quality tools. Less in commodity tools. Very little outside sourcing of components. Very basic highly integrated manufacturing process.
Apparel Includes businesses manufacturing outer-and undergarments for men and women.	Various types and styles of garments. Rapid obsolescence for fashion items. Low obsolescence for under garments and work clothes. Styling is critical as is rapid delivery for fashion items. Typically 5 selling seasons requiring rapid product turnover.	Highly fragmented domestic industry. Very high level of import competition depending on product line. One of the most protected industries in the US. Success of firms is highly dependent on market whims. Shrinkage in domestic employment levels due to rising costs, foreign competition, and higher productivity.	Very basic manufacturing process. Little automation except in fabric cutting. Low scale economies. Batch flow production process with high level of semi-skilled labour content. Raw materials sourced through lextile migs, then cut and sewn in small plants. Flexibility a key priority.

High-tech firms			
Microcomputers Includes businesses producing microcomputers for home and business use.	Standard external apperance. Significant differences in computing power and capabilities. Home and business markets are quite different. Mature stage of PLC in low end of market.	Very competitive IBM compatible market. Numerous makers of IBM PC compatible clones. Number of makers thins out at upper end of scale. Portable/laptop market strengthening. Apple strong in education and home markets. Industry is less than 10 years old.	Assembly of outsourced components. Economies of scale in assembly are low. EOS in components high Repetitive manufacturing processes used. Labour content of product is low depending on stage of PLC and company.
Electrical/electronic connectors Includes businesses producing items to connect various components of electrical/ electronic equipment	Numerous product types depending on application. May be simple one-piece to splice wire or complex microchip carrier for computer use. Often uses high level of precious metals for conductivity. 5–10% of value of electronic products is connectors.	US Industry dominated by AMP Inc.. Numerous smaller manufacturers some are subsidiaries of large conglomerates. Competition depends on products and on the level of patent protection for a product. Also, lease of application tooling to buyer plays a critical role in increasing switching costs. Proximity to buyer also is key. Some foreign competition.	Typically repetitive manufacturing. Low labour content except for custom or semi-custom products. (e.g. military and aerospace). High level of automation for commodities. Success depends on low cost and delivery reliability for most business segements.

according to business unit and industry characteristics and differential emphasis on the competitive priorities of cost, quality, dependability and flexibility.

12.5 IMPACT OF MARKET AND TECHNOLOGY CHARACTERISTICS

12.5.1 Industrial markets

The evidence for the 16 cases suggests that a distinction can be made between the major type of market served and the configuration strategy adopted. For example, there was a strong relationship between serving industrial buyers and the use of a regional manufacturing strategy. Industrial buyers were perceived as more sensitive to quality and delivery reliability than were buyers in consumer markets. Thus, manufacturers serving industrial markets had to support this sensitivity through international location decisions that supported close buyer linkages. Also, the ability of the vendor to offer a broad product line allowed industrial buyers to minimize transaction costs by limiting their interactions to a small number of vendors (Williamson, 1975; Anderson and Gatignon, 1986; Teece, 1986). In the hand-tool industry, this was a key success factor and, as such, heavily influenced not only configuration strategies but also acquisition strategies.

Another important factor was the use of Japanese manufacturing philosophies on the part of industrial buyers. As industrial buyers modified production processes to adapt to the Japanese philosophy of just-in-time (JIT) manufacturing, there has been increased rationalization of supplier networks, with buyers consolidating supply sources to a fraction of pre-JIT implementation levels (i.e. Schonberger, 1982; Hall, 1983).

Business units in the industrial markets category – hand tools and electronic components – tended, in general, to use regional manufacturing configurations to support a customer-focused manufacturing strategy. Customer mandated needs for close coordination of suppliers to ensure product quality and delivery reliability provided incentives to locate in close proximity to major customer markets around the world. This was the predominant mode of operation in the electronic components industry and the hand-tools industry. The two exceptions in these industries, Federated Electronics in the electronic components industry and Valley Tools in the hand-tool industry, adopted strategies that were different from other industry members for the reasons discussed below.

Valley Tools' major competitive advantage was the possession of an internal supply source for a major production input. By integrating vertically they were insulated from disruptions in the supply chain allowing better control over the cost and quality of the final output (Khandwalla, 1974; Kumpfe and Bolwijn, 1988). Also, product characteristics across markets did not change; hand files sold in Brazil were identical to hand files sold in West Germany with only minor exceptions. Because of this, Valley Tools employed a combined regional, coordinated global strategy, sourcing raw material from a large-scale cost-efficient plant in the US and doing final manufacturing locally rather than sourcing regionally as its major competitors did.

Country-of-origin considerations in the hand-tool industry also helped to explain the preponderance of regional manufacturing strategies. Industrial buyers in this industry were very sensitive to the problem of import competition, merchandise trade deficits and their effect on manufacturing employment. It was a matter of pride to use tools that were made in the country in which they were sold. As such, country-of-origin was often used as a component of marketing promotions. One executive at Craftwork Manufacturing emphasized the fact that penetration of the German hand tool market was virtually impossible because the firm had no manufacturing presence there.

This contrasts with the experience of Federated Electronics in the components industry. Federated had little bargaining power with its customers and was under severe cost pressure from both its customers' internal supply sources and from numerous competitors in the Far East. It manufactured an undifferentiated product with a high labour content relative to most other electronic components. Management was forced to capture manufacturing economies through a strategy of coordinated global manufacturing, dispersing labour-intensive operations to low manufacturing cost regions in Mexico and the Far East. Federated Electronics was the only electronics company that truly supported Moxon's (1974) findings that cost pressures precipitated the decision to move manufacturing offshore.

12.5.2 Consumer markets

Business units whose primary focus was consumer markets (apparel and personal computers) were less constrained by customer location in their configuration strategies. There was a greater use of coordinated global manufacturing to meet endogenously generated competitive imperatives. Coordinated global manufacturing implied the disaggregation of the stages of the production process (value chain) and the deployment of each stage to a location with factor endowments most suitable to manufacturing cost minimization of that stage. Outputs at each stage (location) were subsequently coordinated to meet demand requirements at other stages (see Kogut, 1985a).

In the case of the apparel industry, the major market for the end product was the US, but the stages of the production process were dispersed globally according to the cost and quality requirements of the market segments that the business unit was serving. Also important in determining sourcing locations was the Multifiber Arrangement (MFA) and the constraints on supply that this trade framework placed on certain sourcing locations.[4] By using 'arms-length' contractual relationships with apparel subcontractors, apparel business units could acquire the flexibility in volume and product characteristics needed to compete effectively in a highly competitive industry. This illustrated a case in which a non-economic exogenous factor (the MFA) overrode economic factors (and influenced the international exchange process (Toyne, 1989)) in the configuration decision by placing limits on sourcing levels from certain countries.

Interesting exceptions to this rule were found in the cases of Beaumont Jeanswear and Johnson Apparel. These two firms had located highly automated manufacturing facilities in the US to meet the demand for certain products. In both cases the manufactured outputs were atypical apparel products. In the case of Beaumont, jeans had a relatively stable fashion cycle making them more amenable to automation of the manufacturing process. For Johnson Apparel, work clothes fell into the same category: long life cycles and stable designs. Johnson Apparel also was serving industrial markets in the US and, as such, could feature a 'made in the USA' label in its promotional campaigns. Also important is a recent trend in the apparel industry to shorten manufacturing lead times by retaining manufacturing close to the market. This has precipitated increased interest in Caribbean Basin and Mexican sourcing locations.

12.5.3 The technology factor

Technology played a major role in configuration strategies because of its influence on the degree of competition in the industry and the characteristics of the production process and product life cycles. In the two low-technology industries, entry barriers were low – especially in the case of apparel – causing configuration strategies to be designed with manufacturing cost minimization in mind. In the two high-technology industries – especially personal computers – industry technology was in a continuous state of flux. The pressure to be efficient in manufacturing was secondary to the flexibility to support design changes as industry technology advanced and business units strove to maintain technological leadership.

[4] The Multifiber Arrangement is an international trade agreement that establishes export quotas for the major textile and apparel exporting countries in different product categories (see, for example, Beckett-Young, 1989).

In the personal computer industry there was a need to locate manufacturing facilities close to the research and development function so that engineering changes could be incorporated quickly into new products. However, these business units assembled older products in the later stages of their life cycles in regional manufacturing facilities using globally sourced components. In this industry sample, three of the four business units had the US market as their primary focus. The fourth, Technodynamics, was farther along in the process of internationalization and had carefully crafted a coordinated regional/global manufacturing configuration to serve its many non-US markets. Home country manufacturing, however, still dominated for products in the earliest stages of their life cycle. In later phases of the life cycle, as skill requirements decreased and proven manufacturing methods were applied to the production process, corporate planners could safely move the manufacturing process to low labour cost regions.

In the electronic components industry, the role of technology in influencing manufacturing strategy was overshadowed by the market-based need for locations close to industrial markets. Manufacturing of new products, however, invariably took place in the home country because, quite simply, this was where the research and development that preceded new product innovations took place.

12.6 ENDOGENOUS VARIABLES IN THE FRAMEWORK

12.6.1 Competitive priorities

The four competitive priorities – cost, quality, dependability and flexibility – proposed by Wheelwright (1978, 1984b) were of critical importance to the development of effective international manufacturing strategies. Product quality and dependability of delivery were most critical to the industries in the industrial products category: hand tools and electronic components. The cost and flexibility priorities were most critical to firms that were producing products for sale to consumer markets. To be sure, quality – in the sense of being able to provide higher performance products – was still very important to these firms (especially in some segments of the apparel industry), but it did not carry the same weight as cost minimization and production process flexibility.

Cost minimizing strategies, e.g. coordinated global manufacturing, were characteristic of firms that were encountering heavy competition in their industry segments. These firms had as their major imperative manufacturing cost minimization and had designed manufacturing configurations to support this objective. In apparel, the flexibility to source from different regions or to increase or decrease production volumes in response to demand shifts was a critical success factor. Federated electronics and apparel firms in some segments of the fashion industry were under severe competitive pressure and so used coordinated global manufacturing to keep costs low and retain market share. Other apparel firms emphasized high quality in their product offerings and chose sourcing strategies to accommodate this priority. It was widely recognized in the apparel industry that the highest quality manufacturing was available in either the Far East (Hong Kong in particular) or in the US. Time-based competition[5] created configuration patterns that made use of either local production or 'offshore' sourcing supplemented by air shipment. This was very critical for short life cycle fashion items with a finite life.

In sum, cost-based competition encouraged the use of coordinated global manufacturing; the flexibility imperative encouraged a combination strategy of regional and global manufacturing; dependability required regional and home country strategies; while quality as a manufacturing imperative required either home country or regional manufacturing.

12.6.2 International experience

This variable played a role to the extent that it influenced the choice of operating mode. Davidson (1980) and Moxon (1974) both present evidence concerning the relationship of experience to international investment. In this study, firms that had already had some experience of operating internationally were less resistant to increasing international manufacturing when capacity expansion was necessary.[6] However, it appeared that in some cases – personal computers, for example – international experience was a given. This industry became an international industry very quickly and, as such, international sourcing strategies were used extensively. Some business units in the PC industry were, however, better positioned than others to take advantage of capabilities in the international environment. Technodynamics and International Data Technologies (IDT) both entered the PC industry already in possession of extensive international manufacturing assets: Technodynamics had extensive manufacturing assets in Europe and the Far East while IDT had operations in Mexico and Europe. Other firms in the study grew into international manufacturing operations as time passed and their international comfort level rose. One firm, DSR Data Systems, took advantage of the founders' family connections in Taiwan

[5] See Peters (1989) for the role that delivery speed is playing in the design of manufacturing strategies.

[6] This supports Starbuck's (1965) hypothesis that young firms tend to overestimate environmental instability because of a lack of experience in identifying important problems.

and Hong Kong to develop a supplier base for printed circuit boards and other subassemblies.

12.6.3 Product characteristics

Product characteristics had an impact on manufacturing configuration strategies from the perspective of both transportation cost minimization and technology levels. As discussed earlier, higher technology products tended to be manufactured in the home country, e.g. the newer generations of personal computer manufacturers' products. Products with a high value to weight ratio tended to have their manufacturing stages dispersed to lower cost regions, e.g. seasonal apparel products and certain electronics components. Moreover, international apparel manufacturing was highly sensitive to quota constraints imposed by the Multifiber Arrangement (MFA): products made from certain materials could only be sourced from certain locations because of the MFA. In the apparel industry, product characteristics heavily influenced sourcing locations: complex products were made onshore or in a high-cost (relative to other locations) non-domestic location such as Hong Kong. Non-complex products (jeans, work clothes, T-shirts, underwear) were also made onshore because their manufacture could be automated in sophisticated high-volume manufacturing facilities. Intermediate fashion products with medium-term lead times were most amenable to offshore production in low labour cost countries. Thus, a manufacturer competing in this niche should be prepared to use international manufacturing as a sourcing strategy.

12.6.4 Process life cycle

It was argued earlier that as a product moves through its life cycle stages, the production process used to manufacture the product also goes through an identifiable cycle (Utterback and Abernathy, 1975; Abernathy and Townsend, 1975; Hayes and Wheelwright, 1979, 1984; Wheelwright, 1984a). In the earlier stages of a product's life, the production process is low volume and skill intensive. In later stages, the process becomes increasingly rationalized through either the application of higher levels of automation or the use of international manufacturing configuration strategies. This research supports this hypothesis.

Products in the earliest stages of their life cycles (the newest generation of personal computers, for example) were typically manufactured in the home country where there could be close control over the production process. At this stage quality problems were anathema to promotion efforts, so business units desired as much control over the production process as possible before manufacturing was shifted offshore to low-cost locations.

As the manufacturing process became more rationalized, product features were standardized and discontinuities in the process were worked out, production was either moved offshore to low labour cost regions or stayed onshore through the substitution of capital for labour. This was supported by the experiences of all four of the personal computer business units investigated. It was also supported by apparel firms with stable product lines, e.g. jeans and work clothes, that required little season-to-season adaptation to fashion trends. Beaumont Jeanswear and Johnson Apparel were able to produce competitively priced products in the US by automating some of the portions of the production process and rationalizing production in the US.

12.7 CONCLUSION

The major contribution of this study is the creation of confluence between two complementary streams of research from the international business and manufacturing strategy literature through the development and application of a unified conceptual framework. Most prior studies have taken a macro-orientation to the issue of international investment rather than looking at the international involvement of manufacturing business units (MBUs) as the end result of internally generated competitive priorities (see, for example, Hymer, 1960; Kindleberger, 1969; Dunning, 1977; Rugman, 1981).

This study has also shown that business unit competitive priorities, expressed as the differential emphasis placed on four key manufacturing performance priorities (Wheelwright, 1978, 1984b), play key roles in the crafting of effective international manufacturing strategies. Configuration decisions were found to be a rational response to the decision to manage the inherent trade-offs between the four dimensions of manufacturing strategy: cost, quality, dependability and flexibility. Numerous examples are presented in the case studies that provide contrasts between business units emphasizing these different competitive dimensions.

The study has also identified the existence of two contextual factors (or conditioning variables) that must be taken into consideration in the study of international manufacturing strategies. These two industry-specific variables give rise to the contention that the level of industry technology and the type of market being served – consumer or industrial – is relevant to the international configuration decision.

Research on international manufacturing configurations offers considerable opportunities for future research. This study has built support for a linkage between manufacturing strategy and international involvement. What is now needed is further study to better validate the linkages and better specify a model of international manufacturing configuration.

12.8 REFERENCES

Abernathy, W.J. and Townsend, Phillip L. (1975) Technology, productivity, and process change. *Technological Forecasting and Social Change*, **7**(4), 379–96.

Adam, Everett E., and Swamidass, Paul M. (1989) Assessing operations management from a strategic perspective. *Journal of Management*, **15**(2), 181–203.

Anderson, E. and Gatignon, H. (1986) Modes of foreign entry: A transaction cost analysis and propositions. *Journal of International Business Studies*, **17**(3), 1–26.

Anderson, John C., Cleveland, Gary and Schroeder, Roger (1989) Operations strategy: A literature review. *Journal of Operations Management*, **8**(2), 133–58.

Beckett-Young, Kathleen (1989) Anatomy of a price tag. *The Washington Post Magazine*, 10 September, 49.

Davidson, W.H. (1980) The location of foreign direct investment activity: Country characteristics and experience effects. *Journal of International Business Studies*, **11**(2), 9–22.

Davis, W.E. (1979) A primer on locating facilities in foreign countries. *Industrial Development*, May–June, 10–14.

Drucker, P. (1971) The rise in production sharing. *The Wall Street Journal*, 15 March, OpEd page.

Dunning, J.H. (1977) Trade, location of economic activity and the multinational enterprise: A search for an eclectic approach, in *The International Allocation of Economic Activity* (eds. B. Ohlin, P.O. Hesselborn and P.J. Wiskman), Macmillan, London.

Ehrenthal, D. and Newman, J. (1988) Explaining Mexico's maquila boom. *SAIS Review*, **8**(1), 189–211.

Eisenhardt, Kathleen M. (1989) Building theories from case study research. *Academy of Management Review*, **14**(4): 532–50.

Farmer, R. and Richman, B. (1965) *Comparative Management and Economic Progress*, Irwin, New York.

Fitzpatrick, Jim (1983) Why textile and clothing industries are shifting to the third world. *Long Range Planning*, **16**(6), 42–5.

Flaherty, M.T. (1986) Coordinating international manufacturing and technology, in *Competition in Global Industries* (ed. M. Porter), Harvard Business School Press, Boston. 83–109.

Grunwald, J. and Flann, Kenneth (1985) *The Global Factory: Foreign Assembly in International Trade*, The Brookings Institution, Washington, DC.

Gunn, Thomas G. (1987) *Manufacturing For Competitive Advantage*, Ballinger Books, Cambridge.

Hall, Robert (1983) *Zero Inventories*. Dow-Jones Irwin, Homewood, Il.

Haug, P. (1986) U.S. high technology multinationals and Silicon Glen. *Regional Studies*, **20**(2), 103–16.

Hayes, R. and Wheelwright, Steven (1979) Link manufacturing process and product life cycles. *Harvard Business Review*, January–February, 133–40.

Hayes, R. and Wheelwright, Steven (1984) *Restoring our Competitive Edge: Competing Through Manufacturing*, The Free Press, New York.

Hill, Terry, (1985) *Manufacturing Strategy: The Strategic Management of the Manufacturing Function*. Macmillan, London.

Hout T., Porter, M. and Rudden, E. (1982) How global companies win out. *Harvard Business Review*, September–October, 98–108.

Hymer, Stephen (1960/1976) *The International Operations of Multinational Firms*, Massachusetts Institute of Technology Press, Cambridge, Mass.

International Trade Administration (ITA) (1983) *An Assessment of U.S. Competitiveness in High Technology Industries*, U.S. Department of Commerce, Washington, DC.

Jolly, V.K. (1988) Global competitive strategies, in *Strategy, Organization Design, and Human Resource Management* (ed. Charles C. Snow), JAI Press, Greenwich, CT.

Khandwalla, Pradip N. (1974) Mass output orientation of operations technology and organizational structure. *Administrative Science Quarterly*, **19**(1), 74–97.

Kindleberger, C.P. (1969) *American Business Abroad: Six Lectures on Direct Investment*, Yale University Press, New Haven and London.

Kogut, B. (1984) Normative observations on the international value-added chain and strategic groups. *Journal of International Business Studies*, **15**(2), 151–67.

Kogut, B. (1985a) Designing global strategies: Comparative and competitive value-added chains. *Sloan Management Review*, **26**(4), 15–28.

Kogut, B. (1985b) Designing global strategies: Profiting from operational flexibility. *Sloan Management Review*, **27**(1), 27–38.

Kumpfe, T. and Bolwijn, P.T. (1988) Manufacturing: The new case for vertical integration. *Harvard Business Review*, March–April, 75–81.

McGrath, M. and Bequillard, R. (1989) International manufacturing strategies and infrastructural considerations in the electronics industry, in *Managing International Manufacturing* (ed. K. Ferdows), North Holland, Amsterdam, 24.

Moxon, R.W. (1974) Offshore production in the less-developed countries by U.S. electronics companies. *The Bulletin*, Nos. 98–9, New York University, New York.

Moxon, R.W. (1975) The motivation for investment in offshore plants: The case of the U.S. electronics industry. *Journal of International Business Studies*, **6**(1), 51–66.

Peters, Tom (1989) Speed becomes a 'leading edge'. *Industry Week*, 19 June, 12.

Pettigrew, A. (1988) *Longitudinal field research on change: Theory and practice*. Paper presented at the National Science Foundation Conference on Longitudinal Research Methods in Organizations, Austin, Texas.

Porter, M.E. (1980) *Competitive Strategy*, The Free Press, New York.

Rugman, A. (1981) *Inside the Multinationals: The Economics of Internal Markets*, Columbia University Press, New York.

Schonberger, Richard (1982) *Japanese Manufacturing Techniques: Nine Hidden Lessons in Simplicity*, The Free Press, New York.

Shepherd, William G. and Hutchins, Dexter (1988) There's no trade deficit, Sam! *Financial World*, **23**, 28–32.

Skinner, C. Wickham (1969) Manufacturing – missing link in corporate strategy. *Harvard Business Review*, May–June, 136–45.

Skinner, C. Wickham (1978) *Manufacturing in the Corporate Strategy*, Wiley, New York.

Skinner, C. Wickham (1985) *Manufacturing: The Formidable Competitive Weapon*, Wiley, New York.

Starbuck, W.H. (1965) Organizational growth and development, in *Handbook of Organizations*, Rand-McNally, Chicago, 451–522.

Stobaugh, R.G. and Telesio, P. (1983) Match manufacturing policies and product strategy. *Harvard Business Review*, March–April, 113–20.

Teece, D.J. (1986) Transactions, cost economics and the multinational enterprise. *Journal of Economic Behavior and Organization*, **7**, 21–45.

Thompson, R.W. and Harding, F. (1985) Product life and plant location. *Dun's Business Month*, November, 73–93.

Toyne, Brian (1989) International exchange: a foundation for theory building in international business. *Journal of International Business Studies*, **20**(1), 1–17.

Utterback, J.M, and Abernathy, W.J. (1975) A dynamic model of process and product innovation. *OMEGA*, **3**(6), 639–57.

Vernon, R. (1966) International investment and international trade in the product Cycle. *Quarterly Journal of Economics*, **80**, 190–207.

Vernon, R. (1979) The product cycle hypothesis in a new international environment. *Oxford Bulletin of Economics and Statistics*, **41**(4), 255–67.

Wells, L.T. (1968) A product life cycle for international trade? *Journal of Marketing*, **32**, 1–16.

Wells, L.T. (1972) *The Product Cycle in International Trade*, Harvard Business School Press, Boston.

Wheelwright, S.C. (1978) Reflecting corporate strategy in manufacturing decisions. *Business Horizons*, **21**, 57–66.

Wheelwright, S.C. (1984a) Strategy, management, and strategic planning approaches. *Interfaces*, **14**(1), 19–33.

Wheelwright, S.C. (1984b) Manufacturing strategy: Defining the missing link. *Strategic Management Journal*, **5**, 77–91.

Williamson, O. (1975) *Markets and Hierarchies*, The Free Press, New York.

PART FOUR

The Content of Manufacturing Strategy

The model of manufacturing strategy put forward by Terry Hill at the start of this book structures the content of the process in terms of five areas: corporate needs, the market strategy, order winning criteria, process and infrastructure. The successful practice of strategic approaches to manufacture require a depth of understanding of how each of these areas fits into the overall development of manufacturing strategy. This part of the book looks at a number of contributions to our knowledge and understanding of the content of manufacturing strategy in each of these areas.

If we follow the Hill model, the first areas of concern are those of corporate and market strategy. There have been two particular concerns that have received widespread attention over the last decade. The first of these is time-based competition. As competitors develop more rapid response in areas such as new product development, lead times and responding to customer requests, manufacturing is being asked to work together with engineering and develop new ways of competing. The paper by Joe Blackburn addresses this issue in the context of manufacturing strategy. He uses methodologies derived from the marketing area to demonstrate the strategic value of response time. These methodologies are ones that can be used by companies to identify their market priorities and order winning criteria. The second concern is that of flexibility. The call for ever increasing flexibility led to a major focus on developing flexible manufacture. However, as practitioners of manufacturing strategy began to grapple with linking market demands for flexibility to the appropriate manufacturing response, it became clear that flexibility was a complex concept. With many different types of flexibility both in manufacturing and in market terms, making clear connections has proved difficult. The research by Stuart Chambers has led to the development of a framework for analysing flexibility in the context of manufacturing strategy. His paper presents and discusses this framework.

One of the most widely used concepts in manufacturing strategy is that of the **focused factory** developed by Skinner. Focus is a concept providing a key link between the market strategy and process choice, particularly in the context of a wide variety of demands. Despite the widespread use of this concept there has been very little researched or written. Quarterman Lee presents a very welcome review of the concept in the context of its application.

The final area in Hill's model: infrastructure is widely felt to be the least well understood in terms of its strategic linkages. It is known to be important and some areas such as quality management have received much attention in trying to relate decisions to their strategic context. Others such as production control and organization have been neglected. The next two papers in this part of the book focus on infrastructure aspects of manufacturing strategy. Susan Misterek and her co-authors examine the links between manufacturing strategy and organizational culture. Ted Weston examines manufacturing strategy in the context of the theory of constraints. Finally William Doll takes an 'alternate' view looking forward to the evolution of post-industrial systems.

CHAPTER 13

The strategic value of response time and product variety

Joseph D. Blackburn, Terry Elrod, William B. Lindsley and Anthony J. Zahorik

13.1 INTRODUCTION

Time has emerged as a dominant factor in determining competitive advantage. By adopting more efficient design processes, flexible manufacturing systems and faster delivery methods, many firms have been able to reduce substantially the time required to conceive of and deliver products to the market and also to offer greater variety to customers. However, there is still a great deal to be learned about the dynamics of time-based competition and its effects in the marketplace.

The major questions addressed in this study are under what circumstances, and to what extent, do customers value time and product variety. The few studies that have been completed have focused primarily on the supply side of the issue – the effects of time-reducing actions on the firm's costs of production, inventory and shipping. However, these internal actions can be expected to have significant impacts on revenues and demands as well, an area about which very little seems to be known. In particular, those activities which shorten cycle time may also have another beneficial effect on the value of a firm's product line: they often yield a flexible manufacturing system that gives the firm the capability to produce a much wider variety of products at little increase in cost.

This research study sought **quantitative** answers to questions such as

Specially commissioned for *Manufacturing Strategy – Process and content.*
Edited by Christopher A. Voss.
Published in 1992 by Chapman & Hall, London ISBN 0 412 43660 4.

the following. What kinds of rewards accrue to those firms whose strategies are time and/or variety based? Can the firm with the fastest response time extract a price premium in the marketplace? and, if so, how much? Under competitive pricing, does faster response yield greater market share and how much greater?

With respect to variety, the objectives were similarly quantitative. To what extent is the choice of a vendor determined by the breadth of their product offering? How does this differ across industries? Can a supplier with a broad line extract a price premium in the market or do buyers simply 'cherry pick' from among available suppliers?

Ultimately, the research sought to quantify the trade-offs, or cross-elasticities, among all three variables (price, time and variety). For example, how do buyers make trade-offs among these three variables in their choice of a vendor? What price premium can a firm extract when it dominates the competition on the two key dimensions of time and variety? What are the potential limits to time and variety advantages?

13.1.1 The research plan: three key industries

Empirical studies were carried out in three key industries: wholesale book distribution, machine tools and plastic injection moulds. In recent years, time and variety have emerged as significant factors in the establishment of competitive advantage in each of these industries. Specific details on each industry study are given below.

In each of the industries studied, we first carried out a descriptive analysis to determine the identity of the significant players, pricing policies, distribution channels, the effects of new technology and foreign competition and other significant factors characterizing the industry. This was followed by extensive data collection to develop estimates of the price-time-variety trade-offs. Data collection consisted of extensive field interviews and survey instruments using mail and telephone. An innovative form of conjoint analysis was then used to develop estimates of the trade-offs.

13.2 BOOK DISTRIBUTION

In the United States, books are a $14 billion dollar industry, demonstrating strong growth, not only in units sold, but also in the number of titles produced. More than 50 000 new titles are produced each year. In the delivery of the product from authors to consumers there are a number of players, as illustrated in the following graphic:

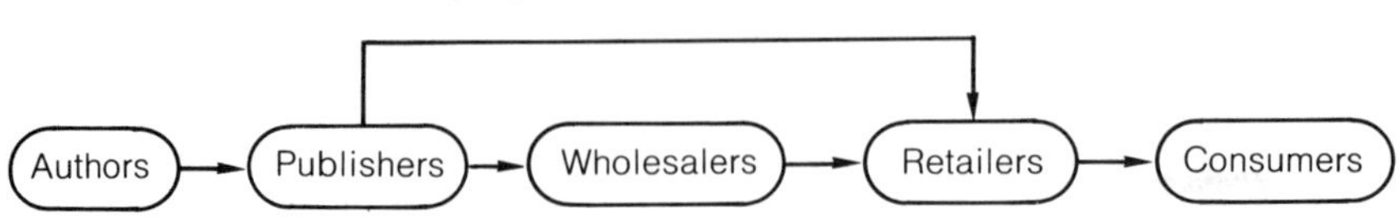

Authors produce the intellectual property which is then put in book form by the publishers. These are distributed to retailers either through wholesale distributors or directly from publishers. Finally, these products are purchased from the retailers by the ultimate consumer, whether they be individual consumers or institutions such as libraries or schools. Although publishers can deliver large shipments of new titles to major retailers quickly, they are not adept at meeting recurring, or reordering, needs of retailers and cannot serve as a single source of supply for a retailer. For this reason, there has been consistent growth in the wholesale book distribution industry.

There are approximately 40 national and regional distributors in the US, although a few major players dominate the national market. The basic reason for the growth of these companies is the service they offer to book retailers. Specifically, they are able to offer a wide variety of titles ordered in smaller quantities and shipped in relatively rapid fashion. Within the distribution industry, the winners have been those who can do this the best. The industry trends have been to guarantee same-day shipping, to increase the number of titles stocked at a warehouse, and to move closer to customers by adding distribution centres. In other words, competition is driving distributors to compete not only on price, but in terms of variety and the speed of service they offer to the retailer.

13.2.1 A survey of customers

For competitive reasons a distributor must know how its customers, the retail bookstores, will react to faster delivery and greater variety. To learn more about this, we conducted a survey of retailers and performed conjoint analysis on the data. This research focused on the retailer's choice of a wholesaler in making the decision to place orders to replenish their inventory. We wanted to know how that selection decision was affected by a distributor's capabilities along three dimensions – time, price and variety. We were able to identify the specific trade-offs buyers were willing to make with respect to price (discounts), variety and speed of delivery, and what changes in these variables would be necessary to cause a buyer to change distributors.

This survey was conducted through telephone interviews of 120 book buyers at randomly selected US book retailers. This paper presents an abbreviated description of the survey and its results.[1]

The study considered five attributes of the services offered by book distributors:

[1] Additional details on the survey and a more extensive industry analysis are provided in Lindsley, Elrod and Blackburn (1990)

1. *Discount* is the percentage unit margin (typically 40–43%) to the retailer reselling the book at list price.
2. *Speed of delivery* is measured by the time in days required for the shipment of books to arrive from the day the order was placed. Delivery time is largely a function of the location of the retailer relative to the distributor.
3. *Delivery reliability* is the percentage of the time that deliveries arrived on time.
4. *Variety of titles (or fill rate)* is represented by the percentage of titles in their typical order that their primary distributor carried in stock.
5. *Loyalty*: under what conditions the retailer would switch from their primary distributor to another?

The typical interview began by establishing the level of service book buyers currently received from their primary distributor. The interview continued with five-choice questions that yielded a quantitative measure of the importance to retailers of improvements in service relative to the current level of service from their primary distributor. The questions asked respondents to choose between improvements in existing service. The importance of the four quantitative attributes was assessed by asking each interviewee three questions, such as:

> Suppose your primary distributor could improve service in one of two ways, either by increasing its discount by one per cent or by delivering one day faster. Which would you prefer?

Loyalty to the existing distributor was assessed by asking two questions, such as:

> Suppose another distributor promised service comparable to your current distributor in the ways we have talked about and in addition promised to deliver one day faster. Would you switch to the new distributor or would you remain with your current distributor?

The importance of the five attributes was assessed by estimating binary logit, which is a regression-like model for the analysis of a binary dependent variable. (The dependent variable in this case is whether or not the respondent chose the first of the two alternatives.) The model allows for tests of statistical significance, as in ordinary regression analysis.

Through our analysis we were able to construct a Customer Satisfaction Index (CSI) for the book-distribution industry focusing on the key variables of interest.[2] This produced the following linear model:

[2] Another variable we examined was reliability. However, this variable was not significant in our model. We concluded that this quality of service was taken as a given, i.e. it was assumed by all those who used distributors.

$$\text{CSI} = 0.64 \text{ Discount Rate} - 0.28 \text{ Delivery} + 0.10 \text{ Variety} + 0.59 \text{ Loyalty}$$

This index shows the relative effect of a change (relative to the current level of service) in these strategic attributes on consumer satisfaction, both in terms of direction and magnitude. The positive coefficients for discount and variety and the negative one for delivery indicate that an increase in the discount rate (i.e., in effect, a price decrease) would raise consumer satisfaction, as would a decrease in delivery time and an increase in variety of titles available. The coefficients also show the magnitude of the effect of such a change from the current level of service.

From this model, the value of a specific change from the existing service can be estimated. Suppose a distributor is contemplating a one-day improvement in average delivery time. (Given that delivery is at present 3.5 days on average, this improvement appears to be feasible.) Such an improvement would increase the CSI to 0.28, given no other changes in service. If, however, the one-day improvement in delivery time were to be combined with a decrease in the average discount, then the CSI might not increase overall. The issue is the relative payoff either to reducing the delivery time or to increasing the available variety of titles. The model can be used to determine payoffs in either increased prices to existing customers or in new business from others' customers. For the former the discount rate variable can be used, while the latter requires the use of the loyalty variable.

13.2.2 Speed of delivery

The payoff from existing customers due to delivery time changes is determined by the discount rate that offsets the decrease in delivery time so as to leave the CSI at zero (and therefore leave the average customers as satisfied as they are at present). This relationship is shown in Fig. 13.1. It indicates that a distributor with a one-day faster delivery time can extract, at the margin, a price premium of about 0.5% – a significant advantage in the highly competitive book industry.

The **loyalty coefficient** shows how much better a competitor must perform relative to the existing distributor if it is to have an even chance of winning away the customer. Seen another way, this variable determines the value of delivery speed in terms of gaining new customers rather than increasing prices. To have the opportunity to gain new customers, a distributor must reduce delivery time by at least 2.2 days, other things being equal. To refer to our example in the preceding paragraph, then, a one-day reduction in delivery time would result in the potential to raise prices to existing customers, but one should not expect it to dislodge many of the competitors' customers.

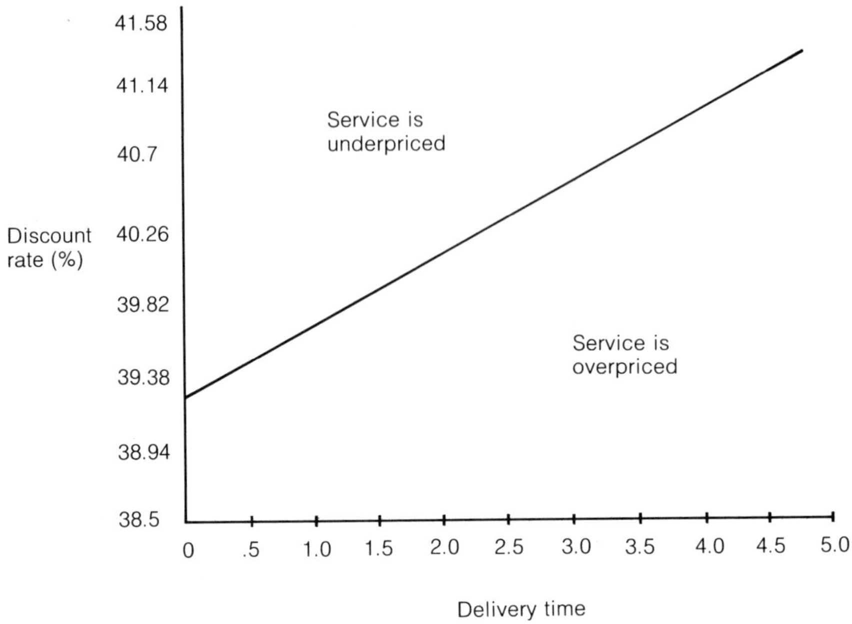

Fig. 13.1 Customer satisfaction trade-off between discount rate and delivery time.

13.2.3 Variety

The payoff due to an increase in variety (that is, an increase in titles maintained in stock) can be determined in a similar manner. The relationship between price and variety is depicted in Fig. 13.2. Figure 13.2 shows that a 5% increase in the fill rate (as 'perceived' by the retailer) would be equivalent to a price change of about 0.8%.

To overcome the inertia of loyalty to a distributor and win new customers, a sizeable increase in fill rate is necessary. Specifically, a distributor must increase the perceived fill rate by about 6% in order to offset the **anchoring effect** of loyalty. Therefore, one should not expect to gain new customers from any increase in the fill rate less than this amount.

The preceding discussion has isolated each of the issues of delivery time and variety. However, a combination of improvements may be equally effective in overcoming the loyalty disadvantage of raising prices. For instance, delivering a day faster and filling 3.1% more of a retailer's title needs also offsets the loyalty advantage of the retailer's current wholesaler. Disadvantages on one attribute can also be overcome by exceptional performance in another. Delivering a day slower can be offset by filling 3.1% more of an order or by offering a 0.44% higher discount.

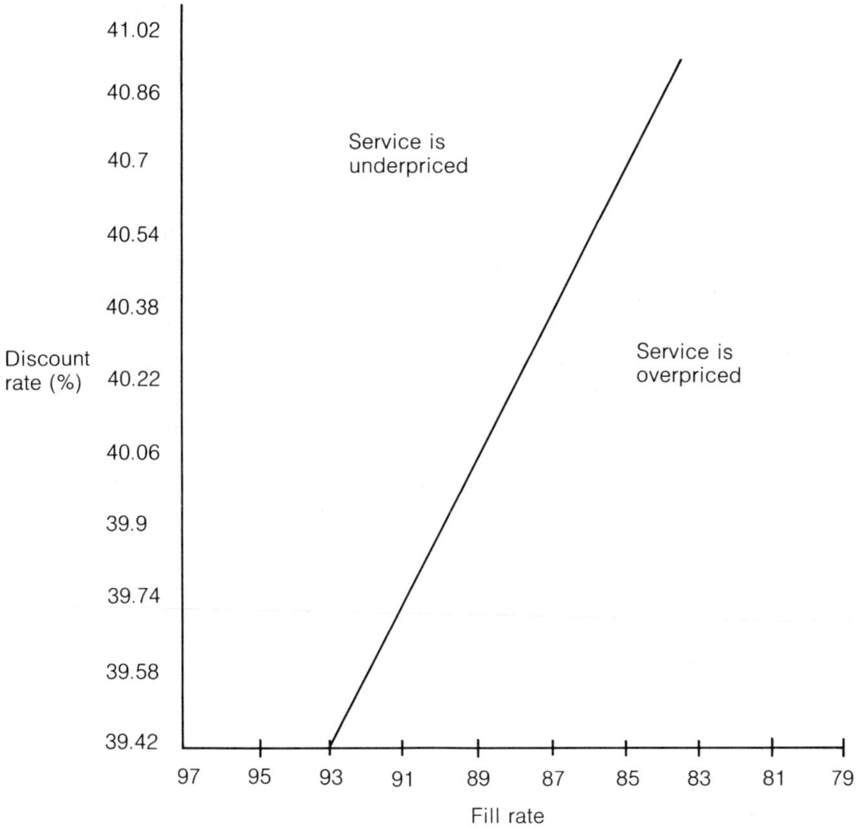

Fig. 13.2 Customer satisfaction trade-off between discount rate and fill rate.

13.2.4 Strategic implications for distribution

How might this model be used strategically by managers? Suppose that a firm wishes to capture a major share in a given market. It can devise a combination strategy based on its own capabilities that would be sufficient to dislodge retailers loyal to a competitor. For example, the firm could increase the discount rate by 0.5%, reduce delivery by half a day, and add inventory to increase fill rate by slightly over 1%.

Another use of the model is in evaluating competitive moves and potential responses. For example, suppose that Firm A is able to reduce delivery time in a certain area by a given amount. From our model, we conjecture that if this amount is less than 2.2 days. Firm B would not see any real slippage in share. If, on the other hand, Firm A is able to reduce it by that amount or more, Firm B has several options to blunt such a move. It can make more titles available or it can increase its discount rate by 1%. Similarly, a manager could evaluate the effect of a competitor offering free freight for retailers. For any given retailer, the value of such a move can be equated to a change in discount rate and then compared to the current level of service (time and/or variety) offered by the wholesaler to determine its ability to overcome the loyalty disadvantage.

In addition, the methodology has great potential in other settings as an analytical tool. In a short time at limited expense, telephone surveys can be conducted to construct similar charts of other industries and these can be used to evaluate strategic options.

Our investigation of the book distribution industry confirmed the importance of both time and variety as competitive dimensions in this industry. There are a number of factors that have contributed to this. On both the supply side and the demand side, forces have been at work to increase the variety of products offered and the speed at which they are delivered. The advent of desktop publishing lowered economies of scale and thus barriers to entry for thousands of small publishers, who consequently have offered a plethora of new titles to consumers. On the time dimension, new information technology and improved logistics systems produced dramatic reductions in customer response time.

This research indicates the segments of the business in which wholesale distributors have made inroads. Specifically, distributors have found an audience mainly among independents, who have focused on variety as a means of differentiation. Also, in terms of product, they have a much larger fraction of lower-priced items. Finally, with the increased ease and speed of ordering, average order sizes have decreased. In this environment, the most profitable wholesale distributors are the ones who achieve the operational efficiencies to compete on speed while matching the variety and price of their competitors. The dual forces of time- and variety-based competition have changed dramatically the structure of the industry.

13.3 THE US MACHINE-TOOL INDUSTRY

The machine tool industry presents a textbook example of how the development of time-based competition can change competitive structure. This $4 billion industry, at the very beginning of the production chain, has always been extremely sensitive to economic cycles, and the standard policy of US machine tool builders has been to smooth out the cyclical demand for its products through the use of extensive backlogs. Change in the machine tool industry in the United States since the 1970s is a story in which Japanese companies rose to preeminence at the expense of United States firms. Although several factors contributed to this shift, it is due in no small measure to the ability of Japanese companies to become time-based competitors – that is, to deliver products much faster than their US counterparts.

As early as the 1960s, machine tool buyers were prodding the industry for faster deliveries. Delivery times for machine tools from many US manufacturers in boom periods often exceeded eighteen months, even though the primary customers, small job shops, had far shorter planning horizons. Consequently, much of the demand for speedy delivery was met by the Japanese, who had begun producing low-quality standard machines in the 1950s but had been steadily improving the quality of their products. The initial strategy of Japanese companies, such as Yamazaki, was to sell standardized products cheaply and from inventory, rather than to match the high-priced customized tools of the US manufacturers. They also participated in Japan's emphasis on the development of computer numerically controlled (CNC) machines.[3]

By the late 1970s, when the oil crisis forced many US manufacturers to retool, the improved quality, low prices, and nearly instant delivery offered by Japanese tool makers sparked a boom in their sales. This move took away forever the low end of the US machine tool market, and eventually led to Japanese suppliers winning over 50% of the entire machine tool market. Since that time, Yamazaki and others have developed flexible manufacturing techniques which enable them to produce an extremely wide variety of products, while continuing to keep delivery times extremely short. The US companies who survived this attack now produce only complex, high-end, customized machines for large customers.

A survey of the principal buyers of machine tools (small, generally unsophisticated tool and die shops) revealed that about two-thirds of the market is time-sensitive to some degree, because of the shorter lead times now given them by their customers. (The one-third of the market which was unwilling to pay any premium to avoid delivery delays consisted primarily of extremely small shops with very unsophisticated equipment.) A telephone survey of shops, employing an efficient form of conjoint (trade-off) analysis designed for the purpose, was used to measure the extent to which potential customers would be willing to pay price premiums to avoid delivery delays.

The study reported here was undertaken to determine the extent to which machine tool customers today will pay more to get products faster. A telephone survey contacted the primary machine tool purchasers for tool and die manufacturers (SIC 3544) in Tennessee. Most of those surveyed were owners of small job shops, which are the primary market for machine tools (lathes, grinders, machining centres,

[3] More details on the evolution of the machine tool industry are provided in Zahorik, Keiningham and Blackburn (1990a)

as opposed to complex transfer lines). Approximately 50% of the firms contacted were willing or able to respond to the survey, resulting in 65 completed questionnaires.

The approach was a conjoint analysis in which respondents were asked to make trade-off choices between waiting times and price premiums that were represented in a series of machine tool buying scenarios. To make the survey fast and easy to answer, a brief and extremely efficient conjoint task was developed. Each of the two attributes, delivery time and price, was varied over three levels, resulting in nine possible scenarios: delivery times used were two weeks, two months, and six months; prices used were a base price, a 10% premium over base, and a 20% premium over base. Of the 36 possible pairs of scenarios, only nine consisted of non-dominated alternatives (under the assumption that a longer wait plus a higher price is always less desirable). Their choices were then analysed to determine the extent to which these purchase decision makers were willing to pay a premium to avoid delivery delays.[4]

The nine paired comparisons are represented as line segments between cells in Fig. 13.3.

Data gathered from each subject was insufficient to determine individual trade-off functions, so the responses were pooled to

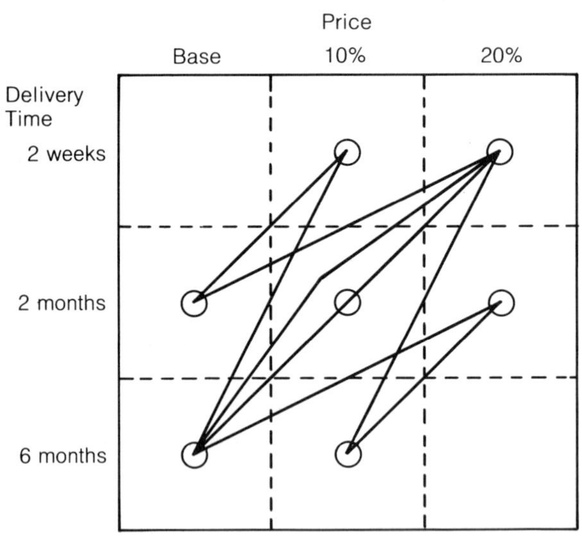

Fig. 13.3 Structure of conjoint study.

[4] Additional details on the interviewing procedure and data analysis are provided in Zahorik, Keiningham and Blackburn (1990a).

determine an aggregate trade-off curve (see Fig. 13.4). In Fig. 13.4, waiting times are compared to their equivalent price premiums. In particular, a two-month wait is considered only as distasteful as a price premium of about 1.5% that is, it is not seen as very burdensome. A delay of six months, however, is considered equivalent to a price premium of about 12%. (Correspondingly, a buyer would prefer to pay a 10% premium than wait a full six months for delivery.)

Our discussions with machine tool buyers, however, had revealed varying levels of sophistication regarding the value of time. Fully 35% of the respondents always chose the alternative with the longer delivery time rather than the higher price. We separated these respondents into a 'price-sensitive' segment. The remaining 65% of the respondents comprised the 'time-sensitive segment' (although some were clearly less time-sensitive than others). The resulting trade-off curves indicate a very different perspective from that provided by the aggregate analysis.

In Fig. 13.5, the trade-off curve for the time-sensitive segment is shown. A two-month delay is equated with a premium of about 4%, and a six-month delay is considered equivalent to a 19% premium. (Many time-sensitive respondents said that a six-month delay was totally unacceptable, and that they would explore other alternatives rather than wait six months.) The curve of the price-sensitive segment, shown in Fig. 13.6, reveals that these respondents were determined not

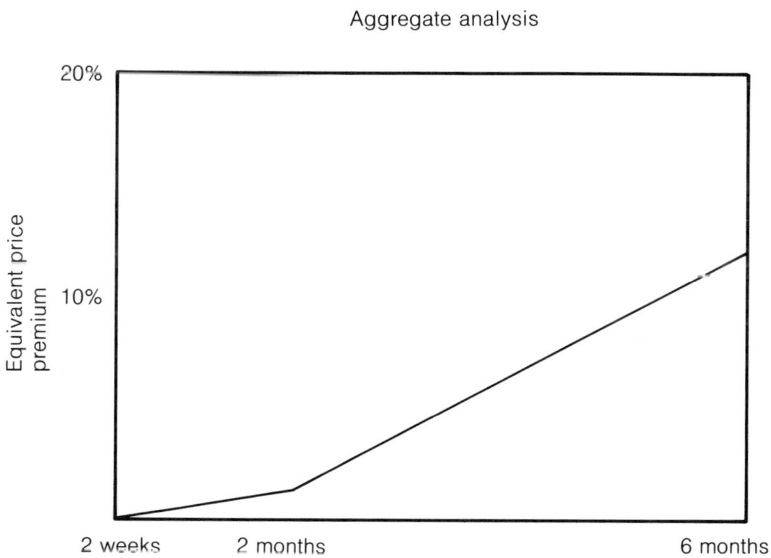

Fig. 13.4 Results of the conjoint study of machine tool buyers; aggregate analysis.

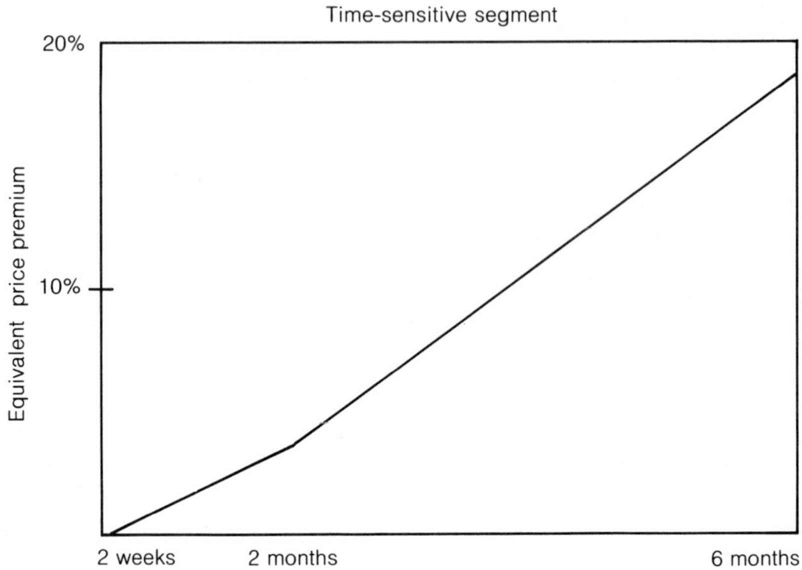

Fig. 13.5 Results of the conjoint study of machine tool buyers; time-sensitive segment.

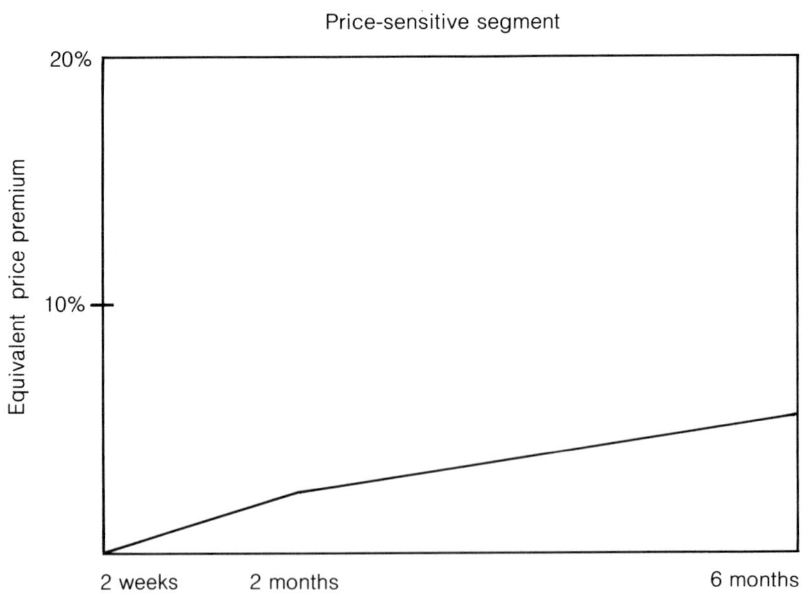

Fig. 13.6 Results of the conjoint study of machine tool buyers; price-sensitive segment.

to pay a premium for faster delivery. The curve is essentially flat, with a six-month wait equivalent to a premium of about 5%.

The results are not unexpected. The metal-working shops represented in this study are fundamental to US productivity and should, therefore, reflect the growing awareness of the importance of time-based competition which has been forced on US industry by its Japanese competitors. At the same time, many of these small shops are owned and run by people whose training has made them highly skilled machinists, but has given them little understanding or appreciation of business planning or strategy.

In light of the study's finding that a large section of US buyers of machine tools are willing to pay hefty premiums for timely deliveries, it is not surprising that the Japanese were able to make such inroads in this market and overcome the reluctance of US machine tool buyers to buy foreign-made products. Once the quality of Japanese machine tools reached acceptable levels, buyers' concerns about retraining workers on the new machines were easily overcome by the promise of rapid deliveries at competitive *or lower* prices and accompanied by excellent service.

The story of the machine-tool industry is one that has become familiar to many US industries during the last few decades. That story holds important lessons for still other industries. Foreign competitors have entered US markets at the low end, using price- and time-based strategies. US competitors have conceded these 'unprofitable' low-end sales to the foreigners, only to find that they quickly leverage their reputations for quality to gain access to the high end of their markets. The unsophisticated attitude shown by purchasers in one-third of the machine shops in this study suggests that other US industries could soon suffer the same fate.

13.4 THE US PLASTIC INJECTION MOULD INDUSTRY

The plastic injection mould industry is fundamental to the manufacturing sector, since it produces the moulds from which all plastic parts are made. The US mould building industry consists of over 3000 machine shops which average fewer than 20 employees and less than \$200 000 in annual sales. Many of these small shops are run by artisans who have little business training and are driven by cash-flow, rather than strategic, concerns. The industry is growing at a rate of about 12% per year, although, because the US Department of Commerce aggregates data on all tool-and-die shops, the actual size and growth rate of the plastics-related portion of the industry are unknown.

The players in the industry consist of end manufacturers who require plastic parts for their products, plastic injection moulders who contract with the manufacturers to produce the parts, and the mould (or 'tool') builders who form the necessary moulds out of metal in machine shops. Despite the increasing demand for faster production by manufacturers, the competitive structure of the industry appears to prevent most tool builders from extracting premium prices for faster delivery.

This study was conducted to determine why few mould builders today use lead time as a competitive weapon and the potential for mould builders to use time for strategic advantage. To further understand the mindset of mould builders regarding time competition, extensive structured telephone interviews were conducted with owners of eighteen mould building shops in the state of Tennessee and the Midwest.[5] In these interviews, owners were asked what they felt were the most important attributes a customer considered when choosing a mould builder. There was general agreement that these attributes were quality of workmanship, price, general reputation for being reliable and easy to deal with, accuracy in forecasting delivery dates, and speed of delivery. In particular, all respondents stated that high quality is considered a requirement to remain in business these days.

Subjects were then asked to rank the five attributes in importance. From these rankings, a Thurstone Case V scaling was performed to determine the average relative importance given to these attributes by the mould builders. (Green and Tull, 1975) From the rankings, it was obvious that quality so far exceeded the other attributes in importance that it cannot be treated as a compensatable weakness, i.e. in most cases, no amount of price discount, early delivery, or other inducement can make up for less than the highest quality. The gap between quality and the next most important attribute is far greater than that between any of the other adjacent attributes. Of particular importance to our study, also, is that lead time speed is considered least important of the five attributes. (Note that this scale represents the aggregate rankings of all respondents. Several individuals placed lead time speed close to the top of the rankings.) In fact, most of the respondents insisted that lead time speed was very important, but just not as important as the other attributes. The perception of these mould builders, on average, is that their customers consider speed the least important of the five attributes, when choosing a mould builder.

This result is corroborated by numerous experiences cited by US manufacturers. A former president of the American Mold Builders Association has written, 'The fastest time to market is a major concern today. Some US shops run three to four weeks late, and seem unconcerned, whereas off-shore shops hustle and at least try' (Meinert, 1989). Blackburn (1990) cites a US manufacturer who had bids for tooling from

[5] A full description of the field study is provided in Zahorik, Keiningham and Blackburn (1990b)

US shops and a Hong Kong firm which were essentially comparable in cost. However, the 'lead time to obtain the injection moulding dies was over three months from US manufacturers; the response in Hong Kong was, "When do you need it?"' Similar anecdotes about the difference in attitude toward time between many US shops and their foreign competitors abound.

13.4.1 Moulders of plastic parts

Moulders of plastic parts play a complex intermediary role between manufacturers and mould builders. In our survey of plastic moulders in Tennessee and Kentucky we found little evidence of plastic moulders who would be willing to pay more for moulds in order to get them faster in general. Exceptions exist, of course. Top mould builders could cite several large, sophisticated plastics moulders in the US who are known to look more for mould performance and fast delivery than for price when soliciting bids.

There was little appreciation of strategic benefits for the moulder of generally faster delivery. Most were content with the current level of delivery of their mould builders. In fact, moulders were more concerned about lead-time variance, so that they could schedule their shops smoothly. A typical response was from a moulding plant owner, who said that the average 14-week lead time promised by his usual mould builders was fine – there would be no value in getting things any sooner than that – but that he would not tolerate late shipments. In fact, if a mould arrived three weeks late, it was likely that he 'would never use that mould builder again'. Another moulder said that he would 'prefer to have a mould delivered as promised in 14 weeks than to be told 12 weeks, with the possibility that it would arrive up to two weeks later'.

Ultimately, we found that some plastic parts moulders do compete on speed, although they do not regularly extract premium prices for this attribute. Plastics moulders tend to pass all tooling costs and lead times through to the end manufacturers and are relatively insulated from issues of lead time and early delivery. There appeared to be little value in conducting a conjoint study among plastic moulders about the economic value of speed from their mould builders, because most of them do not think in these terms. As they see it, the value for faster delivery of moulds is determined and paid for by the end manufacturers. The moulders did not feel that they would reap any strategic benefits that would warrant paying from their own pockets for faster tooling.

13.4.2 End manufacturers

Because moulders do not generally believe that they derive any benefits from faster tooling, except keeping their manufacturing customers happy, research must be focused on the manufacturers to find an appreciation for the competitive value of time. However, this makes the use of formal evaluation models, such as conjoint analysis, virtually impossible. The responsibility for tooling in manufacturing companies is diffused across functions and is at different administrative levels in each company. Therefore, our description of manufacturers' valuations of delivery time for moulds is based upon inference from the data already described and on discussions with acknowledged industry leaders.

One indication that many manufacturers still do not give prime importance to time is the continued use of the standard competitive bidding systems. This system has not encouraged time reductions because the practice of awarding contracts primarily on the basis of price leads to unforeseen delays and disruptions. The lowest bid, particularly if made by a shop which has seriously underestimated the complexity and cost of the mould, may result in a lower quality mould which does not work properly when completed. The important day in the contract for a mould is not the day on which the mould is delivered to the moulder, but the day on which the first successful plastic part is made from it. So, to the extent that a mould requires rework or modifications, the low bid may represent false economy. As in any industry, you get what you pay for.

Not all manufacturers, however, choose their mould builders primarily on the basis of price. We stated above that a few mould builders have progressed beyond the highly competitive area of similar companies to establish niches at the forefront of technology and business sophistication which have allowed them to charge higher prices for faster delivery times (among other things.) But, do they have any regular clients, or are they used only as a last resort when manufacturers face otherwise impossible deadlines? It became clear from talking to some of these forefront mould builders that they do indeed have loyal customers, end manufacturers who are knowledgeable about tooling and are themselves time-based competitors, for example GTE, AT&T, IBM, Xerox, Smith Corona, Kodak, Hewlett Packard, Motorola, DEC, Polaroid and Kohler. Many of these companies face competitors, particularly from Japan, who are able to get products to market much faster, in part because of faster tooling. Estimates of the average advantage of Japanese mould builders over their US counterparts differ, but are all substantial. According to George Stalk (1988), 'In plastic injection moulds, Japanese companies can develop the moulds in one-third the time of US competitors and at one-third the cost.' In its study Xerox found that its competitors were getting tooling in half the time at half the cost, due to 'stable designs/fewer changes, the use of job shop production techniques, 3¥7 shift versus 1 in a typical US shop, NC versus attended equipment, the use of semi-skilled and unskilled labor rather than highly skilled tool makers, and wage rate differences.' (Connorton, Greenslade and Williams, 1985.)

IBM is one such time-sensitive firm which has felt extreme time pressure from the Japanese, and has revamped every stage of the product development process to shorten the cycle. In particular, since tooling is a major part of the development cycle, IBM planners now work directly with mould builders to reduce cycle times, and are willing to pay extra to accomplish it. Some of the company's specific time-reducing methods are described in the next section. According to Bob Vines (1989), Early Manufacturing Involvement Manager at IBM, who specifies tooling and manufacturing processes for the production of printers at the Lexington, Kentucky plant, 'Time is everything. It's well worth it to pay double to cut production time in half.'

Xerox's Plastics Commodity Strategy has led to an increased sensitivity of that company to the factors which have put it at a competitive weakness in procuring plastic parts. The study pointed out that plastic parts constituted 10% of product costs on its large copiers, and up to 30% for small desktop machines, and estimated that annual corporate expenditures would grow by 240% between 1985 and 1989. Given the time and cost disadvantages it faced in securing tooling, as cited above, the company has sought to revamp its approach to choosing suppliers. As part of the strategy it defined **model suppliers** for plastic resins, for moulds and for moulded parts. It described the ideal mould builder as one 'who recognized the current and future threat to their business and demonstrated a willingness to aggressively take action', particularly in the area of 'identifying cost effective tool building processes and technical developments in tool building with emphasis on CAD/CAM utilization and the work process changes CAD and CAM can make.' This requirement for advanced technical sophistication is intended, in part, to allow Xerox to link itself with its tool builders and moulders for better communication and reduced production time (Barden and Williams, 1988). These actions appear to have paid off – Xerox won the Malcolm Baldridge Quality Award in 1989.

Not all manufacturers have this appreciation for faster tooling. Many firms seem willing to live with the state of technology provided by the mould suppliers of their regular plastic parts suppliers, and to choose on the basis of price. The Government, in particular, is said to choose mould builders primarily on the basis of price. A manufacturer of industrial equipment told us that, 'all other things being equal', he would certainly prefer to go with a tool builder who was faster, but, when asked how he would respond if the faster tool builder were a bit more expensive, he frankly didn't have an answer.

13.4.3 A better arrangement emerges

For the sophisticated time-sensitive firms a new structure of collaborative arrangements has evolved (see Fig. 13.7). In many cases these manu-

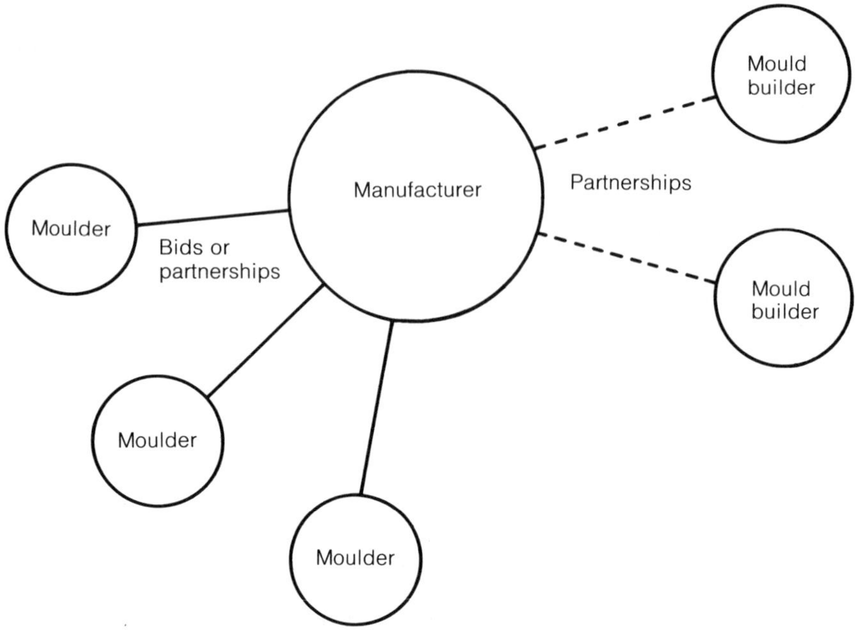

Fig. 13.7 Cooperative alignment used by time-based manufacturers.

facturers now identify fast and highly sophisticated mould builders and establish long-term relationships with them, so that, for example, both can commit to purchasing complementary CAD/CAM systems, and other mutually beneficial time- and cost-reducing expenditures. The manufacturers regard the extra cost of tooling incurred by using these mould builders to be offset by other benefits realized from a long-term partnership. In the new alignment, the manufacturer may request bids from several plastics moulders but will specify its expensive time-based mould shop partner as the mould builder. To capitalize on this new relationship several of the top mould builders we spoke to have vertically integrated by adding their own moulding facilities.

13.4.4 Rewards for time-based competitors

To what extent do those mould builders with fastest average delivery times successfully charge higher prices for their moulds? It is difficult to obtain useful price information from industry data. There are, of course, no published price lists, since each job is unique, the mould builders' circumstances change from one bid to the next, and different sets of competitors vie for each contract. It would be difficult to rigorously assess differences even among average prices without obtaining a great number of, typically proprietary, sets of bids.

Until such a study is undertaken, we must rely on anecdotal information about the industry to see the extent to which the top time-based competitors are well rewarded. Some indicators are these:

1. *Minco Tool* of Dayton, Ohio is widely reputed to charge 20% to 30% more than most other shops. It is one of the largest and most successful shops in the United States (110 employees, sales of nearly $10 million per year) on the basis of extremely high-quality moulds which are guaranteed to work the first time (a time-saving issue), fast delivery times, and 'never having missed a delivery date'. They are used in emergency situations by many manufacturers, but are specified as the mould builder of choice by many time-based manufacturers. They are also vertically integrated, and are capable of producing production runs of small- and medium-sized plastic parts. The company experienced rapid growth from 1976 to 1980, but has capped the growth at its current size in order to insure that quality standards are maintained.
2. *Snider Mold* of Mequon, Wisconsin is a producer of large cavity moulds, and is a very effective time-based competitor. According to the company president, Jim Meinert, the company's capabilities for producing large, extremely high-quality moulds mean that it has virtually no competition, and his business has grown fourfold in just four years. Undaunted by the growth of Asian mould builders, 20% of his output is exported to China. The company works to establish strategic alliances with several regular customers, who collaborate closely on ways to cut production times. For example, when Kohler, a regular customer, is anticipating the need for a new mould, it tells Snider to order the steel immediately and pays for the inventory, saving an important time lag. (Most mould builders do not keep mould steel blanks in inventory.) Snider has also purchased Kohler's design software so that its CAD data can immediately be programmed into Snider's machines.

   According to Meinert, the company never makes the low bid on jobs, although he does feel he must be within 10% of the bottom to be competitive. His advantage is rather in quality, lead time and service. He feels that only a small percentage of his income comes from producing moulds. Most is from engineering support and speed. He intends to integrate vertically into plastic moulding so as to avoid dealing with moulders.

There is an increasing demand for faster tool building for those builders that wish to pursue it, as US manufacturers are forced to react to the Japanese challenge. However, the ability of the US mould industry to meet this demand is limited by the number of shops which have the management skills, aggressiveness and foresight to adopt this new way of thinking. Moreover, those shops which are willing to become time based will need to constrain growth in order to control quality, and may therefore not have sufficient capacity to meet the growing demand for fast tooling. This slack will be taken up readily by a rapidly improving off-shore industry. Only mould builders at the leading edge of technology have been able to set themselves apart from the competition in ways that allow them to charge premium prices for their rapid delivery of moulds. These leading companies are typically able to charge premiums of 20–30% and have achieved rapid growth.

The appreciation for this speed is not universal, however. A survey of plastic moulders found that most see little advantage to themselves in having faster mould suppliers and are content to pass all lead times through to the manufacturers that hired them. Most small mould builders also felt that they had little prospect of charging premium prices for faster mould production. The appreciation for the value of faster tooling time appears to reside with those sophisticated manufacturers who now tend to choose the faster (and more expensive) mould builders, and then direct their plastics moulders to use them. The sophisticated mould builders are starting to move toward vertical integration, acquiring the capacity to mould their own plastic parts. At the same time, the sluggish pace of many mould builders is allowing foreign competitors to make inroads into the market.

13.5 SUMMARY

In all of the industries studied, time proves to be one of the dominant factors in determining market leadership. In some industries, such as plastic injection molding, this is not common knowledge, particularly among the industry laggards. Many of these firms who think that the historically long lead times in their industry will persist are, we think, in mortal danger. Our surveys indicate that certain customers, usually the bigger players, are willing to pay a premium for quick response, and these customers are developing long-term relationships with the time-focused suppliers.

In wholesale book distribution product-line breadth, or variety, is also an important factor in a retailer's decision concerning the supplier of first resort. In this industry buyers are concerned about all three factors – time, price and variety – and are willing to make trade-offs, but competitive pressure is forcing distributors to improve along all three dimensions. We suggest, based on the research presented here, that these observations apply not only to book distribution, but to wholesale distribution in general.

Our overall finding is that there are indeed measurable benefits accruing from reduction in time and increased variety. These results can be quantified clearly, and the trade-offs defined. Managers can then devise appropriate strategies that translate into powerful competitive advantages.

13.6 ACKNOWLEDGEMENT

Funding for this research was provided by The Boston Consulting Group, The Dean's Fund and the Operations Roundtable of the Owen Graduate School of Management, Vanderbilt University.

13.7 REFERENCES

Barden, Jim and Williams, Ed. (1988) *Plastics Commodity Update*, Xerox Corporation Business Studies Products and Systems Group, Webster, NY.

Blackburn, Joseph D. (1990) Time-based competition, in *Strategic Manufacturing: Dynamic New Directions for the 1990s* (ed. Patricia E. Moddy), Irwin, Homewood, IL.

Connorton, Larry, Greenslade, Bob and Williams, Ed (1985) *Xerox Corporation Plastics Commodity Strategy*, Xerox Reprographics Business Group, Webster, NY.

Green, Paul E. and Tull, Donald S. (1975) *Research for Marketing Decisions*, 3rd edn, Prentice-Hall, Englewood Cliffs, NJ.

Lindsley, W., Elrod, T. and Blackburn, J. (1990) Time and variety competition in the book industry. Operations Roundtable research paper, Owen Graduate School of Management, Vanderbilt University.

Meinert, James (1989) *Competing in the World Market: How US Plastic Tooling Companies can get their Fair Share of the Pie*, Snider Mold Company, Mequon, Wisconsin.

Stalk, George, Jr (1988) Time – the next source of competitive advantage. *Harvard Business Review*, July–August, 41–51.

Vines, Bob (1989) Personal communication.

Zahorik, A., Keiningham, T. and Blackburn, J. (1990a) Time competition in the machine tool industry. Operations Roundtable research paper, Owen Graduate School of Management, Vanderbilt University.

Zahorik, A., Keiningham, T. and Blackburn, J. (1990b) Time based competition in the American plastic injection mold industry. Operations Roundtable research paper, Owen Graduate School of Management, Vanderbilt University.

CHAPTER 14

Flexibility in the context of manufacturing strategy

Stuart Chambers

The simplicity of the word 'flexibility' belies the very considerable complexity of the subject. Yet it is perhaps because of its apparent simplicity that the word has been absorbed into the loose everyday language of manufacturing businesses (and increasingly in services, too), to the extent that everything that carries the label 'flexible' must be good. It is not surprising, then, that the marketers of new capital equipment and systems attach great weight to this attribute, regardless of the manner in which flexibility is provided. Yet it must remain the buyer's responsibility clearly to define and select what type of flexibility, if any, is needed. This paper proposes that such clarity can be achieved through manufacturing strategy and provides a simple analytical framework which can be used to link flexibility types with the stages of manufacturing strategy development outlined by Hill (1989).

14.1 THE QUEST FOR FLEXIBILITY

There are many reasons why the issue of flexibility has assumed an increasingly important status in manufacturing over the last decade. These can broadly be categorized into two parts – major changes in market requirements, and new opportunities offered by processing equipment and systems technology. The combination of these changes creates a climate where the acquisition or development of flexibility becomes an 'obvious' strategic solution, but also where costly mis-

This paper was presented at the Annual Conference of the UK Operations Management Association, 25/26 June 1990 at the University of Warwick, England

Specially commissioned for *Manufacturing Strategy – Process and content.*
Edited by Christopher A. Voss.
Published in 1992 by Chapman & Hall, London ISBN 0 412 43660 4.

judgements are made which can damage a business's long-term competitiveness.

14.1.1 Market changes

Almost all manufacturing companies have experienced unprecedented increases in global, national and/or local competition for orders, which has had many repercussions including:

1. Reduced total volumes (often associated with a fall in added value and contribution) result in the need to reduce manpower and plant. Consequently, each remaining employee or item of plant had to be capable of performing a wider range of tasks.
2. Diminishing order volumes required more set-ups, which gradually became a higher proportion of total product cost, and also absorbed capacity.
3. Shortening order lead times required manufacturing to respond with reduced order backlogs and/or shortened processing times.
4. Many customers increasingly exercised their 'right' to vary schedules in terms of both quantity and timing.
5. Product proliferation and customization led to a widening of technical requirements, increased variety and reduced order volumes.
6. Shortening OE product life cycles reduced the viability of hard tooling and of dedicated plant.
7. Economic and policy pressures arose for reductions in inventory, regardless of its function in the manufacturing system.
8. Increasing numbers of customers demanded high levels of delivery reliability to support their efforts to reduce inventory (just-in-time policies included).

14.1.2 Processing equipment and systems technology

The equipment available to most industries has changed radically in the last decade as a result both of the widespread incorporation of programmable control and of redesign to improve features, including consistency of quality, processing speed and set-up time. These characteristics are often marketed in terms of providing the user with enhanced flexibility, but usually at a premium price.

14.2 THE NEED FOR A STRATEGIC APPROACH

The characteristics of all decisions concerning the extent of flexibility should place the subject in the domain of manufacturing strategy, as described by Hayes and Wheelwright (1984):

1. Extended time horizon,
2. Significant long-term impact,
3. Concentration of effort on chosen activities,
4. Series of decisions over time must be supportive and consistent,
5. Pervasive – the activities of all levels of the organization must support the strategy.

Skinner (1985) argues that flexibility should be considered in a strategic context, particularly at the time of process investment. Hayes and Wheelwright (1984) point out that it can be the basis of competitive advantage, and Buffa (1984) comments that there can be a premium price for responsiveness to customer needs. Most papers on flexibility confirm that it should be a strategic issue, encompassing infrastructure (subsystem) developments as well as process aspects.

Changes in flexibility are clearly not just technology issues to be considered or specified by process engineers. The ramifications of proposed changes must be understood and managed by the business and evaluated against an understanding of the needs of its markets. To achieve this, there must be a shared use of language and a sound conceptual basis for the work. Then it should *not* be necessary, and may be highly undesirable, to rely on 'acts of faith . . . from engineers or manufacturing managers' as a basis of 'evolutionary implementations (of flexible manufacturing)' as proposed by Macbeth (1900).

14.3 MANUFACTURING STRATEGY FORMULATION

Most approaches to the development of manufacturing strategy have highlighted the need for manufacturing to orientate towards supporting market needs, and also to be involved proactively in the business strategy development. The works of Hayes and Wheelwright (1984) and Skinner (1985) have suggested that manufacturing priorities should be established from assessments of market requirements, as perceived or analysed from a top-down (or macro) stance. Whilst these insights have been crucial in the advancement of the subject, their practical application by busy executives could easily involve blinkered perspectives and shallow analysis.

Most markets, as served by individual businesses, involve a wide range of volumes and market needs which must be clearly understood as part of the manufacturing strategy development process. This demands a methodology which forces executives to examine their business from the level of orders, products and customers: a macro approach. This was provided by Hill (1985) and refined by him in 1989 (Hill, 1989). In my research I have been able to apply the methodology to some 30 businesses over the last five years, and have found the approach to be a highly pragmatic and effective means of enabling executives to integrate elements of manufacturing strategy with the corporate/marketing strategies and market requirements.

The core of this methodology is a five-step framework (Table 14.1), reproduced below which is used in an iterative manner, allowing feedback from each stage to be used to modify other input until a high degree of coherence is achieved.

What differentiates this approach is the application of the concept of **order winning** and **qualifying criteria**, which form the essential links between marketing, the market and manufacturing. This paper will not attempt to present any detail of this approach, which is adequately covered in other writings, but my research has highlighted certain aspects of the work which must be understood clearly.

1. Order-winning and qualifying criteria usually vary widely across a company's total range of products and customers.
2. Marketing descriptions (segments, market names etc.) are usually not a suitable basis for this work, as the criteria often differ within these categories.
3. Order-winning analysis highlights anticipated changes in market requirements over agreed time periods – hence providing the basis for strategic review; future needs are considered.
4. The methodology provides the basis for an agreed view of emphasis and trends. It provides strategic options rather than prescriptive 'solutions'.
5. It leads to a manufacturing strategy oriented towards those factors

Table 14.1 Five step manufacturing strategy framework

Corporate objectives Step 1	*Marketing strategy Step 2*	*How do products win orders in the market place? Step 3*	*Manufacturing strategy*	
			Process choice Step 4	*Infrastructure Step 5*
Growth	Product	Price	Choice of	Function
Survival	markets and	Quality	processes	support
Profit	segments	Delivery	Trade-offs in	Planning and
ROI	Range	speed	process	control
Others	Mix	reliability	choice	systems
	Volumes	Demand	Process	QA/QC
	Standardization/	increases	positioning	Procedures
	customization	Colour range	Capacity	Payment
	Level of	Product range	size	system
	innovation	Design	timing	Work
	Leader/follower	leadership	ocation	structuring
		Technical	Role of	Organisational
		support	inventory	structure

which generate orders in the selected markets, rather than towards narrowly established goals (e.g. reduced costs) which may, or may not, meet market needs.

The order-winning and qualifying criteria which fall within the scope of manufacturing may include weighted combinations of price, quality, delivery speed and delivery reliability. By a late stage in the development of a manufacturing strategy, executives from each function in the business will have gained a good understanding of the meaning and emphasis expressed by these criteria.

Order-winning criteria define the manufacturing tasks which must be provided effectively by the elements of manufacturing strategy. Where there is a mixture of opposing criteria (e.g. price and delivery speed), one manufacturing option is to divide into units and create focus (Skinner, 1974), resulting in a reduced variety of order winners. Where this is not possible, manufacturing must provide the conditions that best match the dominant criteria and, at the same time, provide sufficient and appropriate flexibility to support the other criteria. Alternatively, manufacturing strategy can offer the business option of narrowing the range of criteria by withdrawing from or by altering the offering to certain incompatible markets.

In almost all businesses, order-winning criteria are not immediately obvious, and there is little data or analysis to confirm their existence or relative importance. For this reason perhaps, the concept is not yet widely used in businesses, and the best focus for manufacturing effort remains undefined. Thus many businesses expect their operations to react to whatever is demanded of them at any time, whatever the consequences in terms of effectiveness or efficiency. It is not surprising, therefore, that major changes in processes and infrastructure are only based on the hard data of product and volume, and the soft analysis of order-winning and qualifying criteria is not even considered. In terms of the manufacturing strategy framework, Steps 1 and 2 are taken into account in determining the process investments (Step 4) but Steps 3 and 5 are avoided, often with disastrous business consequences.

Step 3 is critical if manufacturing and corporate/marketing strategies are to act in unison in an increasingly competitive environment. Moreover, it provides a sound basis for the definition of flexibility types, for their role in competitive strategy, and for a review of the elements of the manufacturing system which provides each type.

14.4 TYPES OF FLEXIBILITY

There have been many attempts to define different types of flexibility. Broad categories are identified by Skinner (1985): process, product and volume flexibility. Other writers highlight types of flexibility that may fall between, or outside, these categories. For example, Gerwin (1987) defines rerouting flexibility as the degree to which the operating sequence through which the parts flow can be changed in the event of machine downtime. This will depend both on product design features (i.e. which processes are required) and on operating features (i.e. the extent of duplication and/or interchangeability of equipment). Slack (1983) introduces delivery flexibility as the ability of the system to shorten or lengthen its delivery time; that is, to bring forward or put back production. This type has not been widely recognized, perhaps because it is determined by factors described by Slack as 'soft' – i.e. difficult to define or specify. Table 14.2 summarizes various categories of flexibility reported by the main writers on this subject.

In order to define clearly what flexibilities are required of, or available from, a manufacturing strategy, it is first desirable to provide unique classification of each flexibility type, which can then be used generally in all strategy developments. Eight unique categories of flexibility have been found to be important in this work, and are redefined below.

14.4.1 Technical range flexibility

This is the ability to manufacture new or potential products that have processing requirements that are different from those of existing products

Table 14.2 Types of flexibility identified by selected writers

Manufacturing strategy books			*Papers on flexibility*		
Skinner (1985)	*Hayes and wheelwright (1984)*	*Buffa (1984)*	*Slack (1983)*	*Macbeth (1985)*	*Gerwin (1987)*
Process		Set-up		Set-up time	
				Routing	Rerouting sequencing material
			Quality level		
Product	Product	Product variety	Novel prods	Product	Modification
			Product mix	Product mix	
					Changeover
Volume	Volume		Aggregate output volume		Volume
				Order size	
			Delivery time		

(e.g. closer tolerances, additional complex features, size and design features).

This is primarily an issue of selected equipment technology, whether provided internally or by sub-contractors, but may also involve process choice (e.g. jobbing uses general purpose equipment and high skill levels to provide this flexibility). Being technology-based, technical range flexibility can be well defined and is often promoted by process engineers.

14.4.2 Volume-related flexibilities

Volume flexibility refers to the provision for changes in aggregate output as demand varies. However, my experience in developing manufacturing strategies has shown that volume changes must be more precisely defined, as each type of change has a different impact on the choices of manufacturing response.

(a) Volume flexibility

Total volume can change over time (incremental or step change) irrespective of the level of short-term or seasonal fluctuations. Manufacturing's ability to handle these changes will be referred to as volume flexibility, and will largely be an issue of capacity management. The choice of process will be a major factor, with jobbing providing the greatest potential for incremental change as plant utilization would not usually be a constraint, and heavily loaded line processes having the least volume flexibility. Vertical integration decisions will be critical; for example, the use of number of sub contractors may provide much greater volume flexibility than alternative in-house capability.

(b) Volume mix flexibility

Having identified changes caused by new product designs (technical range flexibility), all other changes in product mix have volume ramifications. Hence, if the volume of some products increases and that of others decreases, then the mix changes even though the total volume may not change. As with volume flexibility, the system's ability to cope (volume mix flexibility) depends not only on overall capacity, but particularly on key processes. Thus, manufacturing strategy should be concerned about the analysis of actual or potential bottlenecks and about the extent of actual interchangeability/capability between processes and tooling. Product mix flexibility will also depend on scheduling procedures and on the way work is structured (which will dictate the extent to which employees can be provided to support the key processes).

(c) Seasonality flexibility

In many industries demand varies with seasonality such that capacity is alternately overstrained and underutilized, yet with neither significant changes in product mix, nor underlying growth and decline. Whilst many of the manufacturing responses can be similar to those required in (a) and (b), it may increasingly be important to consider work structuring that facilitates the transfer of support staff into productive roles during peak seasonality.

My research has shown that marketing functions are often not aware of the actual extent and impact of seasonality on the cost and effectiveness of operations, and this must, therefore, be part of the manufacturing analysis and 'feedback' in the strategy development. Unlike other forms of volume change, the existence of seasonality should be reflected in changing order-winning criteria. Thus, at periods of maximum demand, price may be seen to be less critical and delivery speed (e.g. ex-stock from finished goods inventory) more important. An agreed assessment of these changes can lead to pricing and inventory policies which maximize profitability and at the same time help to level demand.

(d) Delivery speed flexibility

Delivery speed can be required under two types of condition:

1. Where customer order lead times are less than the sum of order backlog (if any) and process lead time; and
2. Where the customer changes schedules (timing or quantity) within the process lead time.

In both situations, manufacturing must provide the ability to respond with short-term increases in capacity (e.g. overtime), or by rescheduling, or using a combination of both. This delivery speed flexibility is described by Hayes and Wheelwright as volume flexibility: the ability to accelerate, decelerate or juggle orders. Slack refers to delivery flexibility in this context.

Delivery speed flexibility is dependent on many process and infrastructure factors. Not only does it depend on the extent of excess capacity, and on all the factors referred under volume flexibility types, but it also involves all procedures in processing orders from their receipt to final delivery. Thus, manufacturing strategies may justifiably review all order-processing administrative procedures which delay orders reaching manufacturing (in some companies studied, this process absorbed up to half of the customer order lead time). Existing scheduling procedures often group orders together in some way to minimize the number and cost of set-ups, but this also can absorb lead time, thus reducing delivery speed flexibility. The ability to change schedules will be affected by the availability of tools (tool preparation/refurbishment procedures) and of raw materials (inventory policies and purchasing procedures).

Although delivery speed requirements may place severe demands throughout the manufacturing system, most companies do not monitor the extent and trends of these demands, and they are often ignored initially or underrated when order-winning criteria are assessed by marketing staff. The application of the Hill framework, along with appropriate analysis of orders, will identify those customers/products/orders which require delivery speed flexibility. This type of flexibility has little in common with technical range flexibility, which has so often been used as the only basis for process investment decisions.

(e) Set-up flexibilities

Reducing the *duration* of set-up (e.g. by improved method or investment in plant) is normally intended to reduce costs, thus allowing a business to compete for smaller orders without loss of margin. There will be gains in capacity which are often not the primary objective. **Set-up reduction flexibility** supports price as an order winner. The average cost of set-ups can be reduced by scheduling together orders with similar processing requirements (families), but this reduces delivery speed flexibility. Procedures that ensure the readiness of tooling and materials can have a substantial impact on set-up reduction.

Set-up timing flexibility concerns the extent to which changes from one order to another can be conducted *when* required; this will often be constrained by limitations in the capacity of set-up personnel or equipment.

(f) Quality flexibility

This was defined by Slack as the ability of the system to change the quality level of one or more of its products. In the context of manufacturing strategy and order winners, quality flexibility would be required if orders for certain products could be won with differential levels of quality conformance (e.g. zero defects or specially designed QC systems). This may require special quality-control design and different work-structuring arrangements such as plant-within-plant. Additionally, it may require selection of preferred sub-contractors and identification of specific equipment at given workcentres.

14.5 FLEXIBILITY WITHIN THE MANUFACTURING STRATEGY

Manufacturing strategy should be developed in congruence with the corporate/marketing requirements (incorporating product characteristics

Table 14.3 Analysis of flexibility in the context of manufacturing strategy

Development of a manufacturing strategy	*Step number*	*1 + 2*				*3*			
	Description	*Define corporate and marketing strategies*				*Define/evaluate order-winning and qualifying criteria*			
	Detail	*Products*	*Volume of demand*			*Delivery speed /reliability*		*Price*	*Quality*
			Aggregate	*Product mix*	*Product seasonality*				
Relevant type of flexibility to be considered (example)		*Technical range flexibility*	*Volume flexibility*	*Volume mix flexibility*	*Seasonality flexibility*	*Delivery speed flexibility*	*Set-up timing flexibility*	*Set-up reduction flexibility*	*Quality flexibility*
Step 4 (process)	Process choice	*	*	*					
	Equipment technology								
	Set-ups							*	
	Interchangeability			*	*	*	*		
	Range coverage	*							
	Capacity								
	Machines/work centres		*	*	*	*	*		
	Key processes			*	*	*	*		
	Support services (e.g. cranes)		*				*	*	

	Inventory								
	Raw material					*	*		
	Vertical integration	*	*	*	*	*			*
Step 5 (infras-tructure)	Set-up support capacity					*	*	*	
	Procedures								
	Order processing					*	*		
	Scheduling			*		*	*		
	Tool preparation					*	*	*	
	Delivery					*			
	Work structuring/skill	*		*	*	*	*	*	*
	Quality control design					*			*

Note: The asterisks (*) indicate aspects of flexibility/manufacturing strategy considered in a specific company. They should not be used as a universal indicator.

and aggregate volume changes) and with the market requirements, as defined by order-winning criteria. For most companies, all these requirements are in a continual state of flux, and must regularly be monitored and forecast.

Since flexibility is the ability to respond effectively to changing circumstances, it should be considered as an interface between the causes of change and the means of response, both of which are monitored by the steps in the manufacturing strategy framework. Thus in Hill's methodology, Steps 1, 2 and 3 would identify all the types of flexibility requirements and Steps 4 and 5 would review all the factors within the manufacturing system which can provide that flexibility. Flexibility lies at the boundary between competitive priorities and the capabilities of the manufacturing system.

Table 14.3 outlines a framework which shows the alignment of flexibility types against the elements of the Hill strategy model. Some of the flexibility types such as volume flexibility, relate primarily to the corporate issues (Hill's Step 1), whereas others more closely relate to the order winners which have to be supported by manufacturing.

The system design variables for process and infrastructure (Hill's Steps 4 and 5) and arranged at the side of the table, forming a matrix, which can then be completed by *individual* businesses, as appropriate. A typical analysis is shown in completed form. This methodology has been found to be useful in a wide variety of circumstances. It encourages executives to consider all the ramifications of proposed investments in process or infrastructure, and highlights aspects of strategy development that must be examined as changes occur in markets, volumes, and order winners.

14.6 CONCLUSION

The framework outlined in this paper places a full range of types of flexibility into a sound manufacturing strategy context. Its use can help to counter the view that *any* form of flexibility is desirable, and can highlight trade-offs between one type of flexibility and another. It helps to illustrate that a manufacturing strategy is complex and unique to each business. This model should be seen as a foundation on which others should build, and will add to the conceptual base which can be used by an experienced practitioner. It is certainly not intended to be a check-list or packaged solution; but should be a means of stimulating a meaningful debate and analysis concerning the essential details which comprise manufacturing strategy. Its use should reduce the risk of major investment being justified on one flexibility assumption, and subsequently being condemned for the lack of other flexibilities.

14.7 ACKNOWLEDGEMENT

The research work described in this paper refers in part to a recent three-year study funded by the Science and Engineering Research Council (ACME), which further developed the Hill Framework.

The work was carried out by Terry Hill (1989) and myself; several case studies in the 1989 edition of *Manufacturing Strategy* were generated in the process.

14.8 REFERENCES

Buffa, E.S. (1984) *Meeting the Competitive Challenge: Manufacturing Strategy of U S Companies*, Irwin, Homewood, IL.

Gerwin, D. (1987) An agenda for research on the flexibility of manufacturing processes. *IJOPM*, **7**(1), 38–49.

Hayes, R.H. and Wheelwright, S.C. (1984) *Restoring our Competitive Edge*, Wiley, New York.

Hill, T.J. (1985) *Manufacturing Strategy*, Macmillan Education, London.

Hill, T.J. (1989) *Manufacturing Strategy, Test and Cases*, Irwin, Homewood, IL.

Macbeth, D.K. (1985) The flexible manufacturing mission – some implications for management. *IJOPM* **5**, 1.

Skinner, W. (1974) The focused factory. *HBR*, May–June, 115.

Skinner, Wickham (1985) *Manufacturing: The Formidable Competitive Weapon*, Wiley, New York.

Slack, N. (1983) Flexibility as a manufacturing objective. *IJOPM*, **3**(3), 4–13.

CHAPTER 15

Manufacturing focus – a comprehensive view

Quarterman Lee

15.1 INTRODUCTION

Focused manufacturing links manufacturing facilities to the competitive factors of the business. It enables a company to gain greater control of its competitive position and centralizes focus on its relative competitive advantage.

Among the most difficult of manufacturing tasks is responding to many disparate market demands. Manufacturing system complexity often exaggerates the difficulty. Rather than blaming manufacturing for its problems, the company should recognize that manufacturing is complex and has profound influence on corporate strategy. There are ways to reduce this complexity with subsequent benefits for doing so. Focused manufacturing is one such way.

A simplistic view of focus is 'variety reduction' and market sector reduction. This option may not be a wise one. A narrow product or process range alone is not necessarily focus. The key is to concentrate the entire plant on the fundamental tasks demanded by the plant's overall strategy and marketing objective. This is a first approach.

A meaningful level of focus is rarely understood let alone achieved. Examples will be drawn from our consulting experiences.

Focused manufacturing limits activity in an organization to a manageable and consistent set of tasks. These tasks directly support the firm's marketing strategy. Doing this concentrates expertise and promotes superior performance albeit in a narrow range.

Specially commissioned for *Manufacturing Strategy – Process and content*.
Edited by Christopher A. Voss.
Published in 1992 by Chapman & Hall, London ISBN 0 412 43660 4.

15.2 SCOPE AND SCALE

Scale refers to the size of a given plant or manufacturing unit. Scope refers to the number of significantly different products, processes, markets and regions which the facility serves.

Early experience in the basic and mass production industries illustrated the economies of scale. Larger equipment and larger plants appeared to bring lower cost. As Hayes and Wheelwright (1984) point out, there are also diseconomies of scale. Large factories are difficult to manage and control. Overhead costs often rise as factories grow larger. If increased size also brings increased numbers of processes, products or markets, the undesirable effects compound.

Making several small factories from a large one does not necessarily improve performance. If the same scope just exists on a smaller scale, performance is, in fact, likely to decrease.

15.3 FOCUS CRITERIA

Reducing scale is easy; reducing scope is more difficult. It requires some basis for segregating operations. This basis or **focus criteria** may take several forms. The more common are:

1. *Process.* A particular process may require high investment and large volume for economic operation. Alternatively, special skills may dictate a concentration based on a particular process. Process focused plants concentrate technical expertise but are difficult to coordinate and control.
2. *Product.* A dedicated facility manufactures a product or group of similar products. Product focus promotes quality, reduces inventory and improves response to changing demand. It often requires broader skills and may reduce new product flexibility.
3. *Market.* Market segments are a basis for focus when customer demands differ substantially. Quality, delivery, option variety and order size may differ between market segments. If so, these market criteria provide a basis for focus.
4. *Volume.* Similar products (or identical products for that matter) may require separate processes and infrastructure for high and low delivery volumes. Production volume can therefore dictate focus.
5. *Geographic.* Physical proximity to customers, supplies or to special skills may be necessary. Here, geography may be a basis for focus. This often arises where delivery speed is mandatory. It also occurs internationally when import restrictions apply.
6. *Infrastructure.* Support systems are, by definition, peripheral to the manufacturing process. However, they and other dominant site

factors can impact organization structures and the associated facilities requirements.

15.4 FOCUS LEVELS

Although the concept of manufacturing focus originally evolved at the site or factory level, it also applies at other levels. Focus dictates (or should dictate) which site addresses each product, process, market, geography or volume. Each site would specialize along the selected focus dimension.

At the regional or global level, particular countries or regions might specialize. For many years the large automobile firms built their large cars in North America and small cars in Europe. Just below the site level, a facility might have two or more **plants-within-a-plant** (PWP). Each PWP is an independent factory with its own infrastructure.

Focus also applies below the PWP level. A layout cell is a space which contains a set of complementary machines, fixtures, activities or people. Layout cells often (but not always) correspond to organizational departments. Departments or cells may specialize by process, product or other suitable focus criteria. This topic will be discussed later in more detail.

Focus can develop at the workstation level. Individuals may specialize by product, process, customer or (again) any suitable focus criteria.

In planning manufacturing strategy, focus issues should be examined at each level since they often differ. A company may decide, for example, on a geographic focus at the site level. At the cell level it may choose a functional focus.

15.5 FOCUS AND PLANT LAYOUT

The plant layout, site plan or map reflects manufacturing focus (or lack thereof). The special case of a plant or site is particularly important and warrants additional discussion.

Site-level focus decisions may involve the full range of focus criteria. The plant layout designer must eventually arrange machines and equipment. Such arrangements require decisions between product and process focus.

15.6 PRODUCTION MODES

Production mode is the fundamental arrangement and method for manufacture. Fixed location, job shop and mass production were early versions. Plant layout writers of the 1950s recognized the connection with layout and referred to it as **layout style**.

Production mode is characterized by several key features:

1. Physical arrangement,
2. Focus,
3. Product flow,
4. Pattern of resource flow,
5. Task sequence,
6. Task cycle,
7. Task repetition,
8. Task balance,
9. Task synchronization,

While many combinations of key features are possible, only a few have practical consistency. Within a facility, however, several modes may exist simultaneously in separate areas. The usual modes are project, functional, toyota, cellular, line and continuous. The modes and their characteristic features are illustrated in Fig. 15.1, and the flow characteristics in Fig. 15.2.

1. *Project mode* is usually associated with a fixed position layout. The product is completed or essentially completed in one location. Products are usually large, e.g. a building or a ship. Resources are brought to the one point and the product is immobile or relatively so.

	Features \ Mode	Continuous	Line	Toyota	Cellular	Functional	Project
	Arrangement	Sequential	⟵			⟶	Random
	Focus	Product	⟵		⟶	Process	
	Product flow	Simple and direct	⟵			⟶	Complex and mixed
	Resource flow	Fixed	⟵			⟶	Very flexible
T	Sequence	Fixed	⟵			⟶	Very flexible
A	Cycle time	Very low	⟵			⟶	Very high
S	Repetition	Very high	⟵			⟶	Very low
K	Balance	Very high	⟵			⟶	Very low
	Synchronization	Very high	⟵			⟶	Very low

Fig. 15.1 Production modes: characteristic features.

2. *Functional modes* are characterized by groupings of functional specialties. Similar machines and equipment form a functional department. Materials move from one functional area to another. The result is a complex material movement pattern. Labour typically possesses one functional skill, with operators concentrating on their own efficiency. This and inter-functional moves retard material velocity.
3. *Cellular modes* have work cells processing families of parts. In a group technology cell, a small number of work locations each have a primary group of operators. The emphasis is on group effort and individuals may move with the work piece from machine to machine through the cell. Machine utilization is not emphasized. Small batch

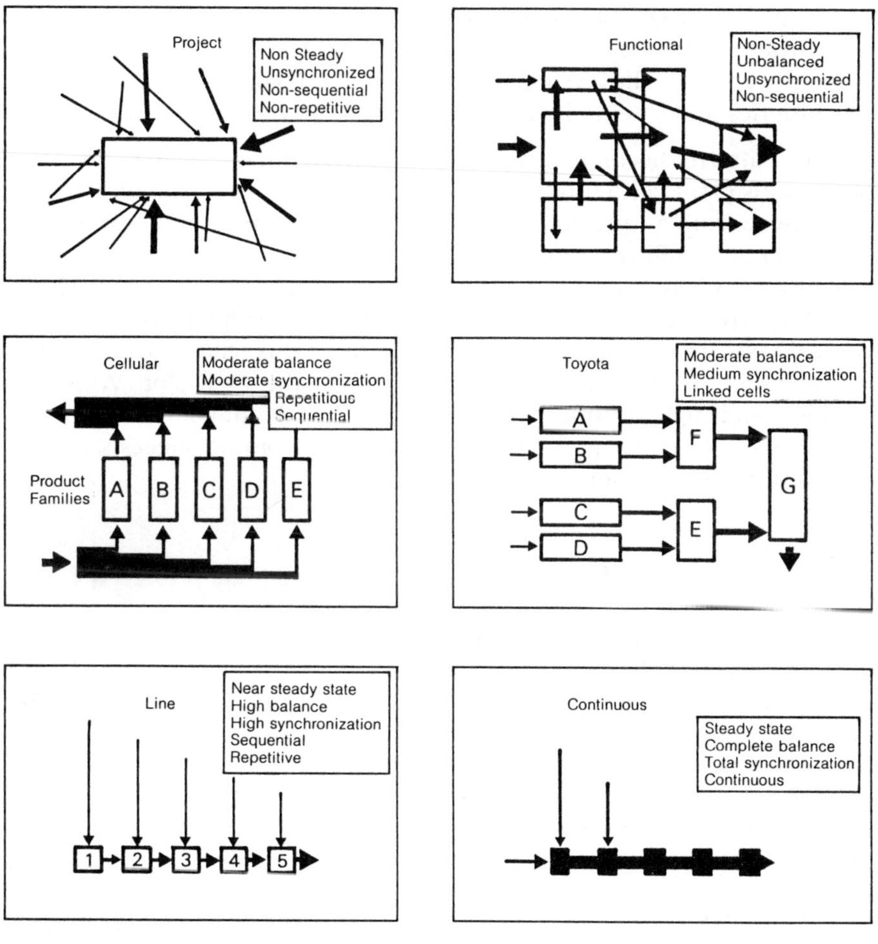

Fig. 15.2 Material flow patterns.

sizes and quick throughput are the benefits. The flow patterns may be straight through, L or U shaped.

4. *The Toyota system* incorporates linked cells with minimum in-process inventory and a pull system of material flow control. Just-in-time inventory, Kanban production control, total quality concepts, scientific maintenance and Vendor families are considered to be part of the Toyota system.
5. *The line or detroit method* is a sequenced flow pattern with fixed work stations and short cycle operations linked by a moving line. Material is delivered to the beginning of the line or to work stations in batches. Products are completed by the time they arrive at the end of the line. Flow line work is characterized by high-volume, repetitive, short-cycle work.
6. *Continuous flow patterns* have few system entry points. Materials flow through a linked sequence of unit processes with fixed means of transfer. The process operates continuously, products are homogeneous, the system is saturated and volumes are high. Oil refineries and steel mills are examples of continuous flow processes.

Flow patterns become less complex with cellular, Toyota, line and continuous plant configurations. Material. velocity increases with movement away from a functional layout. This has important implications for the material handling designer.

Figure 15.3 is typical of many product–volume profiles. On this chart, each bar represents a product or group of highly similar products. The vertical axis represents production volume. This chart overlays the typical range of application of each mode on the *P–V* plot. With few products and high volume, continuous and line production are normal. At the opposite extreme, high variety and low volume suggests project or functional production.

15.7 DEGREES OF FOCUS

Only line and continuous production can achieve pure product focus. Only the functional mode can achieve a pure process focus. Cellular and Toyota modes are intermediate. Line and continuous modes can be intermediate between product and process focus when they produce more than one product. The project mode is a special case which fits neither the product nor process category.

There are many ways to mix focus. Parts and materials may move through a process focused machine shop, for example. Product focused assembly lines might then perform final assembly.

In Table 15.1, the advantages and disadvantages of product and process focus are illustrated.

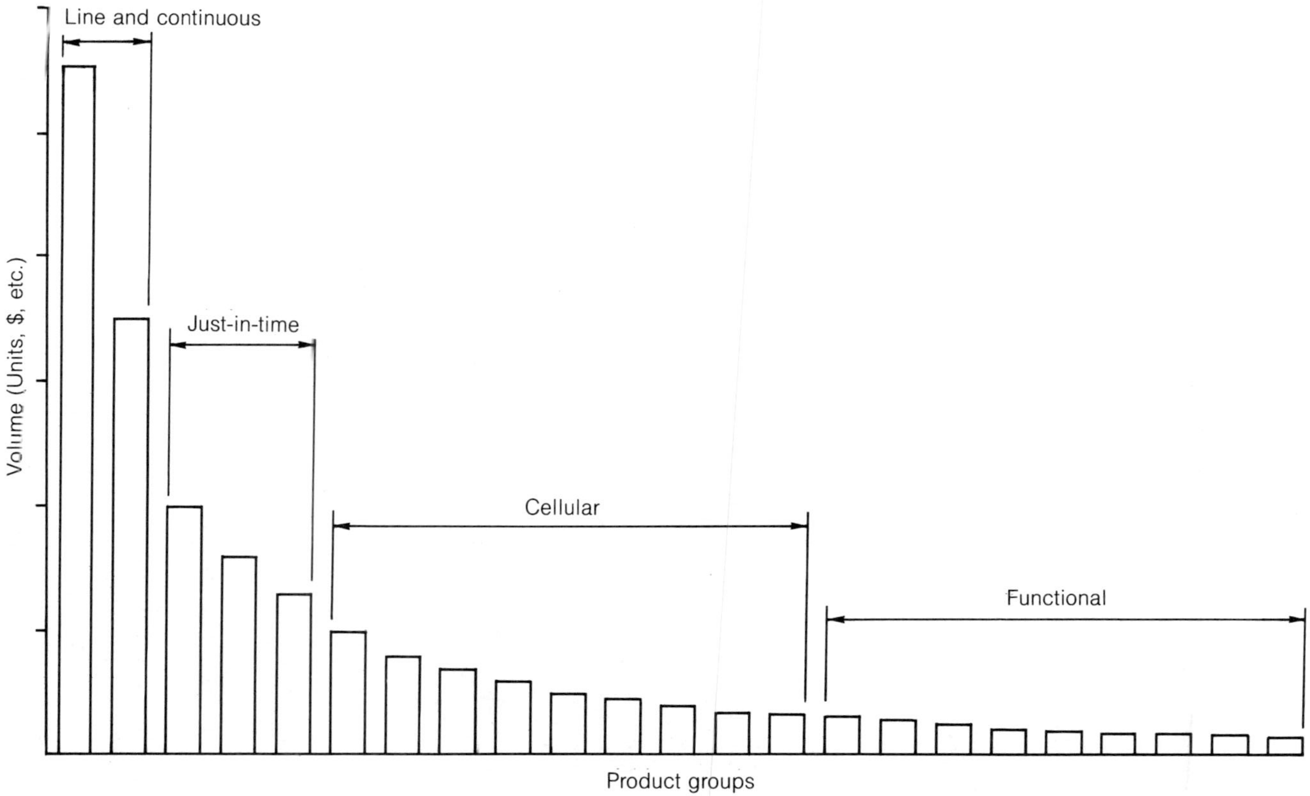

Fig. 15.3 Product-volume analysis.

Table 15.1 Focus comparison

Product focus	*Process focus*
Advantages	
Simplified cost control	Traditional managers comfortable with this approach
Greater cooperation across functional disciplines	Ability to share scarce resources and increase utilization
More responsive purchasing and support services	Typically used in industries like chemical processing
Flatter organization structure	Suited to situations with a heterogeneous labour force
Gearter accountability and responsibility lower in the plant	Usually, the units are cost centres
Simpler plant controls	
Faster product velocity through greater inventory turn, hence, reduced cost	
Simplified and more focused quality improvement (QI) efforts	
Used in situations with lower 'economies of scale'	
More responsive to customer needs	
Overall overhead typically reduced	
More amendable to being managed as a profit centre	
Disadvantages	
Some duplication of key management positions	Increased friction between functional managers
Potential for lower machine utilization	Higher control costs and increased probability of error
Traditional managers are often uncomfortable with this focus	Product costing difficult
Sometimes has trouble handling very different new products	Effective quality improvement (QI) efforts are more difficult
Shop floor 'monuments' must still be shared	Difficult to downsize
	While conventional thought has process focus as being the most flexible, is that it typically is less flexible than a product focused plant

15.8 A FOCUS ALGORITHM

In the author's consulting practice, inappropriate manufacturing focus is frequently seen built into plant layouts. In the usual case, a functional layout is producing high or medium volume products. An example is a manufacturer of minicomputers. This company had five basic models and assembled 3–10 of each per day in a large functional assembly area. Its new plant was arranged with a small assembly line for each basic model. A smaller functional area built prototypes and replacements.

Most manufacturing (and other organizations as well) display a strong bias for the functional mode. The reasons for this bias are unclear but some possibilities are:

1. Functional layouts are often easier to design.
2. Accounting systems do not discourage the large inventory required by functional layouts.
3. Financial policies emphasize high equipment utilization and, in theory, favour functional layouts.
4. Engineers favour high-tech, costly, large-scale equipment which demands high utilization. This, too, favours the functional layout.

Manufacturers misuse the functional mode most often. However, any mode has the potential for misapplication. In one situation, a Detroit-style assembly line builds massive, off-road vehicles at 1.5 per day. The results are poor.

How can the layout designer achieve an optimum degree and mix of manufacturing focus? The algorithm given in Fig. 15.4 can help. The product focused modes, especially line and continuous, offer many advantages in quality, low inventory and efficiency. The Process focused functional mode is most frequently misapplied. For these reasons, the algorithm starts with a pure product focus and line production. It then backs away through the Toyota, cellular and functional modes towards a feasible alternative.

Start with an operation process chart for each distinct product. Any industrial engineering handbook or text shows the conventions for constructing these charts. Line the charts up, side by side as shown in Fig. 15.5.

Each operation on a product is a **product operation** (PO). In the typical situation, a single PO requires too little time and equipment for a dedicated workstation or department. The layout designer must somehow group various POs.

One method is to group all product operations which require the same equipment or type of process. Envelopes 1 and 2 in Fig. 15.5 illustrate this. Such grouping provides a pure Process focus. At the other extreme, a designer might group all operations required by a single product. He would then locate these in a single product department or workstation. Envelope 3 in Fig. 15.5 illustrates this. This is a pure product focus.

In Step 2, the designer examines each product for a trial product focused grouping. There are only two valid reasons for rejecting a pure product focus:

1. The available feasible processes have a large scale with respect to product volume. They cannot economically process a single product; for example, a small turned pin requires only 0.25 machine hours per week for the anticipated product volume.
2. Some element of the infrastructure has a large scale with respect to

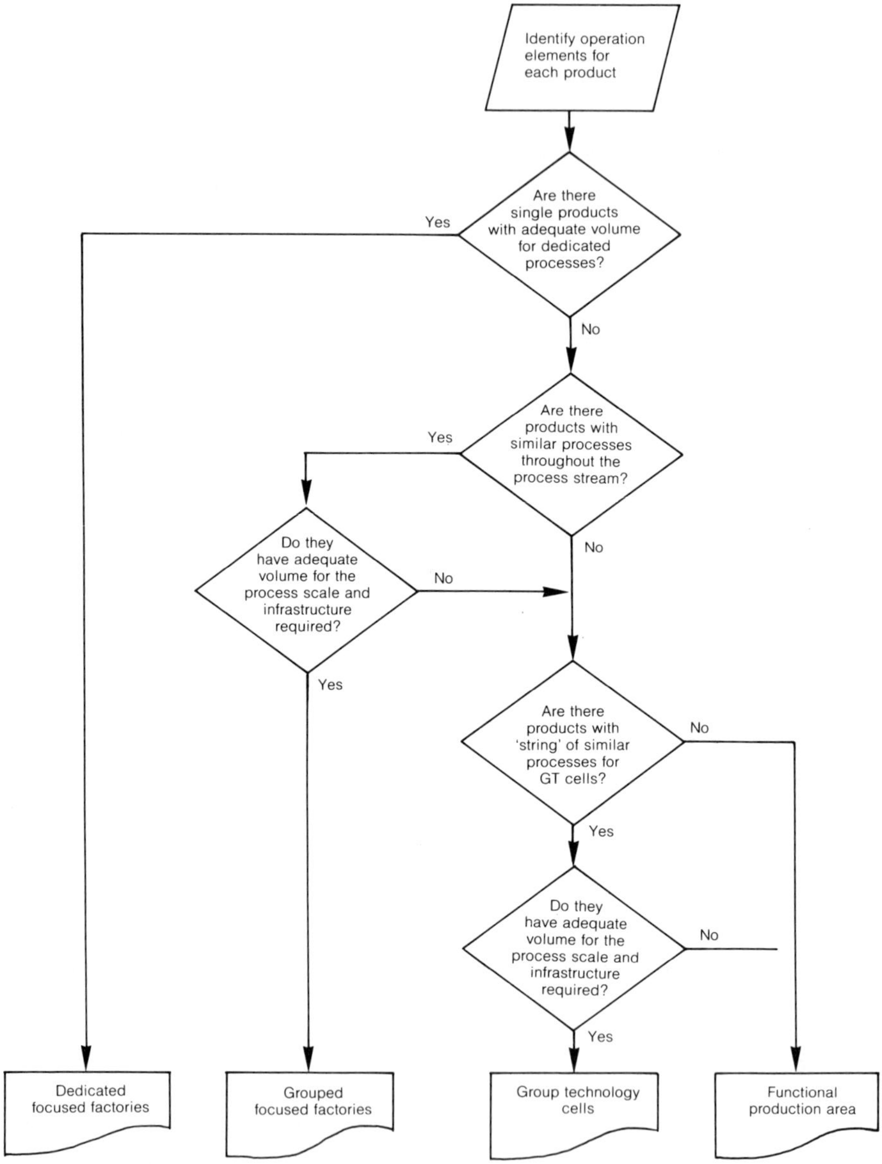

Fig. 15.4 Manufacturing focus algorithm.

product volume. It cannot reasonably serve a single product; for example, a highly skilled electronic technician calibrates a circuit board. Production only requires his skills about two hours per week.

If a trial product passes both tests, it should have its own manufacturing area and possibly its own PWP. Remove such products from further consideration.

In Step 3, the designer examines each remaining product. He identifies subsets of the manufacturing operations (strings) which are common between two or more products. When such a string occurs, the designer tests it for adequate volume *vis-à-vis* process and infrastructure scale. If it passes the test, such operation strings become group technology cells, as illustrated in Envelope 4 of Fig. 15.5.

Step 4 aggregates all remaining operations into functional areas.

The identified areas become basic cells for developing layouts. Such layout cells may be GT cells, functional cells or product cells.

An analyst might perform this focusing task on paper as illustrated. More often, the algorithm structures the thought process and may arrive at a cell definition without formally drawing all operation process charts.

The designer may also use the production flow Analysis technique developed by Burbidge or the techniques of a coding and classification approach. These techniques are particularly useful when dealing with hundreds or thousands of parts or products. Such traditional grouping techniques should supplement rather than displace the above algorithm.

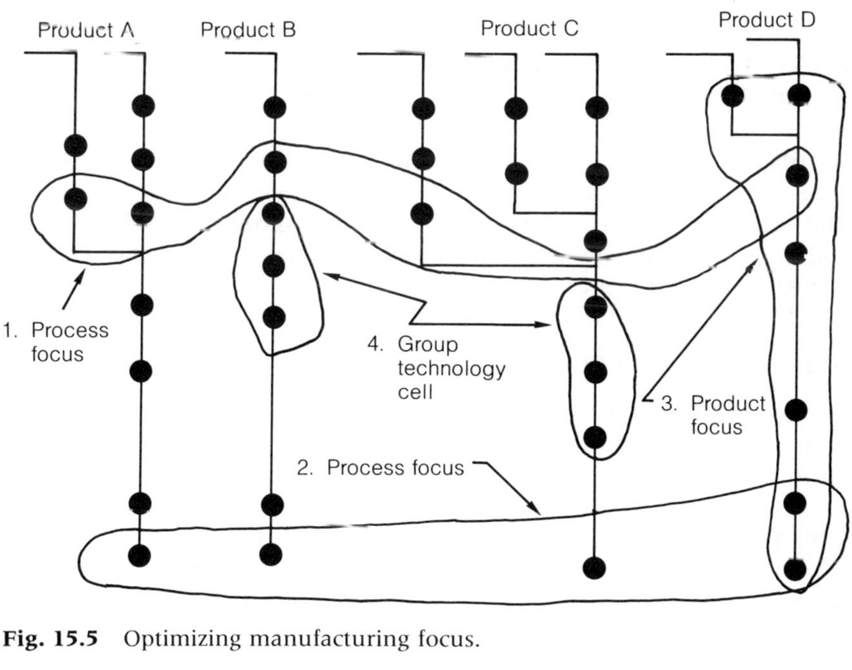

Fig. 15.5 Optimizing manufacturing focus.

15.9 EXAMPLES OF MANUFACTURING FOCUS

The following are examples of manufacturing focus. Names and some facts have been modified to preserve client confidentiality.

15.9.1 Northstar Aluminum Products

Northstar is an aluminium foundry. The company makes about 4400 distinct castings for many customers in several industries. Their castings are used in products ranging from lighting fixtures to decorative bowls and portable pumps. They specialize in medium-precision sand castings using green-sand and jolt-squeeze moulding equipment. Hand ramming produces moulds for prototype or very low-volume work. Northstar offers five alloys to its customers; several alloys may be stress relieved.

Other processes such as die casting, investment, automatic moulding and lost-foam are feasible. These other processes are more labour efficient. However, they also require a high capital investment and have long setup times that are more suited to higher volume than the jolt-squeeze process.

Northstar might also offer other metals. Irons and steels require much higher melting temperatures, and have different moulding requirements. Brasses and bronzes have similar pour temperature and mould requirements, but their lead content requires sophisticated ventilation systems.

Casting metal in sand is an ancient art and is simple in principle. Execution, however, requires skill, experience and close control of many variables. Technically, it is more difficult than many so-called 'Hi-Tech' processes. These technical demands are so severe that few foundries can master more than two or three processes.

Northstar has chosen to focus its plant on a single casting process and a small range of similar alloys. As a result, they have earned a reputation as the lowest cost and highest quality medium production aluminium foundry in the upper Midwest. Process focus has enhanced their expertise in a technically demanding business.

15.9.2 Commercial Fixtures

Commercial Fixtures Corp. manufactures a variety of lighting products which incorporate numerous precision, low-volume castings. These fixtures also require sheet metal parts, electrical components, glass lenses and plastic shrouds. All products are painted to a high-quality gloss finish. The Commercial Fixtures catalogue shows six major fixture groups. Each group has 6–10 styles and 4–8 sizes. Each style is available with high-intensity discharge (HID) or incandescent light, and four voltages.

Until recently, Commercial Fixtures operated an aluminium foundry to supply its casting requirements. Their foundry, however, was too small to purchase supplies effectively. Neither could it support the pattern makers and process engineers required for cost-effective operation. High cost, quality and productivity problems caused Commercial Fixtures to close its casting operation. The firm then contracted with Northstar Aluminum for its cast components.

As a result of this action, Commercial Fixtures reduced the range of their processes. The firm thus achieved a higher degree of process focus at the site level. They concentrated their efforts on assembly, paint and sheet metal forming. Cooperation between the two companies resulted in more cost-effective, focused operations for both of them.

Below the site level, Commercial Fixtures implemented GT workcells in the sheet metal area and dedicated cells for final assembly. The company thus moved towards a higher degree of product focus at the department level.

This combined focus approach has been highly effective. It requires Commercial Fixtures to master only three types of processes, yet it achieves most of the quality, throughput and productivity advantages of a pure product focus.

15.9.3 Inter-Defense Industries (IDI)

IDI manufactures armoured vehicles. During the Cold War IDI supplied many Western armies from its factories in Western Europe.

Through the 1960s, high-variety, low-volume and the unpredictable fortunes of politics and war had led the firm towards a process focus. In the mid 1970s the Mark VIII Heavy Tank was a major success with long-term contracts and stable delivery schedules.

In the first two years of production, IDI encountered difficulties. Despite the high volume and predictable schedules which they had always desired, manufacturing could not cope. High inventories, poor quality, delays and coordination problems threatened an 82-year reputation.

The crises in manufacturing forced an organic restructuring. IDI rationalized manufacturing and rearranged accordingly. The Mark VIII had higher volume and production stability than IDI's other vehicles. It was also more complex and had a higher level of technology. This led to a segregation of the new Mark VIII in its own facility, a focused factory, with a single product.

Within this overall product focus, some components had dedicated processes such as welding, machining and assembly. They were physically arranged in proximity to the assembly areas. Like a river the final assembly line was fed by branches where major components were assembled. These branches were in turn fed by the confluence of smaller production streams.

One major tributary of the final assembly stream was the turret. The turret had guns, missile launcher, fire control, command systems and armour. It is the most complex single sub-assembly with the exception of the hull itself.

Turret production was segregated in a plant-within-plant (PWP). All equipment and operations necessary to make the turret, its sub-assemblies and many components were in this product-focused PWP.

The turret factory was organizationally independent as well. It had its own maintenance, production control and industrial engineering support.

Before the restructuring, production at the workstation level fluctuated wildly. After the changes, day-to-day schedules stabilized as each operation produced only enough for immediate downstream requirements. Inventories shrank and quality problems surfaced quickly. Significant improvements in quality, productivity and delivery resulted.

15.9.4 Kansas Steel Foundry

Kansas Steel grew from a small grey iron foundry which served the repair and replacement needs of the transcontinental railroads. In the early 1980s, Kansas Steel was one of a handful of foundries that could make very large castings. The foundry poured about seven basic alloys from two electric arc furnaces and two coreless induction furnaces.

Several years ago, Kansas steel examined its mix of customers and products. While all of its castings were similar in size, alloy and mouldability, market segments differed in other significant ways. Market segments included:

1. Construction equipment,
2. Oil-field equipment,
3. General industrial equipment, and
4. Military armour.

The largest of these segments was military armour. These castings became part of various armoured vehicles. They included armour pieces as well as mechanical and structural parts. The armour segment represented about 40% of total production but only 20% of profits. This was somewhat disconcerting to Kansas Steel's management. Competition in armour was not aggressive and, pound for pound, armour castings sold for significantly more than the industrial castings.

The plant had great difficulty with the military quality standards as well as the documentation and testing requirements. Rework, repairs and delays were excessive. Management was also concerned about the excessive management support required since it increased overhead for all products.

Kansas Steel decided to split operations into market-focused factories. Fortunately, the site and buildings suited this arrangement. Each factory had separate facilities with the exception of the melt department. Here, two furnaces were dedicated to the new Armor Division (KSAD). The remaining furnaces served the new General Industry Division (KSGID).

The Armor Division enjoys stable production schedules and long-term contracts. Quality standards are very high. The smallest speck of dirt or defect introduced in the moulding department generates massive welding and grinding in the cleaning room.

Those employees who work on documentation with government inspectors have learned to deal effectively with both. The Armor Division hires experienced foundrymen, pays well and has a stable employment policy.

General Industry Division must cope with erratic delivery schedules and economic cycles. KSGID has many new products. However, the quality standards are less stringent and external paperwork is minimal.

The GI Division pays lower wages than KSAD and frequently changes employment levels as production varies. Only a cadre of senior workers can depend on steady employment. Quality standards are lower and procedures different.

As a result of these differences in apparently similar castings, the Armor and GI Divisions evolved different production control systems, quality assurance organizations and suppliers. They have focused two divisions with similar products and volumes around different markets.

15.9.5 Grandma's Pies, Inc.

Grandma's Pies manufactures fresh pies for restaurants. Their drivers deliver many varieties of high-quality pies to medium-sized, upscale restaurants who special order 24 hours in advance. Requirements for freshness and personal contact led Grandma's to a geographic focus bringing all customers within 150 miles.

15.10 INFRASTRUCTURE FOCUS

Equipment, systems or people may support production without processing the product directly. Such elements are part of the infrastructure. Infrastructure may be physical (facilities) or non-physical (people–information–organization). While somewhat unusual, infrastructure can also be a focus criteria.

Specialized and expensive facility requirements may necessitate a facility focus. Aircraft manufacturers (for example) bring together operations requiring high bay facilities. Such disparate operations as assembly, welding, machining and test may require such high headroom facility for specific parts or products. Such a requirement may bring together dissimilar processes, products and activities which share only a common need for headroom.

Non-physical infrastructure is occasionally a basis for focus. A worldwide manufacturer of toiletries concentrated perfume blending at a single site. The specialists there possessed olfactory skills essential to both new product development and production quality control.

As mentioned before, an argument can be made that only market focus is acceptable since customer satisfaction is the primary goal. However, a pure market focus may not account for the real limitations of manufacturing processes and infrastructure. Moreover, it provides little basis for the planning and layout of a production facility.

15.11 MATCHING THE LAYOUT AND ORGANIZATION

Organizations have a focus or lack of focus just as plants and departments do. When the reporting structure groups people with similar job functions, the organization has a functional focus. Product-focused organizations group people according to the products they manage. The criteria commonly used to focus sites and cells also apply to organizations.

Unfocused and inappropriately focused plants, layouts and organizations have handicapped many Western manufacturers. An even more serious situation exists when the focus of the layout differs from the organizational focus. This can happen when, for example, manufacturing departments focus on product lines while the supporting organization has a functional focus.

For facilities planning work a useful tool has been developed to test the consistency between organization and layout focus. The analyst first prepares a detailed and up-to-date organization chart. Every person at the site is represented. The analyst then draws an envelope around those persons whose normal workstation is within a given contiguous area.

The envelopes may be compact and include only people on the same branch of the organization tree. The layout and organization are then consistent. Fig. 15.6 illustrates this. Official channels of communication coincide with the layout and those who communicate have workstations in close proximity.

In other instances, the envelopes may appear disjointed, stretched and

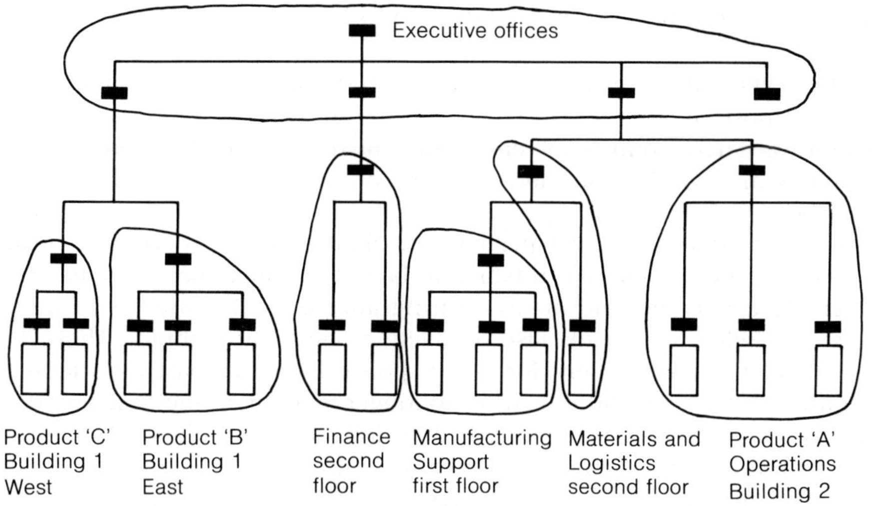

Fig. 15.6 Organization analysis; good organization/layout alignment.

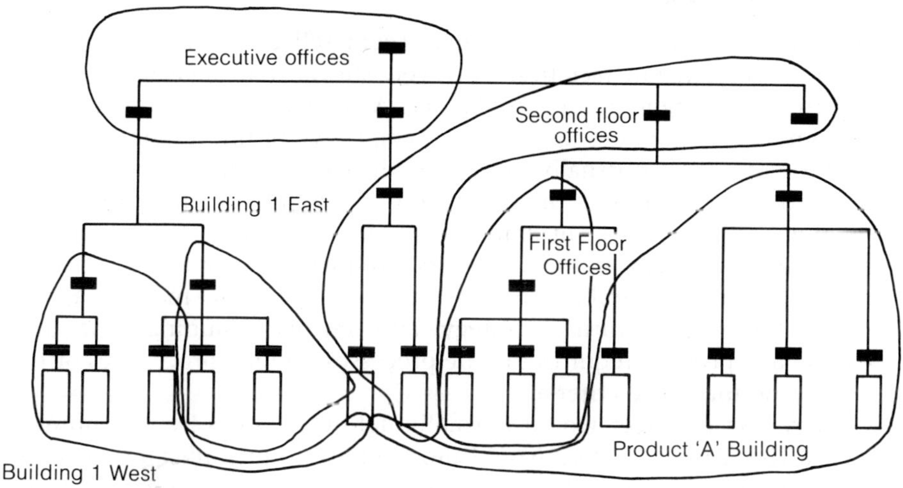

Fig. 15.7 Organization analysis; poor organization/layout alignment.

generally 'messy' as in Fig. 15.7. This indicates a mismatch between layout and organization. People have workstations far removed from their superiors, subordinates and colleagues. Such a situation makes for poor communication along the chains of command.

15.12 FOCUS FOR TIME-BASED COMPETITION

Stalk (1990) and Blackburn (1990) suggest that time-based manufacturers, most of whom are Japanese, are moving away from focused factories. Such companies emphasize flexibility in their production facilities and produce a wide range of products at comparatively low volume.

From a traditional perspective, flexibility and low volume lead towards process focused functional layouts. Functional layouts are, after all, a very flexible manufacturing approach. But process focused factories seldom provide fast or even reliable throughput. They do not meet the requirement for fast response that time-based marketing requires. If the concept of focus is no longer valid in a time-based world, it raises again Roger Schmenner's question, 'How should you organize production?' (Hayes and Schmenner, 1978).

In an attempt to re-answer Haye's and Schmenner's question, additional issues came to light:

1. What is a product?
2. What is flexibility?
3. How can we gain flexibility?

What is a product? The answer may seem self evident. Experience in planning production facilities, however, shows that answers often are controversial, vague and contradictory. A workable answer depends on the purpose of the question and the perspective of the respondent.

Marketing people usually view a product mix by market segment or the product function. A manufacturer of aircraft engine parts views them as GE, Rolls Royce or Pratt and Whitney products. A manufacturer of commercial light fixtures sees its products as either 'specification grade' or 'architectural'. A market-oriented definition may be appropriate for manufacturing strategy and factory planning, but only when each product requires unique processes, unique procedures or has unique performance criteria.

A product engineer would normally define products by part number or drawing number. They would most often group similar products by function or appearance. Product engineers thus think in terms of bearings and bushings or turbine blades and compressor blades.

For manufacturing strategy and factory design, a product definition is suggested based primarily on process similarity. If items can be manufactured in the same sequence on the same equipment they are in a single product group. If, in addition, these items require no significant changeover, they are for manufacturing purposes, considered as identical products.

Note that the above definition states that products are in the same family or group if they can be manufactured with the same process. This implies process standardization. It also implies that a revision of process technology or equipment may change the product definition.

Hill (1985) suggests that only market requirements are a proper basis for focus. But situations requiring identical products to be made in different factories or departments due to different market requirements are unusual. In practice, it has been found that defining products on the basis of their process similarity generally brings one of the following results:

1. The process-oriented definition corresponds to market-oriented definitions.
2. Different markets make similar demands on manufacturing.
3. The differing market demands are not contradictory.

Here are some examples.

Conplastics Corporation extrudes vinyl products, such as vinyl pipe, conduit and guttering, for the building industry. The marketing people classify their markets as 'wholesale' and 'retail'. Wholesale customers operate large chains of home-improvement stores or lumberyards. Retail customers are small lumberyards or home-improvement contractors.

These two markets make quite different demands on the sales department, requiring different personalities, discounts and order procedures. But to manufacturing, these market differences are transparent. The large wholesalers actually order and ship for individual stores, with such orders being similar in size and mix to those for retail stores. Retail contractors usually accumulate several contracts before ordering, and several contractors may combine orders for each shipment. These shipments are similar as well.

For manufacturing, the important product differences are those features which require major changeovers or different sizes of extruder, not who the end customer is.

15.12 FLEXIBILITY

By flexibility, most manufacturing people mean the ability of a factory easily to adapt to varying conditions. Both economy and speed are usually implied. While cost and response time go together, they are in fact different issues. The conditions referred to may include changes in:

1. Product mix,
2. Volume,
3. New products.

While these three types of flexibility are sometimes related, they are also separate issues which should be addressed individually in different ways.

An extrusion plant can easily vary its output but rapid changes in product mix may require expensive setups.

Flexibility and improved response time will be important future issues for many manufacturers. How to achieve them needs to be examined and trade-offs are also involved.

Process focused factories and departments are the long-accepted means to attain flexibility in product mix, volume and often in the introduction of new products. However, they do not perform well on delivery speed or reliability. Process focus is an inadequate strategy for the time-based competitor.

Group technology cells achieve good flexibility within the range of their product families. In addition, new products often fit well in existing families, which eases their introduction. GT cells also have good response, with low inventory and fast throughput.

Another approach to flexibility is process technology (as distinct from process focus). Certain technologies may lend themselves to faster changeovers. This provides high product mix flexibility. Some technologies have lower tooling costs than others, which increase new product flexibility. Numerical control machine tools, for example, are highly flexible for both product mix and new products.

Flexible processes are not necessarily sophisticated or expensive. Manual methods are often quite flexible and should not be discounted. Small sacrifices to direct labour productivity can bring large benefits in productivity of capital, material flow and support requirements.

Toyota approaches product mix flexibility by dedicating simple, inexpensive machine tools to particular products, which seems an anomaly. They then merely turn the machine off if that product is not required. They thus achieve good mix flexibility with only a small trade-off in floor space.

The use of dedicated but small-scale equipment can also bring new product flexibility. A manufacturer of sheet metal products is an example. This firm specializes in sheet metal parts for defence electronic systems. They have many new contracts with low production for a number of years. When terminated, most products do not return. This firm has achieved flexibility by using simple conventional sheet metal equipment arranged in manufacturing cells. Their equipment, however, is not bolted to the floor. Each piece sits on a skid and is connected with flexible rubber cord to a receptacle. It can be moved with a fork truck in only a few minutes. With this concept, one-product workcells are rapidly put together and then disassembled when no longer required. Sophisticated NC equipment is not required since a setup remains for the life of the contract.

15.13 MANUFACTURING STRATEGY AND PLANT DESIGN

Relationships exist between product definition process selection, and production mode. For success these must work together in a coherent manufacturing strategy. For success in a time-based market, they must satisfy requirements for quick response and rapid new product introduction.

The time-based competitor will most likely choose a manufacturing strategy which emphasizes workcells. Group technology cells will combine with flexible process technologies. A time-based competitor might also use simple conventional tooling in a dedicated workcell.

Time-based competition does not negate the advantages of product focus. It simply requires it at a lower level and smaller scale. Defining the product mix in terms of process commonalities clarifies the idea of manufacturing focus and permits a more universal application.

15.14 MANUFACTURING FOCUS AND MANUFACTURING STRATEGY

The focus criteria selected should be consistent with corporate goals, market strategy, manufacturing processes and infrastructure. Developing such focus is one of the most important elements of manufacturing strategy.

Focus is our best means to reduce manufacturing complexity and direct technical and knowledge resources at customer demands. By doing this it makes manufacturing an integral part of the corporate strategy rather than merely being a hapless supplier of commodities.

15.15 REFERENCES

Blackburn, Joseph D. (1990) *Time-Based Competition*, Business One Irwin, Homewood, Illinois.

Hayes, Robert H. and Wheelright, Steven C. (1984) *Restoring Our Competitive Edge*, Wiley, New York.

Hill, T. (1985) *Manufacturing Strategy*, Macmillan, London.

Stalk, George, Jr (1990) *Competing against Time*, The Free Press, New York.

CHAPTER 16

Manufacturing strategy and the theory of constraints

F.C. 'Ted' Weston, Jr

16.1 INTRODUCTION

The subject of manufacturing strategy is concerned with moving a manufacturing enterprise from its current competitive state to a future competitive state. This future state is generally focused on domestic and international markets, long-term profitability, issues of quality and cost control, organizational structure and lines of communication, international competition, information systems and similar considerations.

As is well known, manufacturing strategic planning is often not engaged in at all or is pursued separately from corporate strategic planning. Manufacturing strategic planning must consist of an active programme to insure that conversion resources are available in the proper quantities and in the proper locations, all at the desired times. These resources must effectively and efficiently convert customer requirements into low-cost, high-quality products and services, which are and deliverable in a customer-defined timeframe. Such planning assumes that strategic plans that are developed within the manufacturing area are consistent with both other functional area strategic plans (e.g. engineering, marketing and finance), and corporate strategic plans.

The exclusion of manufacturing strategic planning and/or the failure to integrate the manufacturing strategic planning function with corporate strategic planning, possesses the potential for producing negative long-term organizational results. Credit for bringing manufacturing back to

This paper is a revision of a paper with the same title presented at the Fifth International Manufacturing Strategy Conference held at the University of Warwick, England, June 26–27, 1990.

Specially commissioned for *Manufacturing Strategy – Process and content.*
Edited by Christopher A. Voss.
Published in 1992 by Chapman & Hall, London ISBN 0 412 43660 4.

the attention of top management must be credited to Skinner (1978, 1985), and additional credit must go to Wheelwright (Wheelwright, 1981; Hayes and Wheelwright, 1984). These plus other authors recognize the importance of manufacturing to the economic wellbeing of individual manufacturing firms and the national economies within which the firms operate.

The theory of constraints (TOC) is an emerging philosophy that focuses on where constraints lie within or external to the organization. These constraints prevent the manufacturing firm from moving to higher levels of profit and therefore success in the market place. As offered by Goldratt TOC focuses on identifying physical or policy constraints and then exploiting these constraints in terms of the most efficient use of manufacturing resources. Once exploitation of physical resources has taken place, subordination of other planning and scheduling activities to these constraining resources occurs. Subordination is followed by elevation of the constraining resource(s) so that a higher total level of organizational performance may be achieved. This process is repeated until internal constraints are removed and constraints move outside the direct control of the firm, e.g. in the market place.

Recent literature on strategic planning alludes to the idea of focusing on constraints. Yip (1989) discusses the need for firms to learn to globalize. He defines three steps essential in developing a total worldwide strategy. The essence of these steps is developing a core strategy which includes the basis for sustaining a competitive advantage in domestic markets initially and then expanding to foreign markets. This article suggests that any number of arenas may serve as effective constraints to competing in the world market-place. Constraints may include existing cost structure, market participation, product offerings, location of value-added activities, marketing approach, channels of distribution and competitive initiatives.

Dertouzos *et al.* (1989) defines a set of recommended strategies for industry, labour and government. The first recommendation for industry suggests, 'Focus on the production process, with the objective of improving long-term productive performance'. Long-term productivity performance is precisely the focus on maximizing throughput (T) while minimizing inventory (I) and operating expense (OE) advocated by Goldratt.

Goldratt and Cox (1984) clearly bring home the point of focusing on long-term performance. This was accomplished with both the definition of the goal (e.g. to make money now and in the future) and the focus on a set of realistic measurements that could be used by everyone in a manufacturing environment to gauge whether or not progress against the goal was being achieved. Collectively, throughput, investment in inventory and operating expense, constitute the basis of any set of measurements of success. Other measurements are secondary as linkage to short- and long-term profit is less clear. Mission statements that talk about good customer relations, happy employees, market share and even quality, must eventually focus their impact on the bottom line of the firm.

Unfortunately, many manufacturing firms often permit functional area objectives which are in conflict with larger corporate goals. This may be particularly true in companies where the manufacturing function is dominated by marketing, finance or engineering. In these firms, functional area goals often tend to promote competition between functions. Such competition often has growth of the functional area (e.g. marketing) as a goal, which translates directly to the promotion of the power and authority of the executive heading the functional area. The result of functional competition is decisions and practices that lead to local optimization versus corporate (global) optimization. The example of large finished goods inventories, as may be desired by marketing and sales, versus low finished goods inventories as may be desired by finance, is a standard illustration. Batch sizes represent another.

Competition between divisions of the same company is similar to competition between functions. Often it is non-productive and generally serves individual executives. Such competition is often confusing to customers who are often forced to figure out on their own where products from the same company belong in the customer's environment. The concept that the goal of the company is to make a continuing profit often gets lost in divisional rivalries.

Hazeltine and Baragallo (1990) argue that manufacturing strategy is the key to success in the 1990s. They stipulate that flexibility, integrated technology and a proper understanding of the role of the human factor, are the three strategic elements necessary to compete in a global market-place. They state that a '... it is necessary that a corporate culture for improvement be in place'. The emphasis on the corporate culture for improvement is precisely what Goldratt is saying in the theory of constraints with his emphasis on *how* to bring about desired changes, and the level of executive that must become actively involved in initiating the change process. One of the emerging lessons of the Baldrige Award in the US is the combined effect of corporate culture plus constant personal involvement of senior executives in achieving enhanced quality performance and a refocusing of the company or division.

It may be argued that the literature has always called for top management support when a significant level of refocusing activity is contemplated. However, that has not always happened. Lip service is often rampant and lacking in credibility. Many, if not most, manufacturing concerns today still focus on short-term profits, monthly and quarterly shipments, shop-floor efficiencies per reporting period and stock market reaction to quarterly earnings reports, etc. This author and virtually everyone else close to the manufacturing scene are calling for a focus on the long term. This includes re-emphasis on manufacturing strategy as a means for assessing the manufacturing component in a organization with global objectives. It is suggested that the TOC is a singular philosophy that is attempting to bring together a myriad of potentially conflicting tools and continuous improvement techniques such as MRP II, JIT and TQM. The integration of these continuous improvement techniques with a focus on constraints management and long-term profitability is central to the TOC.

16.2 THE THEORY OF CONSTRAINTS

As stated, an initial perspective to understanding the TOC is first to identify and then to exploit constraints, subordinating related decisions to these constraints, and then to proceed to elevate and systematically remove each constraint. In the TOC, this is a continuous, dynamic process that can never be concluded simply because there will always be a constraint of some type, internal or external, that will prevent the firm from achieving a higher level of profit. Because of this continuous focus on constraints, the TOC virtually negates a traditional programme management approach with specific starting and concluding dates.

16.2.1 What to change

A second perspective of the TOC, parallel to the first perspective which begins with identification of constraints, is consideration of three fundamental questions. The first question concerns what to change. For many firms, this step involves identification of what changes in either strategy or tactics are to be made within the firm. This may include management policies, production capacity and priorities, and process and product changes, as well as identification of marketing, engineering, finance and/or production constraints. In general, the 'what to change' question suggests systematically identifying, and then removing, internal and external constraints, regardless of the functional location of these constraints.

For many firms, there must be a 'felt need' to change, followed by the power, authority and willingness to inaugurate the desired changes. White (1980), explaining the Lewin–Schein theory of change, discusses the idea that people tend to resist change, often out of fear. The need to change must first be perceived as a positive condition by the individual before resistance to change can be overcome. In essence, ownership of the idea that change is needed must be realized by the corporate power structure or the CEO directly. The importance of recognizing the need to change is most likely true for divisions or entire companies as much as it is for individuals.

The Association for Manufacturing Excellence (1988) clearly cites several instances where the felt need to change had to be acknowledged before effective changes were realized by the companies depicted. For many of these companies, symptoms such as loss of market share, costs getting out of control and products that were no longer competitive, became the realization that change must become the norm.

Competition, including loss of market share and/or increased foreign competition, or a cost and pricing policy that is significantly different from the industry, often results in a firm's willingness to undergo either mild or drastic change and restructuring. Changes needed often translate to a focus on the constraints to achieving a higher level of profit and therefore organizational success. Much of the programmatic emphasis given CIM, MRP II, JIT and TQM, does not necessarily stipulate that the focus is on elimination of constraints, but there often is a narrower focus of attempting to reduce inventories and/or expenses in a piecemeal fashion or in a particular area. While many of these programmatic approaches to improvement promote positive changes, recognition of the fundamental constraints that inhibit progress are often unrecognized.

Internal constraints may reside in functional areas such as R&D, engineering, finance or manufacturing. Internal constraints may include direct or *de facto* policies that result in the misuse of cost-accounting concepts or the failure to recognize the importance of information systems as a strategic element in the manufacturing firm. Unfortunately, information systems personnel have generally not teamed up with accounting personnel to uncover misuses of accounting information. The continued practice of focusing on shop-floor efficiencies is a good example.

External constraints may include recognition that markets for products are saturated, that prices are too high or new channels of distribution are blocked, or that new sources of funding are not available or are severely limited. The lack of foreign language skills or knowledge of foreign business practices on the part of senior corporate officials may appear as a barrier to entering foreign markets. Also, corporate understanding of customs and import regulations may be minimal, domestic regulatory constraints may be present, and legal and/or environmental barriers to introducing current or planned products may exist. The TOC would advocate that the firm should do everything possible to remove external constraints with the result that internal *policy* constraints provide the focus of management's attention. For many firms, constraints which are perceived as being external, such as the inability of marketing and sales personnel to acknowledge the presence of foreign markets, and/or the ability to speak a foreign language, are really personal or *de facto* internal policy constraints.

Further, it is argued that many firms have only superficially examined, in a cause-and-effect manner, where the *real* constraints to future progress lie – and therefore have only vague notions as to how to proceed. In these companies, strategic planning either does not exist at all, or the strategic plan is of such a nature as to be virtually meaningless when it comes to providing direction for specific actions to be pursued. Plans prepared that do not result in specific actions are generally vague, poorly prepared, threatening, or so radical that support for the plan is difficult at best. Prepared plans must possess credibility and must be so convincing that support from the required executive core is assured.

In accordance with the TOC, internal constraints, now policy in nature rather than physical, can be managed in such a way that corporate goals of profitability are constantly held in focus by managers in all areas. This is in contrast to the pursuit of functional or local area goals. The pursuit of local goals, such as within marketing *versus* manufacturing, accounting *versus* production, and design *versus* process engineering, produces non-optimal results for the firm as a whole. This lack of a global versus local perspective within the manufacturing enterprise has most likely contributed to the corporate demise of many US firms.

The subject of measurements or, as some would prefer, critical success factors, is included also in the deliberation of what to change. The TOC proposes that throughput (T), inventory (I), and operating expense (OE) are measures which have meaning at the shop floor level as well as in the corporate boardroom. T, I and OE have direct translation to traditional measures such as net profit and return on investment. For example, net profit is simply T minus OE. ROI is net profit (T−OE) divided by I. Understanding how these three measurements work together towards building a focus on both short- and long-term profit is critical if progress towards attaining world-class status is to be achieved.

Those involved in strategic planning in manufacturing must begin to pay more attention to a proper set of measurements that effectively can gauge the success or failure of programmatic activity such as CIM, JIT, TQM and TOC. These measurements must be realistic in that local area improvements must translate into global, e.g. corporate, improvements. All too often, improvements are limited to inventory and operating expense reduction with minimum consideration of the impact on throughput. The use of accounting measures, such as efficiency, are not always instructive in measuring the impact of changes on the corporate bottom line.

The question of what to change must therefore begin with recognition that change is both needed and timely. Senior management must be involved with and lead the change process. Constraints to achieving a higher level of profitability must be addressed. The core problems of the organization must be uncovered in such a manner that the direction in which the organization is to proceed is viewed as non-threatening.

Manufacturing strategy in the 1990s must focus on these core problems and combine the power of the senior executives with the culture of the organization to move forward in a positive, non-threatening, manner.

16.2.2 What to change to

Following consideration of what to change is the question of what to change to. The answer to this second question addresses solution of core problems identified as part of the 'what to change' question. The essence of this new question is recognition of physical and/or policy constraints, internal and/or external to the organization, that must be dealt with before the firm can progress to higher levels of profitability. This second question addresses the negative consequences of continuing to operate in a *status-quo* mode with existing constraints remaining in place. The long-term ability of the manufacturing firm to survive in future markets and under global rules, may depend on the acceptance that change is needed. Also, a willingness to develop a corporate posture that is more sensitive to change, that can bring products to market quicker and that can adopt a culture of constant change, is quickly becoming a prerequisite to competing in this new era of global rules. If strategic planning does not consider the corporate question of what the firm must change to in order to become or remain competitive, then what is strategic planning all about?

For many firms, the 'what to change to' question is a focus on arenas of future competition. Most likely, these arenas will not include technology as technology advances are quickly communicated, if not copied, by competitors worldwide (Clark, 1989). However, the future global competitive arena will include a focused factory concept where processes, including flexible manufacturing layout, parts movement systems and information networks, are responsive to ever-changing market demands. The 'what to change to' question will most likely include a factory that includes supplier involvement in the design of new products, a design for manufacturability philosophy, competitive benchmarking in the areas of design, cost and assembly, plus a focused-factory concept that integrates design with assembly with distribution. All areas of change can be viewed as to their impact on the three measures T, I and OE.

While quality, zero defects, total quality management and statistical process control all focus on a more reliable product, it is hypothesized that this perspective will begin to change in the future. This is *not* saying that quality will become any less important as an expectation by either customers or manufacturers. It is saying that virtually all global competitors will be competing on a basis other than what we *now* define as quality – since quality, like technology, will become a virtual constant. Therefore, traditional quality-improvement programmes will soon begin to be viewed with more scrutiny – which is a return to a focus which examines the marginal operating expense of quality improvements in relationship to the impact on throughput. Marginal enhancements in quality must be offset by a combination of lower operating expenses and/or lower inventory plus product throughput. The era of unlimited spending on quality without a corresponding impact on the bottom line, must cease. To date, this has not occurred.

16.2.3 How to change

The third question to be considered as part of the TOC concerns how to change the manufacturing concern from a current undesired state to a future desired state. This is the most difficult of the three questions posed as part of the TOC and the one that is most often inadequately considered in strategic planning. This last question addresses the psychology of moving the organization through a transitional process. As stated earlier, the felt need to change must first be present before real change can occur. The transitional process of moving the old organization and culture to the desired new organization and culture, may include everything from changing organizational structure, removing policy constraints as to which products are promoted in which markets, a review of cost-accounting procedures, inventory policies to competitive practices. Upper management may even begin to challenge its own role relative to the entire corporation and refocus on the corporate whole versus optimization of the individual parts. In other words, the change process must recognize that business as usual is no longer tolerable when it is clearly evident that competitive pressures in the future will only increase.

One lesson is clear when considering any approach to moving an organization to a future, desired state. This lesson is that a quick-fix, unintegrated, lip-service 'here we go again' approach will not work. Lower level managers, both line and staff personnel and even operating personnel, are tired of approaches to change that involve everyone but management, where training separates top from lower levels of management, or excludes top management altogether, and where one set of rules apply at the beginning of the month but not at the end of the month, quarter or year. Any change process that is not worthy of the time of senior executives simply will not be accepted at lower levels and will not work. Vision and direction must come from the most senior executives, not promoted by consultants and/or advisory bodies. Those that are trying to promote change must understand how the process of change can be blocked by managers at the same organizational level but in other functional areas.

These three questions of what to change, change to what, and how to change, constitute the critical questions that any global manufacturing firm must face – and quickly! When addressing these questions, various facets of manufacturing strategy may be addressed. One common theme prevalent in the manufacturing literature today is computer integrated manufacturing or CIM. CIM is currently construed to consist of a rich mixture of concepts, tools and techniques from the engineering, production management, information systems and human-resource disciplines. For many firms, CIM is the heart of their strategic planning (Weston, 1989). Linkages between CIM and the theory of constraints are only now beginning to be developed (Weston, 1990). However, CIM will suffer the same ills of MRP II or other tools if it is viewed as being the province of a single functional area such as engineering. CIM must be viewed as a continuous improvement theme that is dynamic in nature, that is cost and throughput based, that is strategic in concept, and that has the understanding and support of top management. Top management must make the effort to comprehend the integral features of CIM when viewed with TQM, MRP II, JIT and other continuous improvement themes.

The underlying hypothesis of this paper is that the theory of constraints and strategic planning in manufacturing are deeply intertwined. It is stipulated that the essence of strategic planning in manufacturing is, in fact, the dynamic and systematic process of identification and removal of internal and external constraints. Unfortunately, at least in the United States, management's attention is too often narrowly focused on quality, cost reduction, MRP II, JIT, inventory levels, and even leveraged buy-outs. This is in contrast to the longer term issues of systematic identification, exploitation, subordination and elevation of constraints that are reducing the firm's ability to achieve a higher level of both short- and long-term profit. In essence, this paper is arguing that the subject of strategic planning in manufacturing should focus on constraints to achieving higher levels of corporate profitability. To date, this has not been the case.

Stahl (1989) has presented a five-step approach to developing and implementing strategic plans for firms with both domestic and foreign markets. This approach focuses on:

1. Linking the business plan and marketing strategy to the manufacturing operations;
2. Developing a manufacturing mission statement which identifies which manufacturing resources are critical to the business;
3. Conduct a comprehensive 'where are we' analysis of facilities, proprietary technologies and manufacturing systems;
4. Developing a manufacturing strategy that supports the business plan and manufacturing mission statement; and
5. Planning and implementing long-term manufacturing productivity improvements.

What this article by Stahl does not discuss are measurements appropriate for benchmarking results achieved against the strategic plan following his or similar formats. The TOC addresses the measurements issue directly with the focus on throughput, inventory and operating expense. Goldratt (Goldratt and Cox, 1984; Goldratt and Fox, 1986) defines these measurements which are applicable to virtually any manufacturing organization.

Measurements are becoming a critical issue in manufacturing management. Some measurements, such as share of market, are only symptomatic as to whether products are competitive on a price, quality and availability basis. The loss of share of market, such as by US auto makers, has led to strategies of merging with the competition, refocusing on quality, and a re-emphasis on either knowledge workers and/or automation. All measurements must eventually return to a basis of the cost of producing a product (operating expense), the investment in the operations of the company to design, make and distribute products, and in the volume of product (throughput) flowing out of the factory door and into the hands of customers. In this arena, the theory of constraints has captured the essence of how manufacturing strategy must be evaluated.

16.3 SUMMARY

Strategic planning in manufacturing is often not engaged in at all or is relegated to a minor role in the entire corporate strategic planning function. This is itself a strategic error. Strategic planning must be viewed as the process of moving a corporation from some existing level of success and state of competitiveness to a future desired level. The theory of constraints or TOC is a philosophy that is concerned with moving the corporation into the future by first understanding where constraints to progress lie, and then a focusing process on what to change, change to what, and how to bring about the desired changes. The difference between the TOC and other continuous improvement concepts is that the theory is accompanied by measurements which have meaning at all levels of management.

The theory of constraints is not another quick-fix project effort that is here today and gone tomorrow. It is an evolving philosophy that is admittedly incomplete, but one that is rapidly gaining considerable attention both in the US and in Europe. The principles are beginning to be developed clearly with the result that many firms are beginning to focus on this philosophy as an approach that makes sense to management and knowledge workers alike.

Strategic planning and the theory of constraints are virtually different choices of words for addressing the same subject – how to move the manufacturing (or any other) corporation from the current *status quo* to a future desired state. This is a process that involves change. Top management must not only participate in this change process, but must both lead and direct the change process. However, in order to lead the change process, recognition must exist that change is necessary if competing in a global, highly competitive environment, and is integral to the strategic plan of the firm. In order to direct the process, then, top management must be schooled in the theory of constraints. The ability to compete in the global market-place may depend on it.

16.4 REFERENCES

Association for Manufacturing Excellence (1988) *On the Road to Manufacturing Excellence* (video tape).

Clark, Kim B. (1989) What strategy can do for technology. *Harvard Business Review*, November–December, 94–8.

Dertouzos, Michael L., Lester, Richard K., Solow, Robert M. and the MIT Commission on Industrial Productivity (1989) *Made in America*, The MIT Press, 147–67.

Goldratt, Eliyahu M. and Cox, Jeff (1984) *The Goal*, North River Press, Croton-on-Hudson, NY.

Goldratt, Fliyahu M. and Fox, Robert E. (1986) *The Race*, North River Press, Croton-on-Hudson, NY.

Hayes, Robert H. and Wheelwright, Steven C. (1984) *Restoring Our Competitive Edge: Competing Through Manufacturing*, Wiley, New York.

Hazeltine, Frank W. and Baragallo, Ross J. (1990) The key to competitive success in the 1990s. *P&IM Review*, p. 4.

Skinner, Wickham (1978) *Manufacturing in the Corporate Strategy*, Wiley, New York.

Skinner, Wickham (1985) *Manufacturing: The Formidable Weapon*, Wiley, New York.

Stahl, John E. (1989) Manufacturing strategic planning: key to competitive strength. *CIM Review*, Fall, 14–20.

Weston, F.C. Jr (1989) CIM from a top management perspective. *CIM Review*, Winter, 18–22.

Weston, F.C. Jr (1990) CIM and the theory of constraints. (submitted to *CIM Review*, February).

Wheelwright, Steven C. (1981) Integration of manufacturing strategy business strategy, in *Manufacturing Trends in the 1980s*, Booz, Allen and Hamilton, New York.

White, Edna M. (1980) Implementing an MRP system using the Lewin–Schein theory of change. *Production and Inventory Management*, First Quarter, 1–12.

Yip, George S. (1989) Global strategy . . . in a world of nations?. *Sloan Management Review*, **31**(1), 29.

CHAPTER 17

The nature of the link between manufacturing strategy and organizational culture

Susan D.A. Misterek, Roger G. Schroeder and Kimberly A. Bates

17.1 INTRODUCTION

Recently, there has been a great deal of research interest in the area of manufacturing strategy (Ettlie, Burstein and Feigenbaum, 1990). The research has, however, centred on the content of manufacturing strategy and the relationship of manufacturing strategy to business strategy. There has been very little work done on the linkage between manufacturing strategy and organizational culture.

Both academics and practitioners are concerned with the relationship between culture and manufacturing strategy, or even between culture and business strategy. From a practitioner point of view, a new strategy can require a substantial change in culture. Conversely, culture can constrain the implementation of strategy. The precise relationship between strategy and culture is, therefore, of interest since culture serves as both a determinant and an outcome of strategy.

Karl Weick (1985) asserts that business strategy and culture are essentially the same. He states that culture and strategy both are 'deeply ingrained patterns of management behavior', and that they both '[emerge] out of the cumulative effect of many informed actions and decisions taken daily and over years by many employees.' In contrast,

Specially commissioned for *Manufacturing Strategy – Process and content.*
Edited by Christopher A. Voss.
Published in 1992 by Chapman & Hall, London ISBN 0 412 43660 4.

Saffold (1988) argues that business strategy is an outcome of organizational culture. Whether we start from Weick's or Saffold's viewpoint, the relationship between culture and strategy warrants further investigation from an academic perspective.

Beginning from this perspective, this paper will investigate the relationship between manufacturing strategy and culture. Since our interest is at the plant level, we have restricted our study to manufacturing strategy. We are also interested in world-class manufacturing (WCM) strategy as a type of manufacturing strategy and its relationship with specific types of organizational culture. The particular focus of this study is the relationship between a WCM manufacturing strategy and a collectivist, consistent and egalitarian culture.

The first part of the paper will define the concepts of manufacturing strategy and organizational culture and how we expect them to interact. We will then provide methods of measuring these concepts and report on operational tests of reliability. Finally, we will explore the substance of the relationship between manufacturing strategy and organizational culture and the implications of our results for both practitioners and managers.

This paper is part of a larger study of world-class manufacturing being conducted by the University of Minnesota and Iowa State University. This study includes data from 37 plants in the electronics, machinery and transportation industries in the US. These plants include those which are reported to be managed by world-class approaches, as well as those managed by traditional means (Flynn, Flynn and Schroeder, 1989).

17.2 ORGANIZATIONAL CULTURE AND MANUFACTURING STRATEGY: CONCEPTS

This section presents definitions of the concepts of manufacturing strategy and organizational culture which are based on the literature and past research efforts. Manufacturing strategy has been addressed primarily in the operations management literature, while organizational culture has been described in organizational research, in sociology, and in anthropology.

17.2.1 Manufacturing strategy

Wickham Skinner (1969) was the originator of much of the manufacturing strategy literature. He defined manufacturing strategy as the linkage between business strategy and manufacturing decisions. He also argued that manufacturing should provide competitive advantage to the business and not merely make the company's products or services.

According to Skinner, competitive advantage is created when manufacturing decisions are supportive of the business strategy.

Hayes and Wheelwright (1984) have closely followed Skinner's lead by defining manufacturing strategy as a consistent pattern of decision making in the manufacturing function which is linked to the business strategy. They emphasize consistency of decision making as a key to manufacturing strategy. This approach follows Mintzberg (1978), who has argued that consistency in decision making results in a realized strategy which may or may not agree with the intended strategy. Intended strategy is the result of planning, while realized strategy is the pattern of decisions which actually occurs. This consistency in decision making will be an important ingredient in defining and identifying the organizational culture; it is also a result of culture, which we argue is linked to strategy.

Hayes and Wheelwright (1984) define four stages of progression toward attainment of a competitive manufacturing strategy. Stage 1 (internally neutral) is a fire-fighting, reactive form of decision making where top management is not concerned with manufacturing. In Stage 2 (externally neutral) the rationale for manufacturing strategy is to keep up with the industry, in a sense of industry parity. In Stage 3 (internally supportive) manufacturing is closely linked to the business strategy and becomes a force in supporting the business. In Stage 4 (externally supportive) all of the functions are closely linked together and manufacturing is a basis for competitive advantage. These stages are a convenient way to identify and define the extent of a manufacturing strategy in terms of its contribution toward a sustainable competitive advantage.

Finally, manufacturing strategy implies a long-range orientation toward manufacturing. Hayes and Abernathy (1980) define the importance of managing manufacturing for the long run in order continually to introduce new products and/or new processes. They define innovation as an important basis of competitive advantage. Managing for the short run and for immediate financial gain conflicts with the long-run orientation required for manufacturing strategy to be effective.

The literature cited above has provided a rich source of ideas for defining manufacturing strategy and its relationship to culture. Before providing specific measurements for manufacturing strategy, however, we will make a similar analysis of organizational culture and then discuss the linkages between these two concepts.

17.2.2 Organizational culture

Organizational culture is defined in organization research as a complex set of knowledge structures used to perform tasks and to generate social behaviour (Hofstede *et al.*, 1990). Because organizational culture has its roots in cross-cultural research of societies, we will review both organizational and cross-cultural concepts of culture.

Anthropologists examine many aspects of social behaviour (including speech patterns, modes of interaction, formal rules, taboos, and artefacts) in order to grasp the 'native view' or 'insider's understanding and perspective' (Gregory, 1983) of the people being studied. Traditionally, anthropologists have lived with the peoples being studied and used ethnographic methods of describing the culture being observed. Only recently have anthropologists used survey methods.

Spradley and McCurdy (1975) define culture as 'the acquired knowledge people use to interpret experience and generate social behavior' (Gregory, 1983). Schein (1985) defines culture as:

> a pattern of basic assumptions – invented, discovered, or developed by a given group as it learns to cope with its problems of external adaptation and internal integration – that has worked well enough to be considered valid and, therefore, to be taught to new members as the correct way to perceive, think, and feel in relation to those problems.

Gregory (1983) notes that culture refers to *apparently* shared meanings, because people interact *as if* they share meanings. Members of the culture use communication via specialized meanings to interact with the physical environment, to socialize new members, and to negotiate new meanings between groups over time (Gregory, 1983; Schein, 1985). The use of this definition of culture in organizational research draws on an implicit analogy of the organization as a mini-society.

A shared meanings perspective on organizational culture in the context of a world-class manufacturing orientation describes acquired knowledge which guides members' behaviour in the plant. This includes, according to Schonberger (1986), manufacturing practices such as worker empowerment for quality improvement, supervisors as team leaders, and decentralized decision making. This stands in contrast to a traditional manufacturing culture where supervisors are controllers rather than team leaders, where workers are paid to perform tasks but not to solve problems, and where decisions are made by a hierarchial chain of command. The method of decision making and the behavioural pattern are informally transmitted to new members and strongly embedded in the thinking of all managers and workers.

In addition, WCM manufacturing strategies imply JIT methods of production which fly in the face of traditional practices such as long production runs, specialized workers and materials pushed through plants. WCM strategies also require fewer suppliers and different working relationships with those suppliers. These WCM strategies, it is argued, imply a cultural change in the plant, and it is this change in culture which will ultimately result in different patterns of decision making.

Survey research can be used to capture the shared meanings aspects of culture, but not the deeper-ingrained values, artefacts and underlying assumptions which must be observed to be measured. In this study we will be using survey research, recognizing fully the limitations of the aspects of culture which can be measured by this method.

In using survey methods, we take our lead from Hofstede (1980) who used surveys to study differences in culture between nations. Hofstede identified four dimensions that distinguish the culture of one nation from another: power distance, individualism–collectivism, uncertainty avoidance, and masculinity. We believe that Hofstede's dimensions provide useful dimensions for discriminating across organizational cultures as well as across nation-societies. This is consistent with the work of Hofstede *et al.* (1990) on organizational culture, which reported that dimensions of national culture served as discriminators of organizational culture as well.

Of interest here are the dimensions of power distance and individualism–collectivism. Power distance refers to the extent to which a society accepts the fact that power is distributed unequally, and individualism–collectivism describes the extent to which individuals look out for their own interests versus interests of groups in which they are members. These constructs are reflected in the WCM literature in the discussion of team spirit and egalitarian approaches in organizations respectively. Collectivism is indicated by the emphasis on team orientation (Hayes, Wheelwright and Clark, 1988). Power distance is reflected by the changing roles assigned to supervisors', engineers' and managers' interactions with operators (Schonberger, 1986). Although we have used Hofstede's theoretical concepts, many of the actual measures used in this study were developed specifically for use in manufacturing and are based on manufacturing practices attributed to a WCM orientation.

One commonly used definition of culture is 'shared values' (Deal and Kennedy, 1982). According to this widely accepted study, most successful organizations have strong cultures which are characterized by high commonality of shared meanings, unique company rituals and a distinctive organizational milieu. We think that one of the problems in adopting a WCM manufacturing strategy is the strong culture which must be changed. Our definition of a strong culture will be one that has high homogeneity across the plant and a strong commitment from all individuals.

Wilkins and Ouchi (1983) discuss organizational culture in a more holistic fashion than Hofstede or Deal and Kennedy, outlining three cultural types: markets, bureaucracies and clans. They believe that firms will use market-type cultures when plants can easily replace employees and suppliers from the market at low costs of changing from one to another. Clan-style cultures will be more appropriate when tasks are highly complex and uncertain. As a result, workers and suppliers are not easily replaceable and it is costly to use the market to make changes. A WCM plant environment is highly complex and interdependent, and the clan style of culture should be most appropriate for this environment. Matching Wilkins and Ouchi's framework with Hofstede's dimensions, we believe that a firm with low power distance and a strongly collectivist orientation most closely resembles a clan-style culture.

The WCM literature (Hayes and Wheelwright, 1984; Schonberger, 1986; Hayes, Wheelwright and Clark, 1988; Roth and Miller, 1990) seems to support the desirability of a clan-style culture through its emphasis on interfunctional management teams which involve management, supervisors and workers. By encouraging the formation of groups that cut across functional boundaries and cooperate on the shop floor, the employee is encouraged to contribute to the benefit of the whole organization, rather than to any one functional constituency. The clan style of culture is most consistent of Wilkins and Ouchi's three types with the WCM notion of building competitive advantage over time through retained organizational learning (Hayes, Wheelwright and Clark, 1988).

17.2.3 Integration of manufacturing strategy and culture concepts

Based on our examination of the concepts of manufacturing strategy and organizational culture, we argue that they are closely related, but not essentially synonymous, as Weick (1985) has contended. Manufacturing strategy, as we have defined it, consists of both intended strategy and realized strategy. Intended manufacturing strategy is characterized by manufacturing strategy plans such as a mission statement, goals, objectives and policies (Schroeder, Anderson and Cleveland, 1986), plant charters (Skinner, 1974), and by a long-range orientation on the part of management. It reflects overt action on the part of management. Realized strategy, on the other hand, is a consistent pattern in decision making which is linked to the business strategy. It is the pattern of decisions actually made by manufacturing.

Intended manufacturing strategy is clearly something different from culture. Culture consists of shared meanings at the level of the subconscious. We have argued that a certain cultural type would be reflected in WCM plants by use of teams, lack of hierarchy, and cross-functional cooperation. These cultural dimensions are different from the content of intended manufacturing strategy, which is related to planning and linkages to business strategy. On the other hand, realized manufacturing strategy is more similar to culture, but still not identical. We distinguish the two by claiming that culture refers to people's assumptions which impact the *process* of decision making, while realized manufacturing strategy refers to patterns of decisions or the *outcomes* of decision making.

In Fig. 17.1, we have shown the relationships between intended strategy, realized strategy and organizational culture. We believe that organizational culture, the shared meanings among group members, affects both the realized strategy (or pattern of decisions) and the intended strategy or the plans that are created, since such decisions and plans reflect the assumptions and beliefs of the organization's management. Likewise, an intended strategy may act to change underlying values and assumptions over time and therefore change organizational culture via its impact on managerial decision patterns or realized strategy.

The current study is based on the conceptual framework implied by Fig. 17.1, but does not test all of the relationships. Rather, the study tests for the existence of a relationship between organizational culture and intended strategy. For the present, we are not measuring realized strategy. We believe that research on realized strategy is best accomplished via ethnography due to the complexity of charting many decisions over time.

17.3 RELEVANT ATTRIBUTES OF MANUFACTURING STRATEGY AND ORGANIZATIONAL CULTURE

Based on the literature, it is necessary to identify, define, and discuss the relevant aspects or attributes of manufacturing strategy and organizational culture that will be measured. Our identification of relevant characteristics is based on a world-class manufacturing orientation, and the ideas about manufacturing strategy and organizational culture expressed in that realm of literature.

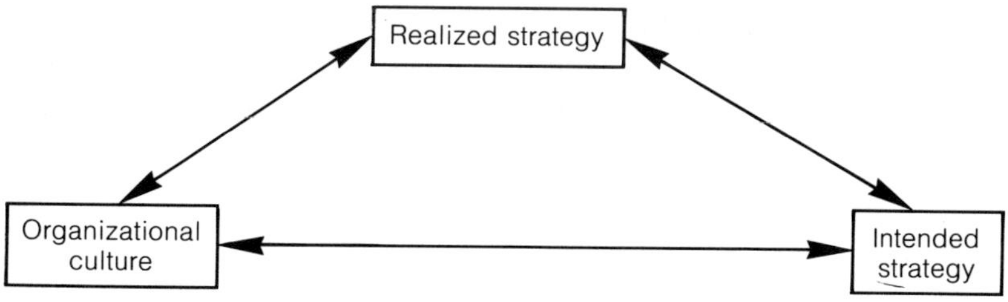

Fig. 17.1 Relationship between strategy and culture.

17.3.1 Manufacturing strategy

The attributes of manufacturing strategy chosen for measurement include the strategic role of manufacturing, the competitive strength of manufacturing, the formality of the planning process, the communication of the strategy throughout the plant, and the extent to which the plant has a strategic long-range orientation. Each of these attributes will be discussed and the content of the measures described.

Strategic role – stages 1 through 4

As discussed above, Hayes and Wheelwright (1984) have described four stages through which the manufacturing function passes as it becomes progressively more important to the competitive position of the firm. Hayes and Wheelwright identify different traits that depict each different stage. The scales describing each stage follow Hayes and Wheelwright closely, and as such four different scales were created to identify the four stages. The Stage 1 scales consider such issues as control systems, fire fighting, short-term performance, and the use of external experts. Stage 2 is characterized by reliance on vendors for equipment, a planning horizon of one business cycle, the use of common industry practices, and the use of industry-wide wage rates. Stage 3 checks for the existence of a manufacturing strategy, and its consistence with the business strategy, and for attention to long-term manufacturing trends. Stage 4 is concerned with the existence of proprietary equipment, integration of functions, and forward visibility with respect to manufacturing capabilities, practices, and technologies. Questions encompassing the issues listed above were created and four scales (one for each stage) were constructed from the questions.

Manufacturing strategy strength

The measure of manufacturing strategy strength is concerned with the strength of manufacturing's contribution to competitive position. Issues included in the scale include the existence of regular systems of plant performance monitoring, a well-developed manufacturing strategy, a well-focused plant, and whether respondents agree that manufacturing provides competitive strength for the business.

Formal strategic planning

World-class plants are assumed to have well-developed planning processes and plant involvement in the formulation of manufacturing strategy. Questions covering these issues are included in this scale.

Communication of strategy

The extent to which personnel know and understand manufacturing's strategy is an indicator of a world-class manufacturer. Questions concerned with whether only managers or whether all plant personnel understand goals, objectives and strategies, and whether employees perceive that they understand the plant's strategy were included in this measure.

Long-range orientation

As discussed above, the existence of a long-range orientation is an important component of manufacturing strategy. Questions concerned with the importance of short-range versus long-range objectives comprise this scale.

17.3.2 Organizational Culture

Three concepts are used to organize the cultural dimensions of a plant with a world-class manufacturing orientation. Individualism/collectivism, or merely collectivism, refers to the extent people work for the benefit of the group rather than the individual. Power distance refers to the extent to which authority is differentiated by position – a slightly different meaning from Hofstede's. Cultural congruency refers to the extent to which attitudes, behaviours, knowledge structures and 'culture' are consistent among employees. Collectivism and power distance were selected from Hofstede's dimensions of national culture because they are visible themes in the world class manufacturing literature. Cultural congruency is closely related to Deal and Kennedy's concept of 'strong culture', but we view a 'strong' culture as one that is homogeneous or internally consistent.

Collectivism

Various aspects of collectivism include how different departments within the plant coordinate and communicate with each other, whether problem solving is done on an individual or a group basis, whether teamwork exists and who participates on the team, and rewards for group rather than individual performance. Four scales addressing these issues were selected to operationalize collectivism (Fig. 17.2).

Power Distance

Authority issues include the extent to which decision making is centralized or decentralized, the visibility of management on the shop floor,

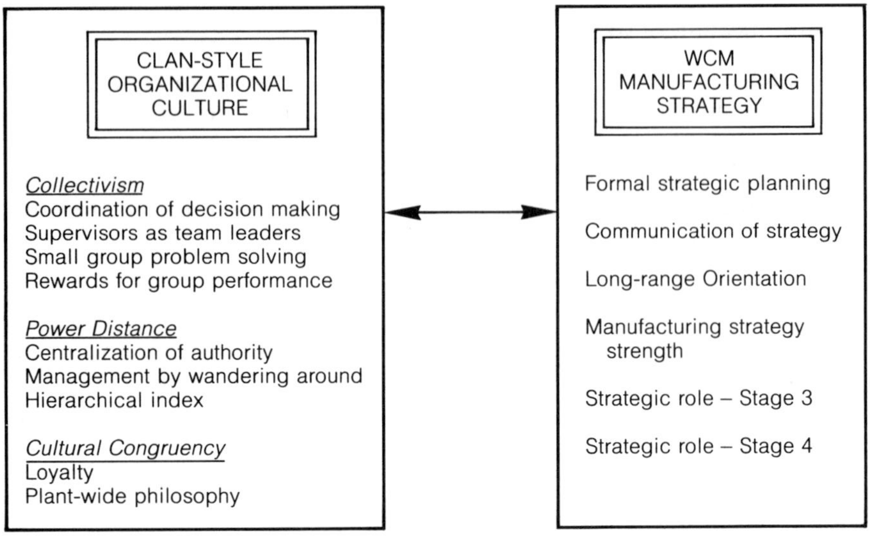

Fig. 17.2 World-class manufacturing strategy and clan-style organizational culture.

and the degree of visible differentiation on the basis of rank. The world-class manufacturing literature has recommended pushing decision making to the lowest level possible, employee involvement of various types, management visibility on the shop floor, and an egalitarian approach to all employees. Such an approach would include common rest rooms, cafeteria, and parking area for management and employees, uniforms for both management and employees, and taking out time clocks. Two scales were selected and an index was created to represent power distance. The scales, named 'centralization of authority' and 'management by wandering around', measure the extent to which decision making is centralized or decentralized and the visibility of management on the shop floor. The hierarchical index is composed of six dichotomous questions that record the existence of egalitarian practices.

Cultural congruency

This concept, which is our interpretation of a strong culture, considers whether employees feel there is a strong, unique philosophy within the plant and also the employee's individual commitment to the organization – how much extra effort would the employee put forth for the organization, whether the employee cares about the fate of the organization, etc. Scales named 'plant-wide philosophy' and 'loyalty' were used to measure this concept.

Even though several of the scales were constructed and validated in other studies, all scales were subjected to reliability and validity analysis to attest to their applicability in the manufacturing environment. All scales did meet the appropriate criteria, with the exceptions noted below. In several cases, it was necessary to delete questions from the scales to meet statistical requirements of reliability.

17.4 ANALYSIS

17.4.1 Data collection and questionnaire design

Data were collected for this study in 37 randomly selected plants from three industries – electronics, automotive components and machinery. The plants included both Japanese-owned and US-owned plants in the US. Each plant was contacted by letter and telephone to request its participation in the study. Of the plants contacted, 60% agreed to participate in the study. A total of 21 questionnaires were sent to each participating plant. Ten of these went to randomly selected workers and the remaining 11 to management. The managers surveyed included the plant manager, plant superintendent, production and inventory control manager, plant accountant, quality manager, process engineering manager, human resources manager, three supervisors and the research study coordinator. Altogether, 722 individual questionnaires were returned.

The questionnaires consisted of questions which belonged to the scales of interest and some quantitative measurements of plant characteristics. Each scale was assigned to at least two different questionnaires in order to assess the consistency of individual responses within a plant. Scales were taken from the literature whenever possible, but many had to be created specifically for this study. As a result, statistical analysis was done to assure reliability of the resulting measurements. The questionnaire was pretested extensively in 12 plants by conducting on-site interviews with managers and workers.

17.4.2 Reliability and validity of the measures

When using questionnaire measures of such concepts as discussed above, it is necessary to determine whether the measures are reliable and valid. Reliability refers to the amount of variation in the measurement instrument – whether the same measures consistently return the same results. Validity refers to whether the measures encompass what they were intended to cover.

Rather than use a single question or response to measure each of the characteristics of manufacturing strategy or organizational culture, a scale consisting of between two and nine questions was used. Most scales consisted of four, five, or six questions. Scales were used because the average of responses to several closely related questions is a more accurate measure of a characteristic than the response to a single question. Multiple responses are assumed to reduce response biases due to variations in wordings, question location, and other factors. Sections (a)–(d) illustrate questions for several selected scales: Strategic Role – Stage 1, Strategic Role – Stage 4, Communication of Strategy, Coordination of Decision Making, Centralization of Authority, and Plant-Wide Philosophy.

(a) World-class manufacturing strategy scales

Stage 1[1] *(scale dropped from final analysis):*

1. We use internal control systems to control manufacturing.
2. Manufacturing at our plant is kept flexible and unfocused.
3. Fire fighting is common at our plant.
4. Short-term performance is emphasized at our plant.
5. At our plant, external experts are used in making important decisions.

Stage 4[1]*:*

1. We actively develop proprietary equipment.
2. At our plant, manufacturing is centrally involved in marketing and engineering decisions.
3. The functions of our firm are well integrated.
4. We pursue long-range programmes, in order to acquire manufacturing capabilities in advance of our needs.
5. We make an effort to anticipate the potential of new manufacturing practices and technologies.

Communication of strategy[1]

1. In our plant, goals, objectives and strategies are communicated to me.
2. Strategies and goals are communicated primarily to managers (reverse scaled).
3. I know how we are planning to be competitive at this plant.
4. I understand the long-run competitive strategy of this plant.

[1] These scales have been copyrighted by Roger G. Schroeder and Barbara Flynn, and may not be used without the written permission of the authors. All other scales were taken or adapted from Price and Mueller.

(b) Individualism–collectivism scale

Coordination of decision making (Patterned after Georgopoulos and Mann (1962))

1. Generally speaking, everyone in the plant works well together.
2. Departments in the plant communicate frequently with each other.
3. Departments within the plant seem to be in constant conflict. (Reverse Scaled)
4. Management works together well on all important decisions.

(c) Power distance scale

Centralization of authority (Aiken and Hage, 1966)

1. I can do almost anything I want without consulting my boss (reverse scaled).
2. Even small matters have to be referred to someone higher up for a final answer.
3. This plant is a good place for a person who likes to make his own decision (reverse scaled).
4. Any decision I make has to have my boss's approval.
5. There can be little action taken here until a supervisor approves a decision.

(d) Cultural congruence scale

Plant-wide philosophy (Price and Mueller, 1986)

1. This plant doesn't have a very definite idea about how things should be done (reverse scaled).
2. This plant produces a very special kind of employee.
3. This plant has a clear view about how to achieve goals.
4. Stories are often told within the plant about accomplishments of past employees.
5. Sayings are often told within the plant which embody organizational wisdom.

Answers to these questions were recorded on five-point Likert scales. The five points were labelled Strongly Agree, Agree, Neither Agree nor Disagree, Disagree, and Strongly Disagree.

Given that scales are used, it is necessary to determine the degree to which the answers to the questions in the same scale are consistent with each other. Where appropriate, questions were dropped from scales to achieve a higher level of internal consistency. The statistical measure used to test for internal consistency or reliability was Cronbach's coefficient alpha (Cronbach, 1951). Nunnally (1967) states that $\alpha > 0.7$ is appropriate for confirmatory research and $\alpha > 0.6$ is adequate for developmental research such as this study. Nine scales exceeded α of 0.7, five scales reported α between 0.6 and 0.7, and two scales were unacceptable ($\alpha \leqslant 0.35$) and were dropped from the final analysis. Overall, the measures of the characteristics of manufacturing strategy and organizational culture that were used demonstrated acceptable reliability, especially since many of these scales were newly developed for this study.

The scales Strategic Role – Stage 1 and Strategic Role – Stage 2 had reliabilities of $\alpha \leqslant 0.35$ and were not used in the analysis. The questions in the scales for Strategic Role – Stage 1 through Stage 4 were based on the descriptions of manufacturing's competitive strength of Hayes and Wheelwright (1984): Stage 1 – Internally Neutral, Stage 2 – Externally Neutral, etc. While the concept of the stages appears to be a useful method of thinking about manufacturing's contribution to business strategy, the questions based on Hayes and Wheelwright's description of Stages 1 and 2 did not show sufficient internal consistency for use in statistical analysis. Stages 3 and 4 did show sufficient internal consistency, and variation in the level of these scores was used to measure the degree to which manufacturing contributed to the competitive position of the firm.

An additional concern is whether the measures draw upon one or multiple attributes. Factor analysis was used statistically to test the unidimensionality of the measures. All scales consisted of a single dimension except Rewards for Group Performance, which captured aspects of rewards for teamwork as an activity separately from rewards given for the performance of the team. Because the scale demonstrated sufficient internal consistency, and because the concept of rewarding for group efforts was important to collectivism, the scale was retained for analysis. Additionally, interviews with practising managers indicate that reward systems are in a state of flux because they are in the process of being revised, thereby leading to the problems encountered with this scale.

The statistical analysis considers whether the organizational culture of the plant and the manufacturing strategy are related. Data were collected from many people in the plant, with some questions (generally the strategy questions) appearing on two questionnaires per plant, and the organizational culture questions appearing on all 21 questionnaires per plant. It is important to see whether there is agreement on responses to these questions between individuals within a plant, to a greater degree than differences between plants. An analysis of variance was conducted comparing the variation in responses within the plant (within-group variance) with the variation in responses between plants (between-group variance). The variation in responses was greater between groups than within groups for all variables, and this difference was statistically significant. This analysis justifies the use of the mean score of the respondents for a given plant as the measure of the plant for the given variable.

Validity was assessed both qualitatively and quantitatively. Since validity refers to the extent to which the measures represent the intended concept, validity is ultimately a matter of subjective judgement. However, evidence from statistical analysis can also be used to support validity. As part of the qualitative assessment, the questions comprising the scales were examined for their consistency with the WCM literature. Also, during the pilot testing of the questionnaire, interviews were conducted with practising managers to determine the clarity of the scales and the manner in which the managers understood the questions.

Factor analysis was used to examine the attributes of the overall concepts of collectivism, power distance, cultural congruency and WCM strategy. Factor analysis shows how the individual questions comprising the scales covary. The questions comprising the four collectivism scales covaried closely, so that a single factor explained over two-thirds of the variation. Likewise, cultural congruency and WCM strategy produced close covariation among the questions comprising the scales so that a single factor explained 62% and 85% of the variance respectively. Power distance questions covaried on two different factors, explaining 51% and 34% of the variance, respectively. The strong tendency of the questions from different scales to covary supports the description of organizational culture using the dimensions of collectivism, power distance, and cultural congruency, and the use of the scales covering various aspects of manufacturing strategy to embody the concept of a WCM strategy.

17.5 ANALYSIS OF CULTURE AND STRATEGY RELATIONSHIP

As discussed above, it is important to determine if there is a relationship between the type of manufacturing strategy in a plant and the type of organizational culture. Organizational culture and manufacturing strategy are each measured using sets of variables, and hence the relationship between the two sets of variables is being studied. Because organizational culture and manufacturing strategy are viewed as reciprocally affecting each other, independent or dependent variables are not specified and causation is not implied.

This approach to research is consistent with the recommendations of Hackman (1985) in his discussion of doing 'relevant' research, i.e. research that is valuable for both theory and practice. Hackman believes that it is difficult to find single causes of good or poor performance; a better approach is to look at the relationships between sets of interdependent practices that together foster good performance. The set of world-class manufacturing practices perfectly fits Hackman's notion of different-but-related conditions that interact, with some redundancy, to bring about certain end conditions.

Canonical correlation is a statistical technique that can be used to study the association or covariation between two sets of variables. The technique first finds the linear combination or weighted set of variables that maximizes the correlation between the two sets of variables, then finds a second linear combination of variables that maximizes the correlation but is independent of the first combination, and so on.

Ordinarily, a 'very large' sample size is required to conduct canonical correlation (Johnson and Wichern, 1988). The sample size for this study is 722 individuals in 37 plants. However, studies such as Hofstede *et al.* (1990) deal with a similar situation – in this case the use of over 1200 individual responses from 20 organizational units as a sample. Hofstede *et al.* argue that 'large' sample sizes are required for the purpose of preventing instability in statistical results. Since the analysis is conducted using the mean of the scores of several individuals within a plant rather than the individual scores, the potential for instability is reduced and a smaller sample size is acceptable.

Canonical correlation was used to determine the degree of association between the six WCM strategy scales and the nine organizational culture scales. Two linear combinations were significant out of six total combinations produced. The two statistically significant combinations, called canonical variates, together explain 89% of the total variance. Also, statistical testing indicated that the probability that the correlations were produced by random error was less than 0.005. Thus we can conclude that a significant relationship exists between manufacturing strategy and organizational culture.

The two canonical variates can be interpreted based on the correlation of the individual variables with each variate (see Table 17.1). The first variate, named the 'World-class Plant', is strongly and positively correlated with the WCM strategy scales, the collectivism scales, and the cultural congruency scales. The second variate, called the 'Traditional, Command-and-Control Plant', exhibits strong, negative correlation with some of the WCM strategy scales and strong, positive correlations with power distance scales.

Our interpretation of this data analysis is that two types of strategy/culture combinations exist in the sample of plants. As expected, the 'World-class Plant' exhibits an organizational culture characterized by strong collectivism or a group orientation, low power distance or a more egalitarian culture, and strong cultural congruency or loyalty and plant culture. This is consistent with WCM literature which calls for increased use of quality circles, work teams, group-oriented reward systems, participation by workers in decision-making and problem-solving processes, and decentralization of authority. Also indicated are a high level of coordination between functional areas, egalitarian practices such as common management and worker rest rooms, cafeterias, parking areas and dress (uniforms), and encouragement of ideas and opinions from

Table 17.1 Correlations between scales and canonical variates

Names of canonical variates	*'World-class plant' variate*	*'Traditional, command-and-control plant' variate*
Strategy scales		
Stage 3	+	−
Stage 4	+	
Formal strategic planning	+	−
Communication of strategy	+++	
Long-range orientation	+	
Manufacturing strategy strength	++	
Culture scales		
Collectivism		
Coordination of decision making	+++	
Supervisors as team leaders	++	
Small group problem solving	++	−
Rewards for group performance	++	
Power distance		
Centralization of authority	−	++
Management by wandering around	+	+
Hierarchical index		+
Cultural congruency		
Loyalty	+++	
Plant-wide philosophy	+++	

Legend:
\+ Correlation between 0.4 and 0.6.
++ Correlation between 0.6 and 0.8.
+++ Correlation between 0.8 and 1.0.
− Correlation between −0.4 and −0.6.
−− Correlation between −0.6 and −0.8.

workers by supervisors. Such characteristics exist concurrently with strong employee loyalty and a 'strong', or homogeneous, consistent plant culture.

The second canonical variate, the 'Traditional, Command-and-control Plant', exhibits strong centralization of authority, a lack of problem solving by small groups, and a negative association with two of the WCM strategy scales. This variate implies that a plant characterized by strong centralization of authority and problem solving by individuals rather than groups will not be a world-class plant. Such plants will not see the benefits of coordination implied by the collectivism scales nor the benefits of strong employee loyalty and strong culture implied by the cultural congruency scales.

Two surprising results emerged from the data analysis. First, it was expected that a traditional plant would not exhibit a long-range orientation. The correlation of the long-range orientation scale with the 'Traditional, Command-and-control' plant combination is +0.39, almost exceeding the traditional cutoff value for significance of 0.40. This implies that a traditional plant may also exhibit a long-range strategic orientation. As such, a long-range orientation may not serve as a discriminating trait of a world-class plant. The second surprise was that management by wandering around correlated positively with both variates. The predictions would have been that management by wandering around, a measure of visibility of management on the shop floor, would be associated positively with a world-class plant but not a traditional plant. The data analysis indicates that management has strong visibility on the shop floor in both world-class plants and traditional, command-and-control plants.

17.6 CONCLUSIONS

Based on the data analysis, there exists a very strong relationship between the type of manufacturing strategy and the organizational culture of a plant. A plant with a world-class orientation – one that was evaluated highly on the WCM scales and contributes strongly to the competitive position of the firm – possesses a collectivist, group-oriented, egalitarian and congruent or consistent culture. A high level of coordination, group problem solving, teamwork, decentralization of authority, loyalty, and a special, plant-wide philosophy characterize such a plant. Likewise, a plant that does not contribute strongly to the competitive position of the firm (a non-world-class plant) is likely to possess a hierarchical, authoritarian organizational culture with decisions being made by individuals rather than by groups. Given that the combinations representing these two types of plants explained 89% of the variance in the sample of plants in three industries, these two plant profiles have strong descriptive power for strategy/culture relationships within the industries in the sample. Also, their potential for describing strategy/culture relationship in plants in other industries appears promising.

We believe that the close fit between strategy type and cultural type occurs as a result of the type of environment encountered by the manufacturing plant in today's society. The manufacturing task is very complex and interdependent, where activities undertaken in one part of the organization have a strong impact on outcomes in another part of the organization. For example, departmental policies in functional areas such as purchasing, inventory control, materials management and maintenance, may have potent effects for the production and quality assurance functions. As such, a hierarchical, authoritarian organizational structure simply cannot manage interdependent relationships between functional areas as well as a flatter, less authoritarian, more group-oriented or collectivist structure. Strong hierarchies are not the best management style or structure for managing today's manufacturing plant.

The results also suggest that the efforts of practising managers to change organizational culture (Ebrahimpour and Lee, 1988) are essential for the implementation of progressive practices in manufacturing environments. This is consistent with the case-study evidence of Maruchek, Pannesi, and Anderson (1990) and Adler (1990). Maruchek, Pannesi and Anderson report many cultural change efforts undertaken in concert with the changing of manufacturing strategies and practices. Furthermore, they indicate that manufacturing executives feel that cultural change is one of the most important factors affecting the implementation of manufacturing strategy. Adler indicates that an organization's culture is a primary influence on the rate of organizational learning and productivity improvement. Adler attributes this to such cultural influences as communications patterns and status hierarchies – notions which are closely related to the concepts of collectivism and power distance used in the present study.

The research also indicates that a long-range orientation on the part of management and visibility on the shop floor are associated not only with world-class plants, but with plants where a more authoritarian, traditional culture exists. An additional explanation for this may be that manufacturing strategy is being formulated at levels higher than top management within the plant. Thus we would be unable to distinguish a long-range orientation given that data were collected within the plant. Also, a long-range orientation may exhibit itself differently in formulation *vis-à-vis* implementation of manufacturing strategy. Long-range plans may have been formulated, but the implementation of such plans may exhibit inconsistency and instability. This may give rise to a retrospective description of a plant as 'short-term orientated' with respect to its 'pattern of decision making' (Hayes and Wheelwright, 1984).

With respect to management by wandering around, or the visibility of management on the shop floor, we believe that it does not serve to distinguish a world-class plant from a traditional plant. We believe that a long-range orientation and visibility on the shop floor are desirable qualities for plant management, but these traits do not serve to aid in discriminating between world-class and traditional plants.

An additional conclusion concerns the research strategy itself. Rather than look for the association between single factors or causes and strategy types, the present analysis looked at the relationships between sets of variables. The philosophy of Hackman (1985), the study of multiple, interrelated-yet-redundant conditions, was used in the construction of the holistic concept of culture. This research strategy proved fruitful in that statistically strong, intuitively believable results were produced. Such a research strategy has much potential for future explanation and description.

17.7 IMPLICATIONS FOR FUTURE RESEARCH

Many questions for research arise out of the results above. The data analysis indicates that a world-class plant is more collectivist and egalitarian, but is that invariably true? To what extent is that result consistent across industries or nations and societies? For example, Japan, the home country of the archetypes of world-class manufacturing, has a much higher level of power distance within its national culture than most other countries (Hofstede, 1980). Would the cultural profile of a world-class plant in Japan exhibit low power distance and hence be similar to that observed in the US?

What are the processes and procedures by which organizational culture can be changed within the plant? Schein (1985) and Deal and Kennedy (1982) both address the issue of cultural change. Schein believes that the ability to change organizational culture and the mechanisms for such change vary with the growth stage of the organization. New organizations have cultures that are relatively easily manipulated; organizations in midlife have more difficulty with cultural change; and mature organizations have extreme difficulty in accomplishing such change. Research into effective and ineffective mechanisms of cultural change in manufacturing plants would add to our knowledge in this area.

As the research results here clearly indicate, there is a strong link between the effects of behavioural variables and strategies. Additional research (both ethnographic and survey research) into the nature of the strategy and behavioural/cultural relationship would be useful for theory and practice. An integrated programme of research using multiple methodologies exploring the issues of strategic initiatives and practices and their relationships with cultural and behavioural variables is recommended.

17.8 ACKNOWLEDGEMENT

This research project was supported in part by a grant from the Japan–US Friendship Commission and a grant from the McKnight Foundation. The authors would like to thank Larry Cummings, Barbara B. Flynn, E. James Flynn, Richard Saavadra and Sadao Sakakibara for comments on earlier drafts of this paper.

REFERENCES

Adler, Paul S. (1990) Shared learning. *Management Science*, **36**(8), 938–57.

Cronbach, L.J. (1951) Coefficient alpha and the internal structure of tests. *Psychometrika*, **16**, 297–334.

Deal, Terrence E. and Kennedy, Allen A. (1982) *Corporate Cultures*, Addison Wesley, Reading, Mass.

Ebrahimpour, Maling and Lee, Sang M. (1988) Quality management practices of American and Japanese electronic firms in the US. *Production and Inventory Management Journal*, Fourth Quarter, 28–31.

Ettlie, John E., Burstein, Michael C. and Feigenbaum, Avi (eds.) (1990) *Manufacturing Strategy: the Research Agenda for the Next Decade, Proceedings of the Joint Industry University Conference on Manufacturing Strategy Held in Ann Arbor, Michigan on January 8–9, 1990*, Kluwer, Boston.

Flynn, E. James, Flynn, Barbara B. and Schroeder, Roger G. (1989) World class manufacturing: theoretical foundations and strategic implications. University of Minnesota Working Paper 89–18, December.

Gregory, Kathleen L. (1983) Native-view paradigms: multiple cultures and culture conflicts in organizations. *Administrative Science Quarterly*, **28**, 359–76.

Hackman, J.R. (1985) Doing research that makes a difference, in *Doing Research that is Useful for Theory and Practice*, (eds. E.E. Lawler, A.M. Mohrman, S.A. Mohrman, G.E. Ledford and T.G. Cummings), Jossey-Bass, San Francisco.

Hayes, Robert H. and William J. Abernathy (1980) Managing our way to economic decline. *Harvard Business Review*, **58**(4), 67–77.

Hayes, Robert H. and Wheelwright, Steven C. (1984) *Restoring Our Competitive Edge*, Wiley, New York.

Hayes, Robert H. and Wheelwright, Steven C. and Clark, Kim B. (1988) *Dynamic Manufacturing*, The Free Press, New York.

Hofstede, Geert (1980) *Culture's Consequences: International Differences in Work-Related Values*, Sage, Beverly Hills.

Hofstede, Geert, Neuijen, Bram, Ohayv, Denise Daval and Sanders, Geert (1990) Measuring organizational cultures: a qualitative and quantitative study across twenty cases. *Administrative Science Quarterly*, **35**, 286–316.

Johnson, Richard A. and Wichern, Dean W. (1988) *Applied Multivariate Statistical Analysis*, 2nd edn., Prentice-Hall, Englewood Cliffs, NJ.

Maruchek, Ann, Pannesi, Ronald and Anderson, Carl (1990) An exploratory study of the manufacturing strategy process in practice. *Journal of Operations Management*, **9**(1), 101–23.

Mintzberg, Henry (1978) Patterns in strategy formation. *Management Science*, **24**, 934–48.

Nunnally, J. (1967) *Psychometric Theory*, McGraw-Hill, New York.

Price, James L. and Mueller, Charles H. (1986) *Handbook of Organizational Measurement*, Pitman, Marshfield, Mass.

Roth, Aleda V. and Miller, Jeffrey G. (1990) Manufacturing strategy, manufacturing strength, managerial success, and economic outcomes, in *Manufacturing Strategy: the Research Agenda for the Next Decade, Proceedings of the Joint Industry University Conference on Manufacturing Strategy Held in Ann Arbor, Michigan on January 8–9, 1990*, (eds. John E. Ettlie, Michael C. Burstein, and Avi Feigenbaum), Kluwer, Boston.

Saffold, Guy S., III (1988) Culture traits, strength, and organizational performance: moving beyond 'strong' culture. *Academy of Management Review*, **13**(4), 546–58.

Schein, Edgar H. (1985) *Organizational Culture and Leadership*, Jossey-Bass, San Francisco.

Schonberger, Richard J. (1986) *World Class Manufacturing*, The Free Press, New York.

Schroeder, R.G., Anderson, J. and Cleveland, G. (1986) The content of manufacturing strata: an empirical study. *Journal of Operations Management*, **6**(3), 136–45.

Skinner, Wickham (1969) Manufacturing – missing link in corporate strategy. *Harvard Business Review*, **47**(3), 136–45.

Skinner, Wickham (1974) The Focused Factory. *Harvard Business Review*, **52**(3), 113–21.

Spradley, James P. and McCurdy, David W. (1975) *Anthropology: the Cultural Perspective*, Wiley, New York.

Weick, Karl (1985) The significance of corporate culture, in *Organizational Culture*, (eds. P. Frost, L. Moore, M. Louis, C. Lundberg and J. Martin), Sage, Beverly Hills, 381–9.

Wilkins, Alan L. and Ouchi, William G. (1983) Efficient cultures: exploring the relationship between culture and organizational performance. *Administrative Science Quarterly*, **28**, 468–81.

CHAPTER 18

The evolution of manufacturing systems: towards the post-industrial enterprise

William J. Doll and Mark A. Vonderembse

18.1 ABSTRACT

The post-industrial manufacturing era is likely to be characterized by increased market diversity, rapid market and technological change, and the world-wide spread of advanced manufacturing technologies. Many firms have had difficulty responding to changing customer needs while improving productivity. Implicitly if not explicitly, top management's responses to these challenges are shaped by two alternative paradigms for managing manufacturing firms – industrial and post-industrial. The industrial paradigm, based on an economic/technical perspective, views manufacturing as a functional area. In contrast, the post-industrial paradigm, based on a customer-oriented sociotechnical perspective, views manufacturing as an enterprise.

This paper presents a stage model of the evolution of manufacturing from craft systems, to industrial systems, to the emerging post-industrial manufacturing enterprise. The attributes of each stage are described and the forces driving the evolutionary process are discussed. The authors contend that industrial and post-industrial enterprises seek innovativeness and efficiency in different ways. This model of the evolution of manufacturing systems is used to provide one explanation of the difficulty firms have had adapting to the new era in manufacturing.

Specially commissioned for *Manufacturing Strategy – Process and content.*
Edited by Christopher A. Voss.
Published in 1992 by Chapman & Hall, London ISBN 0 412 43660 4.

18.2 INTRODUCTION

Manufacturing has entered a new era, created by the convergence of two important forces: 1. increasingly complex, changing, and uncertain markets (Lawrence and Dyer, 1983), and 2. the rapid spread of manufacturing capabilities worldwide (Manufacturing Studies Board, 1986). The result is growing global competition (Young, 1985) that is changing our thinking about the organization of productive efforts. A 'mind set' of mistaken objectives, premises, and patterns of thinking that emphasize a narrow concept of efficiency is hampering the efforts of manufacturers to adapt to this new era. Skinner (1985) contends that this 'mind set', which can be traced to the Industrial Revolution and the work of Frederick Taylor, is now dysfunctional. This mind set leads to a pattern in the evolution of industrial systems that substitutes efficiency for flexibility (Utterback and Abernathy, 1975), leaving firms less able to adapt to market and competitive forces. Piore and Sabel (1984) contend that we are entering a second industrial divide; the industrial model founded on mass production can no longer secure a workable match between the production and consumption of goods.

The market and competitive forces of this new era are changing our concept of manufacturing. Manufacturing is increasingly viewed as an enterprise, i.e. a complex chain or network of interdependent value-adding activities starting with the preparation of raw materials, through fabrication and assembly, to distribution and after-sale service. The successful competitors are becoming service- or customer-oriented (Chase and Garvin, 1989) and the nature of work has become increasingly intellectual (Zuboff, 1984). By reducing product development/throughput time and developing networks of social systems that listen to the 'voice of the customer', these firms seek continuous improvement by coordinating a continuing stream of product and process innovation across their value chain.

Schonberger (1986) identifies innovative firms such as Hewlett-Packard, Harley-Davidson, Honeywell, and Black & Decker that have achieved fivefold, tenfold, and even twentyfold reductions in manufacturing lead times. These 'world-class manufacturers' are achieving continual and rapid improvement in all performance measures (i.e. quality, cost, lead time and customer service) without trade-offs. Schonberger stresses that employee involvement and interaction, both on the shop floor and in the decision making/problem solving process, is the key. Not all US manufacturing firms are making this transition to world-class manufacturing.

Several authors (Toffler, 1970; Naisbitt, 1982; Masuda, 1980; Huber, 1984) contend that we have entered a post-industrial society that will be characterized by increasing complexity, uncertainty and change. Piore and Sabel (1984) content that manufacturing systems are evolving towards flexible specialization, i.e. a strategy of permanent innovation that accommodates ceaseless change rather than attempting to control it. Several authors (Jelinek and Goldhar, 1983; Doll and Vonderembse, 1987) describe the technological innovations such as CAD, CAM and CIM which are thought to characterize this post-industrial approach to manufacturing. Naisbitt (1982) describes an information intensive society where interpersonal relationships (high touch) will play an important role in building effective work groups. In this post-industrial era, fundamental, even radical, changes in managerial and organizational systems may be needed to adapt to more and increasing knowledge, more and increasing complexity, and more and increasing turbulence (Huber, 1984).

A post-industrial manufacturing enterprise is defined as one whose structures and processes are appropriate for a post-industrial society. The post-industrial environment will place new demands on manufacturing enterprises. Increasing market variety and uncertainty, rapid developments in product and process technology, advances in information technology, and increasing global competition will require social, work, and control systems that enable manufacturing enterprises to be innovative as well as efficient.

The authors propose a stage model as a useful way of organizing our thinking concerning the impact of post-industrial society on manufacturing enterprises. The stage model describes the evolution of manufacturing from craft systems, through traditional industrial systems, to the rapidly emerging post-industrial manufacturing enterprise. This paper identifies the attributes that characterize each stage and describes the forces that drive the evolution between stages. The authors contend that post-industrial systems seek innovativeness and efficiency in different ways than do industrial systems.

18.3 STAGE THEORIES

Stage theories provide a useful way of thinking about phenomena that change over time. Kuznets (1965) contends that stage theories provide a framework for prescriptive theory formulation by identifying the attributes of each stage and describing the forces driving the evolutionary process. Stage theories have been prominent during the formative periods of diverse disciplines or areas of study such as economic development (Marx, 1932), information systems (Nolan, 1973), strategy/structural linkages (Chandler, 1962), and organizational life cycle theory (Kimberly *et al.*, 1980).

Stage theories are based on the premise that, driven by environmental forces, elements of a system move through a pattern of distinct stages over time. Each stage is characterized by a set of attributes, relationships between these attributes, and effectiveness criteria. Effectiveness criteria as well as cause and effect relationships are generally stable within a stage but vary between stages (Quinn and Cameron, 1983). Within each stage, shared assumptions about cause and effect relationships shape a common set of premises and patterns of thinking that influence the organization's approach to problem solving.

Transition between stages may not be easy (see Hounshell (1984) and Smith (1977) for a description of the difficulties US firms had in making the transition from craft to industrial systems). It is often associated with crises that must be solved in order for the organization to progress to the next stage. Crises are often caused by the organization's failure to respond in innovative ways to changing environmental factors. Old premises and patterns of thinking may inhibit innovative responses. Organizations that do not successfully resolve the problems associated with transition through stages restrict their growth and may fail.

A number of authors, using a technical/economic perspective, present stage models to describe the changing nature of manufacturing. Wickham Skinner (1985) uses a stage model to describe the Industrial Revolution (i.e. the transition from craft to industrial systems). Utterbach and Abernathy (1975) present a stage model to describe patterns of evolution within the industrial systems stage. They contend that a firm's innovation attempts will vary systematically with differences in the firm's environment, its strategies for competition and growth, and the state of development of process technology used by the firm. Jaikumar (1988) attempts to describe process-control aspects of the post-industrial manufacturing enterprise. He describes six epochs in the evolution of manufacturing process control at Beretta that represent a fundamental shift in the paradigm of production – from a world view of managing material processing to one of managing intelligence. These stage models tend to be historical in nature. With the possible exception of Utterback and Abernathy, they focus primarily inwardly on the technical/economic system and ignore environmental changes. Thus, they tend to view manufacturing as a functional area rather than an enterprise with customers and employees. These stage models give only minimal treatment to the implications of manufacturing system evolution for: 1. value received by the customer, or 2. interrelated changes in the organization's social, work, and control systems (i.e. the organization's values and norms, the changing nature of work, work group organization, and mechanisms used for control and learning).

This economic/technical literature on manufacturing has developed largely independently from related organizational theory literature. Huber (1984) describes the nature of the post-industrial society and discusses the nature of the increased demands that this environment will impose on organizations. However, he does not specifically describe the implications of a post-industrial environment for the design of manufacturing enterprises. Thompson (1967) describes alternative approaches to the design of productive systems (i.e. mediating, long-linked, intensive technologies), but does not relate these patterns of work to the evolution of manufacturing systems. Van de Ven (1986) describes the central problems in the management of innovation, but the insights he provides on the nature of innovative activity are not related to the context of manufacturing. Thus, we know little about the design of innovative as well as efficient manufacturing systems that are required to compete in a post-industrial society.

18.4 A STAGE MODEL OF THE EVOLUTION OF MANUFACTURING SYSTEMS

A stage model seems particularly appropriate for describing the evolution of manufacturing systems and the competitiveness problem facing many industries. The authors contend that manufacturing enterprises, driven by their environment, change in character over time and evolve in patterns roughly correlated to movements between three stages: 1. craft shops employing skilled artisans; 2. long-linked industrial systems using hard automation; and 3. post-industrial enterprises characterized by flexible resources and information-intensive intellectual work. Many firms in the industrial stage are experiencing a crisis in competitiveness. For both market and technological reasons, further incremental changes within the industrial stage are no longer adequate to insure competitiveness. Significant and lasting improvement in competitiveness requires a transition to the post-industrial stage. This transition requires a new paradigm for manufacturing management – a post-industrial paradigm.

This distinction between an industrial and post-industrial manufacturing enterprise is, in large measure, based on Thompson's (1967) conceptualization of long-linked *v.* intensive technologies. Industrial systems correspond to Thompson's long-linked technology; they are characterized by functionally specialized operations arranged in a fixed sequence (Hounshell, 1984) and a hierarchical administrative structure. Post-industrial systems have attributes identified by Thompson as characteristic of intensive technologies. The essential attributes of an intensive technology are customer orientation, flexible resource deployment, and a rich information environment created by direct and continuous feedback from operations. These three characteristics are described by Thompson as necessary to define an intensive technology. They are also necessary conditions to define a post-industrial enterprise. However, they are not sufficient to describe fully the attributes of the post-industrial manufacturing enterprise and the forces that explain its emergence.

The distinguishing differences between craft and industrial systems were technological in nature (e.g. advances in power generation and transmission, mechanical processes, interchangeability of parts, etc.). In contrast, industrial and post-industrial systems may use the same technologies (e.g. CAD, CAM, robots), but apply them in different ways and for different purposes. The distinction between an industrial and a post-industrial organization lies in the values and norms guiding its social system, the organization of its intellectual work, and the methods used to insure control as well as continuous improvement. Table 18.1 contrasts the social, work, and control systems of manufacturing systems.

Values and norms serve an important function in guiding role expectations and resultant behaviours in organizations. Skill-oriented craft systems rely on artisans using simple hand tools to shape products to customary standards of workmanship. In these craft shops, the value placed on the skills of the master workman motivates apprentices to serve many years under the mentorship of the master. In industrial systems, the product is an important aspect of the organization's value system. The social justification of the Industrial Revolution was the improvement in living standards created by mass producing low-cost standardized products of good quality. Product standardization permits functional specialization, including the separation of marketing and manufacturing. This specialization means that most of the work force seldom meet the customer. Thus, the work force associates the value they provide to society with the product rather than the customer (e.g. they view themselves as auto makers, steel workers, refinery workers). In industrial systems, conformance to product specifications rather than value to the customer is the criterion by which quality is assessed. Industrial systems are often market oriented. Marketing serves a boundary-spanning role, buffering the technical core from the customer and interpreting customer needs to other functional areas (Kotler, 1988). Marketing assists in defining customer needs in terms of product specifications that are treated as a given by manufacturing. With the product defined, manufacturing norms focus on improving efficiency/productivity rather than enhancing the value of the product to the customer.

In the post-industrial society, rapidly changing customer requirements and increased competition are forcing manufacturing enterprises to be more customer oriented. The effectiveness of the post-industrial manufacturing enterprise will require continuing, innovative and holistic responses to multiple and sometimes conflicting global performance measures dictated by shifting customer expectations and competitive alternatives. A customer-oriented enterprise has a social system where all employees understand the needs of their customers and how their organization seeks to provide value to customers. Ohmae (1988) contends that being customer-oriented means to search for better ways to deliver value to the customer. Kotler (1988) defines a customer-

Table 18.1 Characteristics of manufacturing systems by stage of evolution

	Craft	*Industrial*	*Post-industrial*
Social system			
Value orientation	Skill	Product	Customer
Norms	Workmanship	Efficiency/ productivity	Product Development and throughput time
Work system			
Equipment resources	Flexible hand tools	Capital intensive – special purpose	Capital intensive – flexible
Nature of work	Skilled manual work	Unskilled manual work/ functionally specialized intellectual work	Information intensive intellectual work
Work group organization	Task oriented groupings	Functionally specialized work groups	Self-organizing and self-directed work groups
Control system			
Perf. measures	Customary standard	Single and task specific	Multiple and global
Information systems	Informal and based on learning from experience	Formal information systems to control task execution and coordinate sequential activities	Formal and informal systems for control, mutual adjustment and learning
Control mechanism	Craftsman	Hierarchical authority structure	Modified market mechanisms (i.e. long-term cooperation)

oriented organization as one in which the customer is the controlling function (i.e. all functional areas are guided by direct contact with the customer). According to Juran (1987), post-industrial organizations are driven by the 'voice of the customer'. As the value to customer orientation encourages a swifter response to customer desires, post-industrial enterprises develop norms focusing on shorter product development (Takeuchi and Nonaka, 1986) and throughput time (Bower and Hout,

1988). This is often referred to as time-based competition (Schmenner, 1988).

The work system describes equipment resources, the nature of individual job assignments, and the way in which job assignments are organized into work groups. In craft systems, groups of artisans planned and executed skilled manual tasks using flexible hand tools. In industrial systems, capital-intensive special-purpose equipment is substituted for labour. Frederick Taylor emphasized a vertical division of labour, i.e. the separation of work planning from work execution. Workers were to execute their tasks in the 'one best way' specified by management. This one best way was often cast in 'technological concrete' by the substitution of fixed capital resources for human effort. In industrial systems, tasks are bifurcated between unskilled manual work (task execution) and functionally specialized work groups performing intellectual work (work planning). The bifurcation of work encourages task-specific innovation but makes innovation across the value chain more difficult.

The work system of the post-industrial manufacturing enterprise is characterized by flexible capital intensive resources, information intensive intellectual work, and self-organizing and self-directed work groups. In post-industrial manufacturing, work is primarily intellectual. To enhance learning, tasks are designed to permit people and groups to plan and execute their own work. The work groups are capable of self-direction because they understand how their activities add value to the customer. The customer rather than the supervisor becomes the mechanism for directing work group activities. These self-directing and self-organizing work groups enhance the organization's ability to implement more integrative (across the value chain) innovations because they are capable of learning.

Values and norms guide the development of role expectations and resultant behaviours, but performance standards, formal and informal information flows, and control mechanisms are necessary for performance improvement. In craft shops, craftsmen are the mechanism for performance improvement; motivated by workmanship norms, they learn from experience by comparing their work with customary standards. In industrial systems, a hierarchical authority structure, using formal information systems, controls task execution and coordinates sequential activities. The functionally specialized nature of work separates the technical core from the customer. Each functional area develops its own task-specific performance measures. Manufacturing, motivated by efficiency/productivity norms, assumes that the product satisfies customer needs and focuses on a single and task-specific performance measure (i.e. cost).

In the post-industrial manufacturing enterprise, changing customer desires for delivery, cost, quality, product performance, and product variety requires the organization to generate creative solutions that enable firms to satisfy multiple and global performance measures. Hierarchical control systems are not effective at generating creative solutions to these complex trade-offs. Driven by the need to shorten product development and throughput time, the post-industrial manufacturing enterprise develops both formal and informal information systems for control, mutual adjustment between self-directed work groups, and organizational learning (Aoki, 1990). A rich information exchange occurs within the firm and with its customers. Informal face-to-face communication between individuals and groups provides a rich medium that enhances learning. This learning occurs within the firm and across the supplier network. The success of products is measured in terms of their intellectual content and timeliness; a key source of this intellectual content is the supplier. To protect this content, the firm develops long-term cooperative relationships with suppliers. This modified market relationship is a control mechanism, replacing the hierarchical authority structure in more vertically integrated industrial systems.

18.5 FACTORS DRIVING THE EVOLUTION BETWEEN STAGES

In stage models, environmental forces drive the evolution between stages by altering the appropriateness of effectiveness criteria, transforming system attributes, and modifying cause-and-effect relationships. Figure 18.1 illustrates the forces that drive the evolution of manufacturing systems. In the Industrial Revolution, technological and market opportunities were the major environmental factors driving the evolution of manufacturing from craft shops to industrial systems using hard automation. The Industrial Revolution increased the intellectual content of work, but the changing nature of work was a consequence rather than a cause of the industrial revolution. The industrial revolution was primarily driven by technological opportunities for mass production; mass markets for standardized products were an important enabling factor. The industrial revolution made possible further environmental changes that are driving another revolution from industrial to post-industrial manufacturing.

The post-industrial revolution is only just beginning; it is quite different from the industrial revolution that has, historically, shaped our premises and patterns of thinking. Whereas the industrial revolution was technology driven and market enabled, the post-industrial revolution is market driven and technology enabled. The intellectual nature of work also plays an important role in the post-industrial enterprise; it is both a cause and a consequence of the post-industrial revolution. Driven by an accelerated rate of market change and increased market variety, post-industrial systems can be distinguished by the way in which social, work, and control systems are designed to enhance the effectiveness of

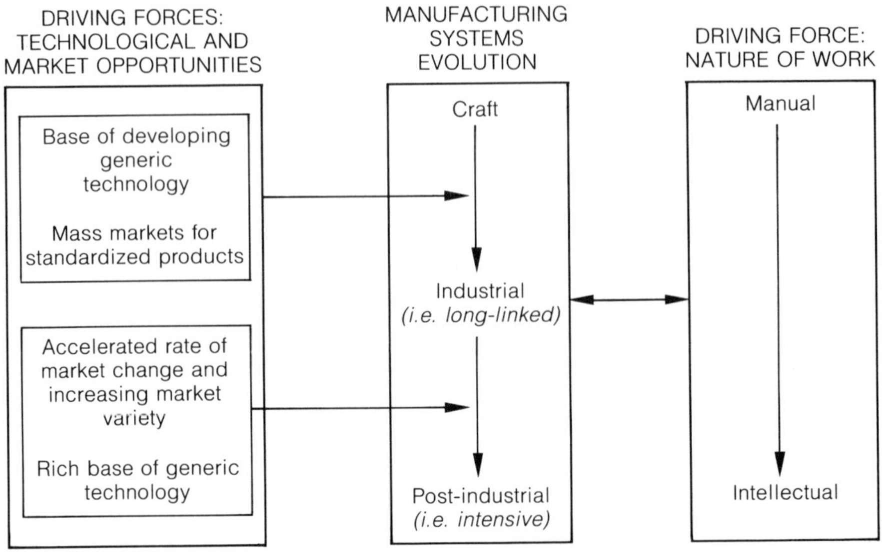

Fig. 18.1 The evolution of manufacturing systems.

intellectual work. Because of the intellectual nature of work and the need for integrative (across the value chain) rather than task-specific innovation, work systems are designed from a sociotechnical perspective (Susman, 1976).

Industrial systems are an efficient means of producing a narrow range of products, but they have problems achieving continuing innovation in products and processes. Post-industrial enterprises can be quite innovative, but they may experience efficiency problems. To remain competitive in the post-industrial environment, manufacturing firms must be both innovative and efficient. The industrial and post-industrial revolutions and associated concerns related to these twin problems of efficiency and innovation are described below in greater detail.

18.5.1 The transition from craft to industrial systems

The Industrial Revolution was driven by non-proprietary (generic) technology advances in power generation and transmission, metallurgy, transportation, and mechanical processes. This emerging base of generic technology had a profound impact on methods of production and distribution. Initially, industrial entrepreneurs with an integrated view of product and process technology organized and built high-volume manufacturing facilities designed to use inflexible resources in a sequential workflow to produce low-cost standardized products. At first, work was primarily manual; hierarchical command and control systems emerged to coordinate this manual work. Developing transportation systems permitted economical distribution over a wide geographical area. These high-volume production facilities required customer acceptance of low-cost, standardized products and increasingly sophisticated marketing and distribution systems.

The US system of manufacturing with its emphasis on interchangeability of parts tended to reduce skill requirements at the workstation. Thus, two kinds of workers developed: those who built, maintained, set up, and improved machines; and unskilled operators who turned out parts. As the Industrial Revolution continued, engineering application of advances in the physical sciences led to increasingly powerful and sophisticated equipment. A specialized staff developed to plan task-specific innovations that substituted capital for direct labour, permitting economies of scale and reducing process uncertainty. The production of goods shifted from skilled artisans, who both planned and executed their own work, to unskilled workers, who used automated machinery to perform work planned by a specialized staff.

In the transition from a craft to an industrial society, product-oriented long-linked technologies, with their superior efficiency in the production of standardized products for mass markets, became the dominant manufacturing pattern. Using a command and control approach, these systems excelled in substituting capital for labour, achieving economies of scale in mass markets, and controlling increasingly complex manufacturing systems through functional separation and hierarchical administrative systems. A pattern of task-specific innovation improved the productivity of labour. However, as direct labour was reduced, further improvements in productivity required more integrative innovations (i.e. innovations that involve simultaneous changes in a variety of activities across the value chain).

Evolution within the industrial stage is described in some detail by Utterback and Abernathy (1975). As industrial systems evolved from uncoordinated, through segmental, to systemic, both product and process innovation became more difficult. The physical processes, methods, techniques, tools, and equipment of production evolved in ways which improved efficiency, but resulted in a more inflexible manufacturing system which could not respond effectively to customer needs. More importantly, management's patterns of thinking and responding to markets and technologies were affected by the very technologies and methods they implemented. Managerial as well as technical systems became rigid, making technological change and continuing improvements in manufacturing performance more difficult.

In these industrial systems, product and process design were approached sequentially. Firms typically employed a technology push approach to process innovation, investing in increasingly automated manufacturing facilities to improve productivity. As demand for product variation increased, long-linked industrial systems produced more products with an infrastructure, both administrative and technical, that was designed to manufacture standard products for mass markets. As customer requirements shifted, the mismatch between customers' needs and manufacturing capabilities increased. Organizations that continued to employ long-linked industrial systems lost market share as competing firms segmented the market. Organizations that responded with product innovation added complexity to a system that could not maintain its efficiency.

Innovations in the technical core are essential for organizational adaptation to a changing and increasingly complex environment, and for continuing efficiency improvements. Seemingly isolated innovations may require adjustments throughout the value chain. A realistic assessment of the type of innovations that will be successful, and how they should be introduced, depends upon an understanding of the productive process that will receive them. Line workers understand the production process but do not possess the technical skills to design innovations. Staff engineers possess the technical skills but may not have the insights that come from experience of implementing innovations that cause change throughout the process.

An equally important consideration is the value of an innovation to the customer. A central problem in the management of innovation is assessing the instrumentality of an innovation from the perspective of shifting, multiple, global performance requirements dictated by changing customer expectations and alternatives. Long-linked industrial systems, with their emphasis on task-specific efficiency measures and hierarchical management systems, have difficulty initiating, coordinating, and assessing the instrumentality of innovations in material inputs, processes, and product tailoring across the value chain. The post-industrial enterprise, with its 'value to customer' perspective, is better able to assess the instrumentality of innovations.

18.5.2 The transition from industrial to post-industrial

In many ways, the evolution of long-linked industrial systems can be viewed as an underlying cause of the changing market, labour force, and technological opportunities that are creating a favourable environment for the post-industrial enterprise. The Industrial Revolution precipitated environmental changes in markets and technology that are continuing and, perhaps, accelerating. These changes are driving a continuing evolution from industrial to post-industrial paradigms as the basis for designing competitive manufacturing systems.

Over time, the industrial revolution has had a profound effect on markets. As industrial systems evolved, workers became more productive and were paid higher wages. Affluent and discriminating customers began to demand greater choice. Customer demands for new and different products increased product variety and accelerated the rate of market change. Changing customer expectations and competitive alternatives also made markets more uncertain. Customer purchase decisions were increasingly based on a combination of global criteria with customers adjusting the importance they place on price, product performance and features, quality, and delivery. Continuing competitiveness required a more holistic response to shifting performance measures dictated by changing customer preferences (i.e. a more customer-oriented approach to manufacturing).

The Industrial Revolution also created conditions favourable for the development of a large base of generic technology for improving products and processes. Firms engaged in research to develop proprietary technology that provided only temporal advantages. Over time, this proprietary technology was added to the stock of available generic technology. Efforts of research institutes and universities also expanded the generic science and technology base. A science of manufacturing was developing. Early developments focused on process control in continuous-flow industries and the automation of specific tasks (i.e. islands of automation) in discrete-part manufacturing. Over time, the emphasis has shifted to the design of flexible and integrated systems for discrete-part manufacturing.

This development of a rich base of generic technology helped to shift the basis of competition from technology driven and market enabled to market driven and technology enabled. When a rich base of generic technology is available to all competitors, technological skill is a requirement for being in business, but it seldom provides sustainable competitive advantage. When technology drives competition, you can't continue to compete on technology alone. Your research scientists may excel, but they are no guarantee that your engineers will succeed in rushing innovative products to market. Nor can you count on technologists to know what customers need and want (Competing Beyond Technology, 1989). Success in a post-industrial environment depends upon the enterprise's ability to anticipate markets and respond quickly and efficiently with products that provide high value to customers. The quality of collaborative efforts is a critical factor in the creation and operation of responsive manufacturing systems that design and produce products with high value to customers. If manufacturing systems are to be designed and operated to be responsive to the 'voice of the customer', collaborative efforts among a variety of individuals with different backgrounds and training are necessary. Collaborative efforts are facilitated when each employee has a holistic view of the manufacturing enterprise and the sophistication to understand how interdependent decisions across the value chain affect performance measures dictated by customer preferences. Employees must be capable of seeing the whole in the part (Van de Ven, 1986) and making decisions that add value to the customer.

Responding quickly to changing markets would be difficult without enabling technology. If applied correctly, computer and communications technology can help create an information-rich environment for reducing product development and throughput time. Advanced manufacturing technology is permitting the design of flexible manufacturing systems that can respond quickly and efficiently to shifting customer requirements. In applying either of these technologies, the intellectual talents of the work force are critical resources.

To be competitive, the post-industrial enterprise must be innovative as well as efficient. Indeed, it must be innovative to be efficient. Innovation must be customer oriented (i.e. driven by a value system that emphasizes value to the customer). This requires a pattern of integrative (across value chain) innovation where: 1. decisions require trade-offs between multiple performance measures; and 2. work groups or task forces have a shared understanding of customer preferences. The performance criteria are global rather than task specific. They describe how the whole value chain is performing for the customer on a set of measures such as cost, delivery, product performance and features, quality, and product variety.

The post-industrial enterprise seeks efficiency quite differently than does an industrial system. First, it focuses on economies of scope (Goldhar and Jelinek, 1983) rather than economies of scale. Second, it encourages learning and improvement in self-directed work groups across the value chain rather than task-specific cost savings. Learning is enhanced by combining work planning and work doing in the same individual or work unit. For example, original equipment manufacturers who previously designed parts and then subcontracted their manufacture are shifting design responsibility to the vendor. Innovation focuses on collaborative efforts within the firm and with suppliers to identify interdependent activities across the value chain that can be changed in ways that add value to the customer and/or reduce cost. A shared understanding of customer needs and applications enables participants to consider the value of the activities to the customer as well as their cost. Trade-off decisions are made in accordance with customer preferences.

Customer-oriented sociotechnical systems of the post-industrial era tend to adopt a technology pull approach to innovation where an assessment of changing customer needs drives the selection of product and process technologies. In the post-industrial enterprise, product design and process selection are often approached simultaneously to shorten product-development time and permit a more holistic response to multiple and global performance measures dictated by customer preferences. In this post-industrial environment, organizations can often compete effectively by focusing on customers and satisfying their needs by successfully implementing readily available generic technology. However, this technology pull strategy depends upon the work force's intellectual ability to identify and implement technologies that enhance the value of the firm's activities to the customer or reduce cost.

18.6 SUMMARY AND DISCUSSION

The transformation from an industrial to a post-industrial society will change the paradigms we use to manage manufacturing systems. In the industrial system paradigm, manufacturing is viewed as a functional area where technical transformation processes convert raw materials into finished products. Within this industrial paradigm, models of the changing nature of manufacturing (Skinner, 1985; Utterback and Abernathy, 1975; Jaikumar, 1988) emphasize technical/economic changes, and largely ignore corresponding patterns of change in social, work, and control systems.

In the post-industrial paradigm, manufacturing is viewed as an enterprise with customers and employees (i.e. a complex sociotechnical system in which people use technology to satisfy customers). Manufacturing becomes a complex customer-oriented network of value-adding activities, primarily intellectual in nature, that are performed by people organized into interdependent sociotechnical subsystems. The intellectual nature of work means that learning at individual and work group levels is necessary to achieve continuing productivity improvements. Social, work and control systems are designed to enhance learning and encourage innovation across the value chain.

These paradigms present two alternative visions of manufacturing system design (i.e. they assume different effectiveness criteria and cause-and-effect relationships). The industrial paradigm focuses on efficiently producing standardized products. In contrast, the post-industrial management paradigm emphasizes value to the customer and seeks to improve product development and throughput time. Implicitly, if not explicitly, top management's strategies for adapting to the challenges posed by the new era in manufacturing are often influenced by their choice of paradigm. As we move towards a post-industrial society, these challenges are expected to continue, even accelerate.

The stage model described in this paper provides, in part, an explanation for the difficulty many firms have had in adapting to the new era in manufacturing. A lack of understanding of the need for a shift in management paradigms between the industrial and post-industrial stages has left firms with a set of mistaken objectives, premises and patterns of thinking that hinder efforts to respond to changing market, technological and competitive conditions. When top management does not recognize the need to shift to a post-industrial paradigm, organizations typically respond to competitiveness problems by emphasizing efficiency improvements (Skinner, 1986). The organization continues to respond using the premises and patterns of thinking of the industrial paradigm until a crisis occurs. In many industries, a competitive crisis is occurring.

By identifying attributes and effectiveness criteria of a post-industrial manufacturing enterprise, Table 18.1 provides a framework for designing the social, work and control systems of a world-class manufacturer. Encouraging better employee understanding of the value of your firm's products to your customers, reducing product development and throughput time, emphasizing flexible technology, using information technology at a local level to leverage intellectual work, combining work planning and work doing to encourage learning, setting up self-managed work groups, and emphasizing multiple and global performance measures are some of the ways that a firm can move towards a post-industrial paradigm. These prescriptions are consistent with the direction of social system changes that are occurring in world-class manufacturers (Schonberger, 1987).

The problems of getting an entire company to make this transition from an industrial to a post-industrial paradigm should not be underestimated. The changes described above may run counter to widely shared objectives, premises and patterns of thinking concerning how organizational activities should be organized and directed. One problem is a narrow concept of productivity that inhibits innovation (Skinner, 1986). Hierarchical boundaries that maintain the more intellectually substantive tasks within managerial ranks can have a chilling effect on opportunities for learning and innovation. Finally, management's concerns about its own prerogatives may inhibit the transition towards a more customer-oriented approach to manufacturing. Teams of workers who understand customer needs have a basis for directing their own activities; however, management may resist the idea of self-managed work groups.

18.7 REFERENCES

Aoki, M. (1990) Toward an economic model of the Japanese firm. *Journal of Economic Literature*, **28**(3), 1–27.

Bower, J.L. and Hout, T.M. (1988) Fast-cycle capability for competitive power. *Harvard Business Review*, **66**(6), 110–18.

Chandler, A. (1962) *Strategy and Structure*, MIT Press, Cambridge, Mass.

Chase, R.B. and Garvin, D.A. (1989) The Service Factory. *Harvard Business Review*, July-August, 61–9.

Competing Beyond Technology (1989) *Harvard Business Review*, **67**(6), 410.

Doll, W.J. and Vonderembse, M.A. (1987) Forging a partnership to achieve competitive advantage: the CIM challenge. *MIS Quarterly*, *11*(2), 205–20.

Goldhar, J.D. and Jelinek, M. (1983) Plan for economies of scope. *Harvard Business Review*, **61**(6), 141–8.

Hounshell, D.A. (1984) *From the American System to Mass Production: 1800–1932*, The Johns Hopkins University Press, Baltimore.

Huber, G.P. (1984) The nature and design of post-industrial organizations. *Management Science*, **30**(8), 928–51.

Jaikumar, R. (1988) *From Filing and Fitting to Flexible Manufacturing: A Study in the Evolution of Process Control*, Working Paper 88-045, Division of Research, Harvard Business School, February.

Jelinek, M. and Goldhar, J.D. (1983) The interface between strategy and manufacturing technology. *Columbia Journal of World Business*, Spring, 26–36.

Juran, J.M. (1987) *On Planning For Quality*, The Free Press, New York.

Kimberly, J.R., Miles R.H., *et al* (1980) *The Organizational Life Cycle*, Jossey-Bass, San Francisco.

Kotler, P. (1988) *Marketing Management*, Prentice-Hall, Englewood Cliffs, NJ.

Kuznets, S. (1965) *Economic Growth and Structure: Selected Essays*, W.W. Norton, New York, 213–16.

Lawrence, P.R. and Dyer, D. (1983) *Renewing American Industry*, The Free Press, London, 1–16.

Manufacturing Studies Board (1986) *Toward a New Era in U.S. Manufacturing: The Need for a National Vision*, National Academy Press, Washington, DC.

Marx, K. (1932) *Capital, The Communist Manifesto and Other Writings*, Random House Publishing, New York.

Masuda, Y. (1980) *The Information Society*, World Future Society, Bethesda, Md.

Naisbitt, J. (1982) *Megatrends*, Warner Books, New York.

Nolan, R.L. (1973) Managing the computer resource: a stage hypothesis. *Communications of the ACM*, **16**(7), 399–405.

Ohmae, K. (1988) Getting back to strategy. *Harvard Business Review*, **66**(6), 149–56.

Piore, M.J. and Sabel, C.F. (1984) *The Second Industrial Divide*, Basic Books, New York.

Quinn, R.E. and Cameron, K. (1983) Organizational life cycles and shifting criteria of effectiveness: some preliminary evidence. *Management Science*, **29**, 33–51.

Schmenner, R.W. (1988) The merit of making things fast. *Sloan Management Review*, Fall, 11–16.

Schonberger, R.J. (1986) *World Class Manufacturing: The Lessons of Simplicity Applied*, The Free Press, New York.

Schonberger, R.J. (1987) *World Class Manufacturing Casebook: Implementing JIT and TQC*, The Free Press, New York.

Skinner, W. (1985) The taming of lions: how manufacturing leadership evolved, 1780–1984 in *The Uneasy Alliance: Managing the Productivity–Technology Dilemma* (eds. Clark, K., Hayes, R. and Lorenz, C.), The Harvard School Press, Boston, 63–110.

Skinner, W. (1986) The productivity paradox. *Harvard Business Review*, **64**(4), 55–9.

Smith, M.R. (1977) *Harpers Ferry Armory and the New Technology*, Cornell University Press, Ithaca.

Susman, G.I. (1976) *Autonomy at Work: A Sociotechnical Analysis of Participative Management*, Praeger, New York.

Takeuchi, H. and Nonaka, I. (1986) The new new product development game. *Harvard Business Review*, **64**(1), 137–46.

Thompson, J.D. (1967) *Organizations in Action*, McGraw-Hill, New York.

Toffler, A. (1970) *Future Shock*, Random House, New York.

Utterback, J.M. and Abernathy, W.J. (1975) A dynamic model of process and product innovation. *Omega*, **3**(6), 639–56.

Van de Ven, A.H. (1986) Central problems in the management of innovation. *Management Science*, **32**(5), 590–607.

Young, J.A. (1985) Global competition: the new reality. *California Management Review*, **27**(3), 11–25.

Zuboff, S. (1984) *In the Age of the Smart Machine: The Future of Work and Power*, Basic Books, New York.

PART FIVE

Research Needs

The preceding chapters have given many perspectives on the area of manufacturing strategy. These have been derived from the literature, from experience and from research. A common theme has been that manufacturing strategy is still in its early days and as a result there are many unanswered questions; these questions provide an agenda for future research. The final chapter of this book is a review by Everett Adam Jr and Paul Swamidass of current writing on manufacturing strategy and of the future research needs.

CHAPTER 19

Assessing operations management from a strategic perspective

Everett E. Adam, Jr and Paul M. Swamidass

Manufacturing strategy research, together with service operations and strategy research, form the background for an examination of operations management. Although much of the assessment and critique is directed at operations strategy, traditional operations management issues such as technology, productivity and quality also come under scrutiny as to their role in strategy. This review identifies three common themes and seven missing themes in the literature, contrasting the 'content' variables so frequently investigated within operations management – quality, cost, delivery and flexibility – against strategic 'process' variables. Conclusions include the partial indictment of operations management literature through the missing themes, an indictment that can be shored-up through increased empirical research in operations strategy that focuses on content, process and performance.

19.1 BACKGROUND

Operations strategy (OS) is attracting increasing attention from researchers and practitioners within the operations management (OM) area. The strategic view of operations management dates back to Skinner's pioneering work (1969, 1974) in which he advanced these major ideas: 1. the manufacturing function can and should be employed as a competitive weapon; 2. cost and efficiency are inadequate goals for manufacturing; 3. 'a factory that focuses on a narrow product mix for a

Source: Adams, Everett E. Jr and Swamidass, Paul M. (1989) © 1989 by the Southern Management Association, previously published in the *Journal of Management*, **15**(2), 181–203. Reprinted with permission.

particular market niche will outperform a conventional plant' (Skinner, 1974, p. 114); and 4. the successful formulation and implementation of manufacturing strategy is a top-down process that requires manufacturing to conform to a manufacturing task derived from competitive strategy.

This article critically assesses operations management from a strategic perspective, with a review of manufacturing strategy (MS) and the literature on service operations, especially as it relates to strategy. We use the themes commonly found in the operations strategy literature (including manufacturing and services) and the themes missing from it to structure our review. The significance of using missing themes in literature as a yardstick for review is that this approach serves as both an indictment of the literature as well as guideline for the direction of future research.

19.2 MANUFACTURING STRATEGY RESEARCH FROM 1969 TO 1988

19.2.1 A thematic approach

Several possible frameworks could be employed in reviewing the last 20 years of manufacturing strategy research. We have chosen to evaluate manufacturing strategy research on the basis of progress made against key themes. What, then, is the source of these themes? The themes used in this review fall into two groups. The first group consists of commonly found themes in the manufacturing strategy literature. The second group consists of important themes that are found in the business strategy literature or in the operations management literature, but ignored or inadequately developed in the manufacturing strategy literature. The two groups of themes are labelled, 'common themes' and 'missing themes' respectively. The common themes in the manufacturing strategy literature reflect manufacturing strategy **process** (i.e. the process of formulation and implementation) and **content**.

19.2.2 Themes commonly found in the manufacturing strategy literature

Common theme 1: The process of formulating and implementing manufacturing strategy is important for guiding manufacturing

Skinner (1969) and Hill (1985) stressed the importance of the process of manufacturing strategy (MS) formulation and implementation and made specific recommendations concerning it. One identifiable theme is attributed to Skinner's work, which suggests a top-down approach for defining and implementing MS. His theme may be summarized as (a) develop a manufacturing task based on business strategy, (b) conform manufacturing infrastructure policies and efforts to the manufacturing task, and (c) give a substantive role for manufacturing managers in defining and implementing MS.

Table 19.1 summarizes the process variables identified in the manufacturing strategy literature. The process variables identified in Table 19.1 are defined in the books or papers cited in the table; however, a brief description of the less familiar variables are included here. The variable 'Manufacturing Infrastructure' is the 'internal systems' including 'organizational levels, wage systems, supervisory practices, production control and scheduling approaches, and job design . . .' (Skinner, 1978, p. 45) through which actual production is carried out. 'Manufacturing Task' is a clear and concise statement of 'what the manufacturing function must accomplish,' which includes specific goals and priorities for the manufacturing function in terms of cost of production, quality, etc. (Skinner, 1978).

The variable 'Order Winning Criteria' was proposed by Hill (1985) as essential to manufacturing strategy process. These are manufacturing criteria surrounding product cost, quality, etc., that help manufacturers to win orders against competitors. Readers may notice much similarity between the two concepts 'Manufacturing Task' and 'Order Winning Criteria'. The variable 'Role of Manufacturing Managers' in Table 19. 1 refers to the importance of the role of manufacturing managers in strategic decision making (Skinner, 1978; Swamidass and Newell, 1987).

There is no apparent pattern in the frequency of variables in Table 19.1, indicating limited consensus among the various writers. Further, a wide variety of terms is used by writers to describe process variables in manufacturing strategy. In isolating the process variables in Table 19.1, we used our judgement to group together several different terms into a single variable. For example, in the table, the term **manufacturing task** proposed by Skinner (1969) is substituted for the terms **manufacturing mission, manufacturing objectives** and **manufacturing criteria** used by other writers. There appears to be a tendency to create new terms without materially adding to the conceptual inventory.

Particularly lacking in the manufacturing strategy process literature is the twin focus on strategy formulation and implementation found in the business strategy literature (Andrew, 1971). Additionally, although business strategy literature investigates strategic planning, planning processes, and planning models, manufacturing strategy literature does not. There is an urgent need to develop a body of literature on the manufacturing strategy planning process. It should be noted that the effectiveness of strategic planning in corporate strategy has been mixed. Ackoff (1981) observes, 'Most corporate planning is like a ritual rain dance: it has no effect on the weather that follows, but it makes those

Table 19.1 Manufacturing strategy process variables in the literature

	Authors					
Process variables	*Hill (1985)*	*Miller and Roth (1988[1])*	*Schroeder* et al. *(1986[1])*	*Skinner (1978)*	*Swamidass and Newell (1987[1])*	*Wheelright (1984)*
Infrastructure	Yes			Yes		
Manufacturing task			Yes	Yes		
Order winning criteria	Yes					
Role of Manufacturing managers				Yes	Yes	
Defining manufacturing strategy		Yes		Yes		
						Yes
Consistency between						
Manufacturing and business strategies	Yes					Yes
Manufacturing and other functional strategies	Yes					Yes
Manufacturing task and infrastructure				Yes		Yes

[1] Empirical studies.

who engage in it feel that they are in control. Most discussions of the role of models in planning are directed at improving the dancing, not the weather' (p. 359). Even so, there are enough reported benefits from management's involvement in strategic planning that the need for the formal study of strategic planning for manufacturing strategy is not diminished.

If business strategy literature is an indicator, there are several variables and issues relating to strategy process that have not yet caught the attention of operations strategy researchers. The research relating to manufacturing strategy process should include, but not be limited to: 1. planning, 2. implementation, 3. decision aids, 4. structure, 5. modelling, 6. cognition and perceptions, 7. political and bureaucratic factors, 8. leadership, 9. organization, and 10. integration.

In the list above, issues such as perceptions, leadership, and organization have a rich body of knowledge in the social sciences, organizational behaviour, and organization theory disciplines. The manufacturing strategy scholars need not duplicate, but must integrate relevant items from sister disciplines to make their research broader, richer and more complete. The work of Swamidass and Newell (1987) is an example of how organization theory, business strategy, and manufacturing strategy literatures can be bridged and integrated.

Common theme 2: The core content of manufacturing stratcgy includes cost, quality, flexibility, and technology

Strategy content 'focuses on the specifics of what was decided, and process addresses how such decisions are reached in an organizational setting' (Fahey and Christensen, 1986, p. 168). Historically, the *Harvard Business Review* has been a leading outlet for manufacturing strategy writings. In the 1980s, however, the topic is addressed more widely in several journals and books.

In reviewing the manufacturing strategy (MS) content literature, the works of Hill (1985), Miller and Roth (1988), Schroeder, Anderson and Cleveland (1986), Skinner (1969), Swamidass (1986) and Wheelwright (1984) were chosen as a representative sample of the current body of manufacturing strategy literature. Table 19.2 summarizes 'content' variables identified in these writings. (The reader is referred to the original sources for definitions of each content variable in Table 19.2.) Although we have carefully examined the studies, our interpretation of an author's intended content variable could be a potential source of error in our synthesis.

Although more than a dozen content variables have been identified in manufacturing strategy literature, Table 19.2 reveals that some variables are more frequently stressed in empirical as well as conceptual writings. The variables most frequently identified are quality, cost, flexibility and technology. We are interpreting the terms **manufacturing process** and

Table 19.2 Manufacturing strategy content variables in the literature

	Authors						
Content variables	*Hill (1985)*	*Miller and Roth (1988[1])*	*Schroeder et al. (1986[1])*	*Skinner (1978)*	*Swamidass (1986[1])*	*Wheelwright (1984)*	*Number of times mentioned*
Capacity			Yes			Yes	2
Cost/price/productivity	Yes	Yes	Yes	Yes	Yes		5
Delivery	Yes	Yes		Yes	Yes		4
Product design and engineering	Yes				Yes		2
Distribution		Yes					1
Employee relations			Yes				1
Facilities						Yes	1
Flexibility	Yes	Yes	Yes	Yes	Yes		5
Focus	Yes			Yes			2
Infrastructure	Yes						1
Quality		Yes	Yes	Yes	Yes	Yes	5
Return on investment				Yes			1
Service		Yes	Yes	Yes			3
Standardization	Yes						1
Technology-process	Yes		Yes	Yes	Yes	Yes	5
Vertical integration						Yes	1

[1] Empirical studies.

manufacturing technology to be mutually interchangeable. Thus, in Table 19.2 and Fig. 19.1, we use the term **technology process**.

Using concentric circles, Fig. 19.1 places the most frequently mentioned variables at the core, and the less frequently mentioned variables are placed in concentric rings farther away from the core.

In Fig. 19.1, immediately surrounding the core is the ring showing delivery, service, and focus variables, which are second only to the core strategy content variables in the frequency of their mention in literature. We are somewhat surprised that dependability (schedule) was not more frequently mentioned as it often is stressed as an operating objective. A number of variables that appear inconsistently in the literature are shown in the outermost circle.

Advanced technologies and their justification are fast developing into an important manufacturing strategy issue. The most common themes in advanced technology acquisition literature are 1. the costing and jus-

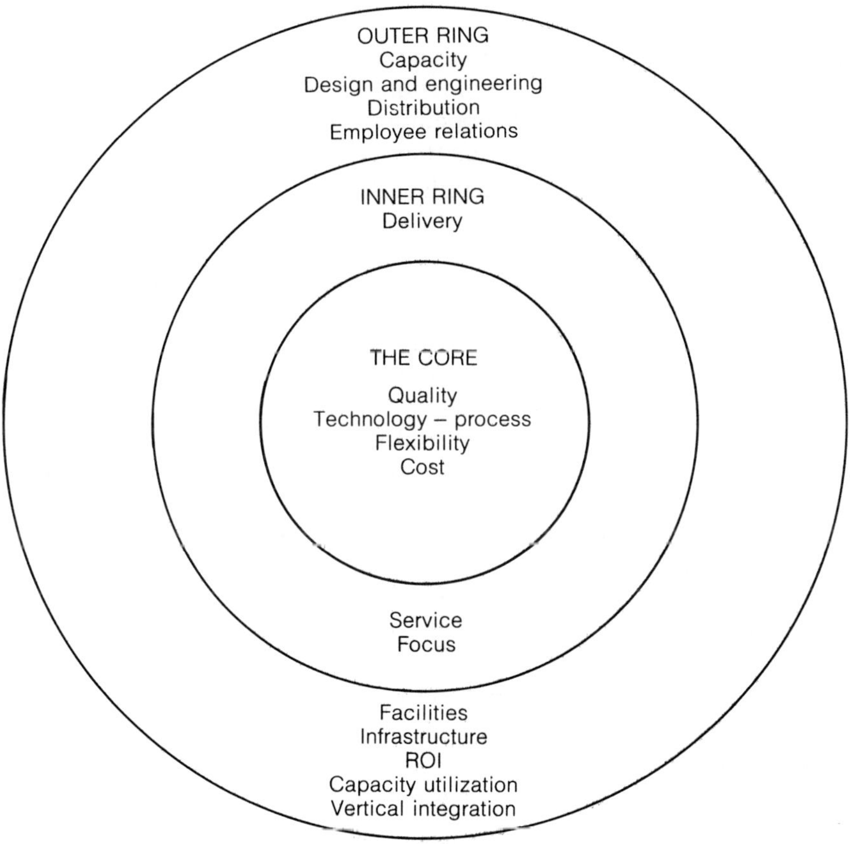

Fig. 19.1 Manufacturing strategy content variable making up the core and surrounding rings.

tification methods for new technologies, and 2. incorporating strategic benefits of new technology in cost/benefit analyses (Bennett and Hendricks, 1987; Blank, 1987; Engwall, 1988; Kaplan, 1986; Meredith and Suresh, 1986; Primrose and Leonard, 1987; Skinner, 1984; and Suresh and Meredith, 1985).

The shortcoming of manufacturing strategy (MS) content research is that it lacks empirical investigations of relationship among its variables. Particularly needed is research that investigates the effect of cost, flexibility, technology/process, quality, and dependability on business performance individually and the effect of their mutual interaction on performance. Swamidass and Newell (1987) provide a rare example of the much needed investigation of relationships between MS variables and performance.

19.3 SERVICE OPERATIONS

The formal study of service operations as a subset of operations management is a relatively new phenomenon. The discussions in this section starts with a definition of service operations, followed by a discussion on the key issues of service operations that have a bearing on strategy.

Services, as contrasted with manufacturing, involve the conversion of resources into an *intangible* output. The distinguishing characteristics of service operations are manifested in many typological or classificatory studies of service operations (Adam, 1983; Albrecht and Zemke, 1985, pp. 3–12; Heskett, 1986, pp. 5–25; Lovelock, 1983; Mills and Marguiles, 1980; Mills and Moberg, 1982; Snyder, Cox and Jesse, 1982). Service operations are distinguished by the following major characteristics: the intangible nature of output, immediate consumption of output, conversion processes that require a great deal of labour and little equipment, direct customer contact, frequent customer participation in the conversion process, and rather simple (or elementary) measures for the conversion activities and outputs. Based on these characteristics, the following are typical service sector industries: transportation, communications, utilities, wholesale and retail trade, finance, insurance, real estate, business services, personal services, and nonprofit services.

There exists a wide range of labour–capital–technology blends that are found in service sector organizations. This diversity in the range of service operations is reflected in the diversity of operations strategy appropriate for the service sector. For example, there is a tremendous difference between a utility delivering nuclear power and a small business providing a haircut. This diversity in labour–capital–technology blend lends itself to a broad variety of strategies in service sector firms, a variety yet to be developed in the formal literature.

The service operations management literature is not abundant, a situation scholars have noted (Chase, 1978, 1985; Collier, 1985, 1987; Heskett, 1986; Hostage, 1975; Lovelock, 1988; Mills, 1986). Even though service operations management problems or issues are repeatedly addressed, in most instances a great deal is yet to be known about them. Some of the established issues in the service operations literature will be addressed as strategy content variables below.

19.3.1 Strategy and service operations

Manufacturing literature has for most of the last century focused on rationalization at the production core – increasing direct labour productivity, process design, inventory control, and computerized scheduling and control. Only recently has our attention in manufacturing been directed at the broader strategic business issues such as quality and automation, which extend beyond cost reduction. In general, the literature on service sector operations lags behind the literature in manufacturing. Currently, in the process of its evolution, we see an emphasis in the rationalization of service operations, rather than an emphasis on broader strategic issues.

In the following sections, we review service operations strategy literature under the heading of strategy content, strategy process, and adapting from manufacturing strategy. It should be noted that the development of the service operations strategy literature has not followed along these well defined headings, which are used to make the review of service strategy literature consistent with the rest of this paper and to offer a guideline for future research in services.

Strategy content. An analysis of service operations strategy content literature reveals these variables: capacity, automation, quality, queuing, staffing, customer awareness, and multi-site operations. Table 19.3 provides the context of each content variable in service operations. Were we to select core content variables from frequency of presentation, the variables would be customer awareness, capacity, quality, and staffing decisions (Chase, 1978; Chase and Prentis, 1987; Collier, 1985, 1987; Heskett, 1986; Lovelock, 1988; Mills and Marguiles, 1980; Mills and Moberg, 1982; Snyder, Cox and Jesse, 1982).

Albrecht and Zemke (1985) explain that an effective service strategy content is arrived at by positioning one's service in the marketplace. Many analogies are made in their writings to 'product positioning', which is reminiscent of the pioneering thinking of Alfred P. Sloan as he developed automobiles for market segments in General Motors' formative years. Their description of service operations strategy parallels a typical forced choice model (Adam and Ebert, 1989), where markets and the environment are balanced against financial and operating capabilities of the firm, forcing the development of strategic choices.

Table 19.3 Service operations content variables in the literature

Content variable	*Service operations usage*
Capacity	Because no inventories are available to buffer demand, capacity defines the capacity to meet demand. Capacity has temporal (time of week, day, etc.) dimensions. Incorrectly established capacity can strangle a service business. On one hand, too high a capacity defines a high fixed cost structure or, on the other hand, too low a capacity can end the business because of inadequate service delivery, resulting in customers who never return.
Automation	Establish the level of technology, mechanization, and automation – often computer-based – that is economically and technically feasible.
Quality	Consistency in service delivery.
Queuing	Managing waiting lines. Use of flexible labour staffing, signs making apparent all stages of the process, and fair queues (first come-first serve). Customer participation in the service process.
Staffing	People-oriented staff given 'customer contact', task oriented staff 'in back office', cross-training, part-time employees, and so forth. Service worker skills are the focus.
Customer awareness	Commitment to the customer as reason for existence. Understanding customer interactions. Proximity to customer.
Multisites	Capability to repeat a winning service business. Standardization of processes and services.

Strategy process. Writers on the subject of service operations strategy process have considered variables such as the service concept, markets, and the service delivery system (Albrecht and Zemke, 1985; Heskett, 1986). Consider Albrecht and Zemke (1985, pp. 39–41, 63–76), who visualized a service triangle that places the customer at the centre. Corners of the triangle are formed by the service strategy, the service system and the employees. Their theme is that a service concept and strategy set the stage for service systems (processes) and people (employees) to interact with and serve the customer.

A representative work in the area is that of Heskett (1986, pp. 5–25), who developed an integrative strategy process for managing services that moves successively from the specification of target market segments, to the development of a service concept, to the development of an operating strategy, and then to the service delivery system. His concept of operations strategy includes specification of the firm's overall strategy and its relation to marketing, finance, operations, and other functional areas. Heskett suggests a position be taken on where management effort is to be focused; either investments, quality and cost control, or on specifying the results to be expected against competition (in terms of quality, cost, productivity, and morale/loyalty of service staff). Because service operations strategy literature seldom distinguishes between content and process, it is sometimes difficult to sort out the variables and often there are overlaps between the works of authors stressing process on the one hand and those stressing content on the other.

Adapting from manufacturing strategy. Much promise for research lies in the transfer of ideas from manufacturing strategy literature to service operations strategy. In the opinion of the authors, an example of a promising approach to service strategy may be adapted from the work of Chris Voss (1986). Based on his experiences and study in Great Britain and Europe, Voss presents a view of manufacturing strategy that links the policy decisions associated with operations to the marketplace, the environment, and the company's overall goals. This view is readily adaptable to services, requiring but minimal understanding of the differences between service and manufacturing operations (Adam and Ebert, 1989, pp. 36–39).

One feature of Voss's approach that is crucial to competitiveness is his market-based view of strategy development. In every area the company chooses to compete, there is a set of specific **market-based criteria for success**. His approach allows the organization to choose the strategic dimensions such as **efficiency**, **price**, **effectiveness**, **quality** and **flexibility** as demanded by the market. An appealing feature of Voss's suggestion is the choice managers must make, given a specific operations mission established from market-based criteria for success. These choices involve facilities, aggregate capacity, choice of process, vertical integration, operations integration, and operations interface with other functions.

In service strategy formulation, because of the closeness of the service business to customers, service operations must be extremely sensitive to customers and markets. The real key to Voss's approach is expressed in the following quote concerning the firm's choice of strategy:

> [Operations] strategy must be changed and adapted to maximize the market criteria for success, or the chosen markets should be changed to match more closely operations capability in terms of market criteria for success. (p. 68)

Voss's statement on setting operations strategy is built on the simple idea of a customer (or market) orientation for operations: match operations to markets or select markets to match existing operations capability.

Common theme 3: Technology offers strategic advantages to service operations

Technology is frequently discussed as a major strategic content variable in service operations (Collier, 1985). The literature on the subject stresses automation to replace labour and automation as a means to provide flexible capacity, but has not paid enough attention to the strategic benefits of technology. The message of this stream of research is that great opportunities exist for technology, not only to replace labour in services, but also to gain strategic advantages in the form of improved delivery speed, increased quality and reliability, and increased new services that the customer could not envision.

Collier (1985) addresses well the revolution taking place in the automation of services and the management of new technology with good documentation. Mills and Moberg (1982) suggest that technology can buffer service operations from customers, yet not seal operations off as in manufacturing. Operationally, this allows for improved efficiency, yet strategically it does not totally isolate customers and operations, an interaction often demanded by customers (especially for personalized services).

It is well recognized that, lacking the benefits of inventories, service operations have major problems with capacity planning, scheduling, and managing demand (Lovelock, 1988). Therefore, new technologies – especially computer processes and light equipment – are proven, flexible approaches for adding capacity inexpensively and for scheduling. It has been shown that with technology **capacity** and **demand** can be managed much better (Collier, 1985; Lovelock, 1988). Through automation and mechanization we can adjust capacity quickly by turning on and off equipment. With computer evaluation of demand patterns we can more effectively manage demand through pricing.

Mills and Moberg (1982) suggest that a stream of research that focuses on technology choices at the client–worker interface is needed. Finally, we agree with many service sector scholars who suggest services be studied separate from manufacturing (Chase, 1978; Chase and Prentis, 1987; Heskett, 1986). The technologies *are* often so different that this seemingly simple suggestion is worth repeating.

19.4 MISSING THEMES IN OPERATIONS STRATEGY

Business strategy and manufacturing strategy (MS) literatures had almost contemporary beginnings. Business strategy literature, which can trace its origin to Andrews (1971) and Lawrence and Lorsch (1967), has since made rapid progress in the research. Growth in the research of business strategy may be attributed to several factors including the contribution of PIMS database to strategy research in the early 1970s, the growth of a strategy movement among management scholars and executives, and the creation of the Strategic Management Society and its journal, the *Strategic Management Journal.* The growth of MS strategy research during the same period has been extremely spotty and severely lags behind business strategy research. The advantage of using themes found in business strategy literature as yardsticks to evaluate MS strategy research is that the themes permit us to see how far the research in MS has yet to go in reaching important milestones already reached in a related discipline.

The missing themes considered in this paper are the need for distinct research streams for strategy content and strategy process, strategic planning, effect of operations strategy (OS) on performance, empirical theory building, taxonomic research, international strategies, and operations management (OM) themes missing in OS literature. The missing theme called 'OM themes missing in strategy literature', which is the last theme discussed here, is a catch-all for several themes that are well developed in the operations management area but have not been successfully absorbed into the operations strategy literature.

Compared to the research on traditional operations management topics such as inventory management, scheduling and methods analysis, operations strategy research is going at a crawling pace. One of the purposes of the missing themes approach is to expose the diversity of research needs in the area in the hope of attracting increased interest and research activity. The missing themes and accompanying suggestions for research apply equally to manufacturing and service operations strategy literature.

Missing theme 1: Operations strategy research needs distinct research streams investigating strategy content and strategy processes

The discussions on the formal distinction between strategy process and content can be traced to Andrews (1971), Ansoff (1965), Chandler (1962) and Schendel and Hofer (1979). Researchers in the business strategy area have established a dichotomy: 'researchers study content, or they study process' (Huff and Reger, 1987).

Although business strategy research proceeds along two distinct paths, one pursuing strategy content and the other strategy process, such is not the case with manufacturing and service strategy literatures. In operations strategy literature, it is common to study the two together (Schroeder, Anderson and Cleveland, 1986; Miller and Roth, 1988). One of the few explicit discussions of the distinction between manufacturing strategy content and process variables in the operations strategy literature is in the work of Swamidass and Newell (1987). Often operations management authors address process variables along with content variables without acknowledging them as such.

Establishing two distinct streams for the content and the process of manufacturing strategy will likely aid the development of the discipline. For example, in the business strategy literature several major research streams on the subject of strategy process have emerged, including strategic planning, strategy formulation, and strategy implementation. The many-pronged research effort in the area of business strategy processes has resulted in rapid growth and maturity of the discipline.

Missing theme 2: Strategic planning is an important strategy process tool for operations management

Strategic planning deals with the structured as well as the unstructured process of strategy formulation. Steiner (1969) and Ackoff (1971) are credited with early work on strategic planning. The literature on strategic planning, which is based on rationality and logic and founded on case illustrations, has been criticized for being an overly heroic approach to strategic management (Huff and Reger, 1987). Yet, regardless of its limitation, strategic planning is vital to the process of strategy formulation.

Hill (1985) and Skinner (1969, 1985) are two advocates of formal planning for manufacturing strategy. Their work, grounded in case studies, stands alone in the area of strategic planning for manufacturing strategy. One of the serious weaknesses in the operations strategy literature is the lack of progress and research interest in the area of structured, formal planning. In contrast, one can point to hundreds of research and scholarly writings in the area of business strategy planning (Huff and Reger, 1987).

The benefits of strategic planning are well summarized in Greene, Adam and Ebert (1985). The benefits of strategic planning are that it offers systematic procedures for identifying upcoming strategic issues and problems (Ansoff, 1965; King, 1981), it formalizes the leadership role of top executives in strategic planning (Hunsicker, 1980), and it provides specific analytic tools and techniques for formulating strategy (Dutta and King, 1980; Hoffman, 1985; Mokwa, 1986).

Swamidass (1986) reported that executives and manufacturing managers show a systematic lack of understanding for manufacturing strategy, resulting in poor implementation. The role of systematic planning for operations may be one important solution to this problem. The need for more vigorous research in the area of strategic planning for manufacturing and service strategy cannot be overstated.

Missing theme 3: The real test of operations strategy is its effect on operating and overall performance

In our judgement, the greatest weakness of operations strategy research becomes evident when one searches for research that studies interrelationships among variables, particularly the effect of strategy content and process variables on performance.

Twenty years after Skinner's pioneering work, manufacturing strategy research is still predominantly one of either identifying strategy process variables and their importance to practitioners (Miller and Roth, 1988; Schroeder, Anderson and Cleveland, 1986; Swamidass, 1986), conceptualization of the variables (Wheelwright, 1984), or axiomatic prescriptions based on case studies (Hill, 1985; Skinner, 1969, 1974). What is glaringly absent is a body of studies that would investigate the interrelationship among variables or the effect of a subset of variables on performance.

In contrast, business strategy research has matured. 'The major thrust of this research has been to identify linkages among environmental conditions, corporate- or business-unit strategic decisions, and economic performance' (Fahey and Christensen, 1986, p. 167). The basic paradigm for business strategy, according to Fahey and Christiansen, is shown in Fig. 19.2. This model enables the investigation of questions of the genre, 'What performance results arise from following specific strategies under different conditions?' To reach maturity in research, this paradigm must become a part of operations strategy research. Swamidass and Newell (1987) provide one empirical investigation of relationships among manufacturing strategy variables that conforms to this research paradigm. In particular, their study investigated the effect of environmental uncertainty on selected manufacturing strategy content and process variables and the latter's effect on plant performance.

The study by Schoeffler, Buzzell and Heany (1974) was a forerunner to numerous business strategy studies that have adhered to the above paradigm. In their study. Schoeffler, Buzzell and Heany, investigated the impact of market share, investment intensity, and company factors on performance. Such studies reveal the effect of strategic choices on performance. They provide evidence for the numerous claims by strategy writers and material for theory building. Clearly, the effect of strategic choices on performance must be a part of all strategy-related research.

A prerequisite for studies that investigate the effect of manufacturing

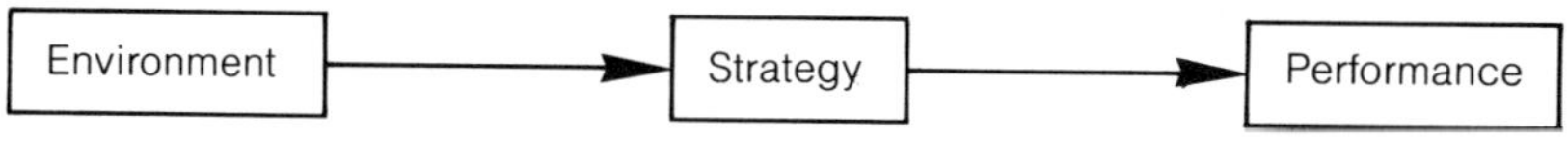

Fig. 19.2 Typical research paradigm in business strategy literature.

strategy variables on performance is an agreement in the discipline on suitable measures of performance. One study that investigated the effect of manufacturing strategy variables on performance used performance measures commonly found in the business strategy literature – growth in sales, growth in return on assets, and growth in return on sales (Swamidass and Newell, 1987). If the volume of work in the business strategy literature is any indication, developing appropriate measures for performance is extremely important to the study of strategy (Bowman, 1980; Kudla, 1980; Montgomery and Singh, 1984; Montgomery, Thomas and Kamath, 1984; Schendel and Patton, 1978). Taking this argument further, one stream of manufacturing strategy research should concentrate on performance measures that are appropriate for investigating manufacturing strategy, but are not explored by business strategy researchers.

The consequnce of the failure to research the impact of manufacturing and service operations on performance is that we are not any wiser about the axiomatic claims and prescriptions about operations strategy that proliferate in the literature.

Missing theme 4: Operations strategy theory development should use empirical research as building blocks

In the 1980s, empirical studies based on surveys have appeared (Miller and Roth, 1988; Schroeder, Anderson and Cleveland, 1986; Swamidass, 1986; and Swamidass and Newell, 1987). The continuing survey of manufacturers concerning their strategies by the Manufacturing Futures Project, Boston University, is a welcome contribution to the development of manufacturing strategy literature.

There is much to be understood about operations strategy from scientifically designed field studies that are generalizable. The lack of field data is a deterrent for empirical studies because prescriptive writings on operations strategy remain untested. Thus, theory building and validation of prescriptive ideas have been arrested by insufficient empirical research.

Empirical studies of operations strategy based on objective data are very rare. Operations strategy research based on PIMS-type (Schoeffler, Buzzell and Heany, 1974) databases are needed to move forward and approach maturity. The lack of debate and discussion on research methods and research strategies within the discipline accentuates the limitations of operations strategy research.

Missing theme 5: The development of taxonomies and classificatory schemes would improve operations management and strategy

Taxonomies provide a mechanism through which both operations management content and strategies (or patterns of strategies) could be classified for better and more detailed investigation. Business strategy researchers have devoted considerable effort in identifying and empirically validating strategy taxonomies (Dess and Davis, 1984; Galbraith and Schendel, 1983; Hambrick, 1983; Porter, 1980).

Although there has been some effort within the broader context of operations management to develop taxonomies (Adam, 1983), operations strategy literature is weak on taxonomical research, both at the conceptual and empirical level. Conceptual development of strategy taxonomies usually precedes empirical research. For example, Porter's generic strategy types were confirmed by the investigation of Dess and Davis (1984). Sometimes taxonomies are revealed by sound empirical research. When using PIMS data, Galbraith and Schendel (1983) found four 'distinct, internally consistent patterns of strategic behavior,' which they labelled **efficiency, service, product innovation**, and **brand identification**.

In the logical progression of manufacturing and service operations strategy research, the development of well rounded and well respected manufacturing strategy taxonomies is needed. The most commonly stated manufacturing strategy taxonomy identifies cost, quality, dependability, and flexibility (Buffa, 1984; Skinner, 1978; Swamidass and Newell, 1987; Wheelwright, 1978) as the strategic manufacturing options. However, as Fig. 19.1 shows, there are many more strategic options cited in the literature. Taxonomic studies could reconcile empirical findings. The typology of manufacturing flexibility, which is a manufacturing strategy core variable, proposed by Swamidass (1988) is an example of a typology related to operations strategy that may stimulate more research.

Kotha and Orne (1989) provide just the typology we are suggesting. They offer a conceptual framework that links manufacturing to business unit strategy and focuses on developing the notion of generic manufacturing strategies. Kotha and Orne relate Porter (1980, 1985) to Hayes and Wheelwright's (1984) traditional manufacturing classification. Their eight generic manufacturing strategies reflect process structure complexity, product line complexity, and organization scope.

Missing theme 6: Operations management needs to reflect the international context of business

Manufacturing and service firms seemingly share the desire for internationalization or globalization in distribution and consumption of products and services. Sometimes, the globalization of operation is a strategic necessity for competitive survival (Buffa, 1984; Hayes and Wheelwright, 1984; Jolly, 1987; Starr, 1988).

The work on international operations dating back to Stobaugh (1969), Vernon (1971, 1974), Vernon and Wells (1976), and Wells (1972) must be integrated into contemporary operations management literature. Established themes in international operations management that have been around for over two decades are vertical integration (both forward and backward), prolonging product life cycles by manufacturing in developing nations, rationalization through international operations, defensive and offensive strategies in international plant locations, hollowing-out the US manufacturing core, operating cost trade-offs, and technology transfer.

Global competitiveness. Within the production and operations management literature, 'global competitiveness' or 'manufacturing and global competiveness' is an important emerging theme. Examples are studies on product profiling (T. Hill, 1987), the role of the operations manager in internationalizing manufacturing (M. Hill, 1987), speed of change in product-market posture (Hirano, 1987), and factory level practice comparisons between nations or cultures. Among the broader works on this theme of global competition are Buffa (1984), Hayes and Wheelwright (1984), Jolly (1987) and Starr (1988). Key issues in these writings seem to centre on 1. gaining access to markets, 2. designing products (or services) by tailoring for a specific market versus standardization, 3. providing low-cost/high-quality manufacturing locations and 4. meeting national requirements for manufacturing, such as value added in the purchasing (consuming) country.

The strategic issues concerning technology, political environment, currency, geographical distances, culture, and so forth have been addressed in the international business area literature since Vernon (1971). Yet, because the manufacturing and service strategy literatures have hardly benefited from it, international business literature must be selectively integrated into international operations management literature.

Given the current state of development of the literature, it would be helpful to have a typology for international production and operations management to map the area, identifying key variables and their interactions. Such a typology would be helpful for the presentation of what we already know and would serve as a driver for future investigations. The typology of Adam (1983) might be extended to include an international dimension.

Emerging research interest. Most recent research in international production/operations management tends to fall into one of two categories: 1. an across-the-board comparison of industries, countries, or regions on productivity, performance, and managerial behaviour, or 2. an evaluation of an operations problem/issue in the international context.

The studies by Miller and others emanating from Boston University's Manufacturing Roundtable provide a good example of the first category, where manufacturing managers are annually surveyed in Japan, North America, and Europe on a variety of issues (DeMeyer *et al.*, 1987). Similarly, Rho and Whybark (1988) compare the People's Republic of China and South Korea, focusing on the non-fashion textile and small machine tool industries. This stream of research typically investigates production and inventory practices in various countries, which may include sales forecasting, production planning and scheduling, shop floor control, purchasing and materials management (Rho and Whybark, 1988). Further, Australia's manufacturing practices are reported by Harrison (1987) as are the United Kingdom's by New and Meyers (1987).

Typical of the second category is the in-depth comparison of product development processes within North American and Japanese manufacturing firms (Clark and Fujimoto, 1987).

Missing theme 7: Major themes in the operations management literature such as JIT, productivity, and quality are not integrated into manufacturing and operations strategy

The literature that is identified with manufacturing strategy *per se* has generally limited its focus to issues flowing from Skinner's work. However, there are a number of strategic decision issues in operations management that have been excluded from OS discussions. For example, though the literature on plant location, technology, design, just-in-time (JIT) and total quality control (TQC) production have strategic implications, they have not received direct attention in the strategy literature. In the context of this review, we present JIT, plant location, productivity, and quality as some of the OM themes that should receive more direct attention in operations strategy literature.

JIT production. Just-in-time (JIT) production poses an opportunity and challenge to the manufacturing strategy literature. Its benefits to manufacturers are reported to be very high, ranging from 500% to 2000% improvement over pre-JIT levels of performance (Schonberger, 1987). Although the benefits of JIT are yet to be documented systematically in large sample studies, the benefits of JIT so frequently reported by practitioners qualifies JIT for consideration as a strategic variable. By any measure, the impact of JIT on manufacturers has been strategic: for individual manufacturers, the impact has been long-term gains in competitiveness that is anything but marginal (Schonberger, 1987). The competitive advantage gained through JIT and TQC appears in cost, quality, flexibility, and dependability – all recognized strategic goals. Yet, manufacturing strategy literature has not formally embraced JIT for its strategic value to manufacturers.

Location. A distinctive feature of the operations management literature is the use of management science to model problems. Although location decisions are legitimate strategy issues, and management science models on location and capacity expansion abound, the respective literatures stand apart from each other. For example, field studies (such as the one by Schmenner, 1982) and location models must be brought together in the operations strategy literature.

Productivity. Operations management strategies are reflected in objectives that stress schedule (delivery), quality, cost, and flexibility. Cost is a key strategic variable (Porter, 1980; Skinner 1978). A common surrogate for cost is productivity because it more accurately measures inputs to and outputs from the conversion (transformation) process.

Productivity can be measured broadly (at the national level), measured narrowly (at the level of employees), or measured at many levels in between. At the level of the firm or division, measurement of productivity is not practised widely. Experts estimate that less than 2% of firms do so, even though several measurement techniques do exist (Kendrick, 1986; American Productivity Center, 1986).

Can the strategic value of productivity be unleashed by the appropriate use of productivity measurement? This is an area for fruitful investigation. There is little progress in using well known productivity measurement procedures as a method of performance measurement. One exception is the progress made in the modification of total factor economic productivity measurement to be consistent with accounting records (American Productivity Center, 1986).

Another area of strategic importance is the measurement of the productivity of indirect and professional employees. The performance of shop floor employees is the most closely measured, the performance of service employees is measured less and the performance of indirect, professional, and management employees is measured even less.

Productivity measurement leads to competitive advantages. Consider, for example, banking deregulation. Increased competition and increased cost of money to banks has brought about the pricing of services previously free to bank customers. This forces banks to identify costs and get involved in costing labour and technology on a service-by-service basis. The gains made by individual banks in measuring productivity translates into a strategic advantage for banks that now understand their costs and can price accordingly in a competitive environment.

One group of employees whose output has not been measured effectively consists of professionals: computer systems analysts, engineers, managers, physicians, professors, and so forth. Managers, in particular, need to have their productivity measured as a basis for improvement. Estimates suggest there are 10 million managers and supervisors at work in the United States alone. The strategic value of developing measures for these employees is worthy of investigation. Productivity measurement is a meaningful and fruitful research topic for operations strategy scholars because it is data rich, offering a great deal of opportunity for intensive analysis.

Quality. Quality is frequently mentioned in the operations strategy literature, yet it fails to receive the attention given in the traditional operations management literature. In this discussion, selected issues surrounding competitive quality that have a bearing on operations strategy are addressed.

Profitability, to a larger extent, is driven by quality (Adam, Hershauer and Ruch, 1986; Heyl, 1987). Many US manufacturers view quality in the context of the rather obsolete classical cost–quality trade-off. It is common knowledge that quality related costs are prevention costs, appraisal costs, internal failure costs, and external failure costs. The key to attaining competitive quality (i.e. quality with profitability) lies with prevention. A proven assumption of total quality assurance is that an increase in prevention expenditures will generate a proportionately greater savings in the other, downstream cost categories. Clearly, prevention will result in the reduction of internal and external failure costs because fewer defective items will be produced. Plus, less appraisal will be required because the products will be made correctly the first time (Evans and Lindsay, 1989). Strategically, improved quality results in increased market share. With increased demand *and* lower costs due to quality improvements, price can be reduced to increase market share and profits. The strategic use of quality derives from the understanding of the above cause-and-effect sequence, which we refer to as the value of competitive quality.

Among those most influential in the quality improvement literature are Deming (1975, 1982, 1985), Juran (1982), Crosby (1979) and Feigenbaum (1983). Each has a unique, yet outstanding managerial approach to quality. Manufacturers adopting their work have demonstrated the strategic value of quality not only to individual firms, but also to entire industries and the nation. Japan is a case in point. Their prescriptive work covers issues pertaining to operations strategy formulation and implementation. The empirical research on this prescriptive literature is yet to develop in earnest. The potential of a large body of studies on the formulation and implementation of a strategy for attaining competitive quality based on the various paradigms recommended by these writers is yet to be exploited.

Deming's method has had a great international impact. In the early 1950s, Deming began spreading statistical thinking and continual improvement to Japanese top management. He outlines his framework in terms of 14 points, the final one stressing a never-ending process of improvement (Gitlow and Gitlow, 1987).

Juran (1982), founder of the Juran Institute, provides another paradigm for the effective use of quality. Juran, like Deming, has a programme of quality improvement that is broken down into a series of 16 steps. He notes that statistical quality control is a useful and necessary tool in quality improvement programmes. However, he says, many zealous managers regard the tool as a panacea to their quality problems. Juran blends statistical analysis with rather classical managerial action: planning, organizing and control. His theme is that quality improvement is a long-term process (Juran, 1982).

Crosby (1979) expresses a 14-step process for quality improvement in four absolutes of quality. The first is that quality means conformance to requirements. Next, defect prevention, not inspection, is the only way to achieve true quality. The third absolute proclaims that the only acceptable performance standard is zero defects. Finally, Crosby states that the cost of quality is the cost of non-conformance – the cost of doing things wrong. He claims that non-conformance costs can add up to 25% of selling price for manufacturing firms and 40% of operating costs for service companies (Crosby, 1979). However, critics say that it is a good programme for changing attitudes, but it is not a good basis for specific actions (Fierman, 1985).

Total quality control (TQC), a term that was coined by Feigenbaum (1983) in the late 1950s, is another major theme found in quality-improvement literature. This concept denotes that in any given organization effective quality management must be a total, company-wide effort aimed at the avoidance of problems that will inevitably arise. Continuous correction of these problems leads to continuous improvement of quality (Evans and Lindsay, 1989; Ishikawa, 1985).

Several recurring themes in the quality-improvement literature are continuous improvement, quality treated as everyone's job, the relationship between quality and productivity, and the role of management. These themes primarily stress the process of attaining quality. Although operations strategy literature is weak on strategy process research in general, the literature is rather extensive and diverse in the area that prescribes the process leading to world-class quality levels in products (Adam and Barker, 1987; Heyl, 1987). However, most of the literature on the topic is rather prescriptive, leaving much room for empirical theory-building work that is yet to develop. Further, there is a need for longitudinal field studies that more clearly identify relationships between quality, productivity and profitability. Any investigation in this area will be incomplete without a fair share of the research focusing on the processes associated with the Japanese experience in attaining strategically effective quality levels in products.

The root of Japan's success with quality points to human factors. The Japanese process for attaining competitive quality has somehow found the secret of harnessing the energy of its people more effectively than others (Ishikawa, 1976, 1985). There are many reports depicting the Japanese production worker as a robot who is completely dedicated to the well-being of his or her company. A study by Weiss (1984) dispels three myths associated with Japanese manufacturing: compared with US workers in an electronics manufacturing firm, Japanese workers are not significantly less absent, are not less likely to quit, and do not work harder.

An exploratory study by David Garvin produced some evidence explaining the differential between Japanese and US manufacturers in the process used to gain competitive quality. In Japan, quality problems are attributed to external factors such as poor materials or purchased parts. In the US, the problems stem from internal sources – workforce, maintenance or adjustment of equipment or processes. In addition, the Japanese strategic objective is to produce high-quality products. US manufacturing managers tend to subordinate quality objectives to the drive to meet production schedules (Garvin, 1986).

19.5 CONCLUSION

Three common themes and seven missing themes were used as a framework to review operations management literature from a strategic perspective. The missing themes in operations strategy literature identified here form logical indictments against the literature while automatically pointing to important directions for future research themes. In reading this review, the reader may rightly come away with the feeling that the strategy research in the area is spotty and lacks dedicated research effort.

The prescriptive work on manufacturing strategy that has gone on for nearly two decades and the growth of more prescriptive work on the process of attaining world-class quality levels are encouraging occurrences because many variables and constructs are being generated for eventual empirical theory-building, which unfortunately is not happening with any regularity.

Two concluding remarks are offered about the state of operations strategy literature. First, theory building seems to be endlessly stalled for want of high quality empirical research. The reasons for this problem are many, including a lack of interest among operations management researchers for empirical research and a lack of training in empirical research methods among operations management scholars. Operations strategy is not readily subject to a management science model which is the preferred research tool for operations management researchers. Empirical methods are more appropriate for the investigation of strategic issues in operations.

Secondly, the missing themes illustrate another important problem hindering the advancement of research in the area: the failure of operations management scholars to integrate and build upon themes developed in the business strategy area. Even though many would contend that business strategy research is soft or weak, by comparison, operations strategy research is anaemic. This is a condition that is made worse by its isolation from relevant research occurring in related disciplines.

19.6 REFERENCES

Ackoff, H.I. (1971) On the use of models in corporate planning. *Strategic Management Journal*, **2**, 353–9.

Adam, E.E., Jr. (1983) Towards a typology of production and operations management systems. *Academy of Management Review*, **8**, 365–75.

Adam, E.E., Jr and Barker, E.M. (1987). Achieving quality products and services. *Operations Management Review*, Winter, 1–8.

Adam, E.E., Jr and Ebert, R.J. (1989) *Production and Operations Management: Concepts, Models, and Behavior*, 4th ed, Prentice-Hall, Englewood Cliffs, NJ.

Adam, E.E., Jr, Hershauer, J.C., and Ruch, W.A. (1986) *Productivity and Quality: Measurement as a Basis for Improvement*, Research Center. College of Business and Public Administration. University of Missouri, Columbia, MO.

Albrecht, K. and Zemke, R. (1985) *Service America! Doing Business in the New Economy*, Irwin, Homewood, IL.

American Productivity Center (1986) APC performance measurement system, Houston, TX.

Andrews, K.R. (1971) *The Concept of Corporate Strategy*, Irwin, Homewood, IL.

Ansoff, H.I. (1965). *Corporate Strategy*, McGraw-Hill, New York.

Bennett, R.E. and Hendricks, J.A. (1987) Justifying the acquisition of automated equipment. *Management Accounting*, July, 39–46.

Blank, L. (1987) Total return requirements for technology investments. *Journal of Cost Management*, Fall, 46–8.

Bowman, E.H. (1980) A risk return paradox for strategic management. *Sloan Management Review*, **21**(3), 17–31.

Buffa, E.S. (1984) *Meeting the Competitive Challenge: Manufacturing Strategy for U.S. Companies*. Irwin, Homewood, IL.

Chandler, A.D. (1962) *Strategy and Structure: Chapters in the History of American Industrial Enterprise*, MIT Press, Cambridge, MA.

Chase, R.B. (1978) Where does the customer fit in a service operation? *Harvard Business Review*, **56**(6), 137–42.

Chase, R.B. (1985) The 10 commandments of service system management. *Interfaces*, **15**(3), 68–72.

Chase, R.B. and Prentis, E.L. (1987), Operations management: a field rediscovered. *Journal of Management*, **13**(2), 351–66.

Clark, K. and Fujimoto, T. (1987) Overlapping problem solving in product development. Paper presented at Research Symposium: Issues in International Manufacturing, INSEAD, Fountainbleau, France.

Collier, D.A. (1985) *Service Management: The Automation of Services,* Reston Publishing, Reston, VA.

Collier, D.A. (1987) *Service Management: Operation Decisions,* Prentice-Hall, Englewood Cliffs, NJ.

Crosby, P.B. (1979) *Quality is Free,* McGraw-Hill, New York.

DeMeyer, A., Nakane, J., Miller, G. and Ferdows, K. (1987) *Flexibility: The Next Competitive Battle* (Manufacturing Roundtable Research Report Series), Boston University, School of Management, Boston, MA.

Deming, W.E. (1975) On some statistical aids toward economic production. *Interfaces,* **5**(4), 1–5.

Deming, W.E. (1982) *Quality, Productivity and Competitive Position,* MIT Press, Boston, MA.

Deming, W.E. (1985) Transformation of western style of management. *Interfaces,* **15**(3), 6–11.

Dess, G.G. and Davis, P.S. (1984) Porter's (1980) generic strategies as determinants of strategic group membership and organizational performance. *Academy of Management Journal,* **27**, 367–88.

Dutta, B.K. and King, W.R. (1980) Metagame analysis of competitive strategy. *Strategic Management Journal,* **5**, 301–18.

Engwall, R.L. (1988) Investment evaluation methodologies. *Journal of Cost Management,* Spring, 40–4.

Evans, J.R. and Lindsay, W.M. (1989) *The Management and Control of Quality.* West Publishing, St Paul.

Fahey, L. and Christensen, H.K. (1986) Evaluating the research on strategy content, *Journal of Management,* **12**(2), 167–83.

Feigenbaum, A.V. (1983). *Total Quality Control,* 3rd edn, McGraw-Hill, New York.

Fierman, J. (1985) Why enrollment is up at the quality college. *Fortune,* **111**, 170.

Galbraith, C.S. and Schendel, D. (1983) An empirical analysis of strategy types. *Strategic Management Journal,* **4**, 153–73.

Garvin, D.A. (1986) Quality problems, policies, and attitudes in the United States and Japan: an exploratory study. *Academy of Management Journal,* **29**(4), 653–73.

Gitlow, H.S. and Gitlow, S.J. (1987) *The Deming Guide to Quality and Competitive Position,* Prentice-Hall, Englewood Cliffs, NJ.

Greene, C.N., Adam, E.E. and Ebert, R.J. (1985) *Management: For Effective Performance,* Prentice-Hall, Englewood Cliffs, NJ.

Hambrick, D.C. (1983) Some tests of the effectiveness and functional attributes of Miles and Snow's strategic types. *Academy of Management Journal,* **26**, 5–26.

Harrison, N. (1987) *Manufacturing Futures Survey: Australia.* Paper presented at the Research Symposium: Issues in International Manufacturing, INSEAD, Fountainbleau, France.

Hayes, R.H. and Wheelwright, S.C. (1984) *Restoring our Competitive Edge: Competing through Manufacturing,* Wiley, New York.

Heskett, J.L. (1986) *Managing in the Service Economy,* Harvard Business School Press, Boston, MA.

Heyl, J.E. (1987) The strategic role of quality and productivity: building competitive strength through operations, in *Proceedings of Fourth Annual OMA Conference, Phoenix, Arizona* (ed. J.E. Heyl).

Hill, M. (1987) Internationalization of manufacturing across political and economic systems: the role of the operations manager. Paper presented at the Research Symposium: Issues in International Manufacturing, INSEAD, Fountainbleau, France.
Hill, T. (1985) *Manufacturing Strategy: The Strategic Management of the Manufacturing Function*, Macmillan, London.
Hill, T. (1987) Product profiling: developing concepts and creating language in manufacturing strategy. Paper presented at the Research Symposium: Issues in International Manufacturing, INSEAD, Fountainbleau, France.
Hirano, M.J. (1987) Dynamic implications of speed of change in product-market posture. Paper presented at the Research Symposium: Issues in International Manufacturing, INSEAD, Fountainbleau, France.
Hoffman, R.G. (1985) Economic scenarios for project evaluations. *Journal of Business Strategy*, **5**, 66–74.
Hostage, G.M. (1975) Quality control in a service business. *Harvard Business Review*, **53**(4), 98–106.
Huff, A.S. and Reger, R.K. (1987) A review of strategic process research. *Journal of Management*, **13**(2), 211–36.
Hunsicker, J.Q. (1980). Can top managers be strategists? *Strategic Management Journal*, **1**, 77–83.
Ishikawa, K. (1976). *Guide to Quality Control*, Nordica International Limited for the Asian Productivity Organization, Hong Kong.
Ishikawa, K. (1985) *What is Total Quality Control? The Japanese Way*, Prentice-Hall, Englewood Cliffs, NJ.
Jolly, V. (1987) Competition global strategies. Paper presented at the Research Symposium: Issues in International Manufacturing, INSEAD, Fountainbleau, France.
Juran, J.M. (1982) *Juran on Quality Improvement*. Juran Institute, New York.
Kaplan, R.S. (1986) Must CIM be justified by faith alone? *Harvard Business Review*, **64**(2), 87–95.
Kendrick, J.W. (1986) *Improving Company Productivity: Handbook in the Cases*, Johns Hopkins University Press, Baltimore.
Kenworthy, H.W. (1986) Total quality concept: a proven path to success. *Quality Progress*, **19**(7), 21–5.
King, W.R. (1981) The importance of strategic issues. *Journal of Business Strategy*, **1**(3), 74–6.
Kotha, K. and Orne, D. (1989) Generic manufacturing strategies: a conceptual synthesis. *Strategic Management Journal*, **10**, 211–31.
Kudla, R.J. (1980) The effects of strategic planning on common stock returns. *Academy of Management Journal*, **23**, 5–30.
Lawrence, P.R. and Lorsch, J.W. (1967) *Organization and Environment: Managing Differentiation and Integration*, Division of Research, Harvard Business School, Boston.
Lovelock, C.H. (1983) Classifying services to gain strategic marketing insights. *Journal of Marketing*, **47**, 8–20.
Lovelock, C.H. (1988) *Managing Services: Marketing, Operations, and Human Resources*, Prentice-Hall, Englewood Cliffs, NJ.
Meredith, J.R. and Suresh, N.C. (1986) Justification techniques for advanced manufacturing technologies. *International Journal of Production Research*, **24**(5), 1043–57.
Miller, J.G. and Roth, A.V. (1988) *Manufacturing Strategies. Executive Summary of the 1988 Manufacturing Futures Survey* (Manufacturing Roundtable Research Report), Boston University, Boston.

Mills, P.K. (1986) *Managing Service Industries*, Ballinger, Cambridge, MA.
Mills, P.K. and Marguiles, N. (1980) Toward a core typology of service organizations. *Academy of Management Review*, **5**, 255–65.
Mills, P.K. and Moberg, D.J. (1982) Perspectives on the technology of service operations. *Academy of Management Review*, **3**, 467–78.
Mokwa, M.P. (1986) The strategic marketing audit: an adoptive utilization perspective. *Journal of Business Strategy*, **6**(4), 88–95.
Montgomery, C.A. and Singh, H. (1984) Diversification strategy and systematic risk. *Strategic Management Journal*, **5**, 181–91.
Montgomery, C.A., Thomas, A.R. and Kamath, R. (1984) Divestiture, market valuation, and strategy. *Academy of Management Journal*, **27**, 830–40.
New, C. and Meyers, A. (1987) Management of manufacturing: UK. Paper presented at the Research Symposium: Issues in International Manufacturing, INSEAD. Fountainbleau, France.
Porter, M. (1980) *Competitive Strategy: Techniques for Analyzing Industries and Competitors*. Free Press, New York.
Porter, M. (1985) *Competitive Advantage: Creating and Sustaining Superior Performance*. Free Press, New York.
Primrose, P.L. and Leonard, R. (1987). Performing investment appraisals for advanced manufacturing technology. *Journal of Cost Management*, Summer, 34–42.
Rho, B. and Whybark, D.C. (1988) *Comparing Manufacturing Practices in the People's Republic of China and South Korea*. Unpublished manuscript, Indiana Center for Global Business, Indiana University School of Business, Bloomington.
Sasser, W.E., Olsen, P. and Wyckoff, D.D. (1978) *Management of Service Operations: Text, Cases, and Readings*. Allyn & Bacon, Boston, MA.
Schendel, D.E. and Hofer, C.W. (1979) *Strategic Management: A New View of Business Policy and Planning*, Brown, Boston.
Schendel, D. and Patton, G.R. (1978) A simultaneous equation model of corporate strategy. *Management Science*, **24**, 1611–21.
Schmenner, R.W. (1982) Multiplant manufacturing strategies among the Fortune 500. *Journal of Operations Management*, **2**(2), 77–86.
Schoeffler, S., Buzzell, R.D. and Heany, D.F. (1974) Impact of strategic planning on profit performance. *Harvard Business Review*, **52**(2), 137–45.
Schonberger, R.J. (1987) *World Class Manufacturing: The Lessons of Simplicity Applied*, Free Press, New York.
Schroeder, R.G., Anderson, J.C. and Cleveland, G. (1986) The content of manufacturing strategy: an empirical study. *Journal of Operations Management*, **6**(3), 405–16.
Skinner, W. (1969) Manufacturing – missing link in corporate strategy. *Harvard Business Review*, **47**(3), 136–45.
Skinner, W. (1974) The focused factory. *Harvard Business Review*, **52**(3), 113–21.
Skinner, W. (1978) *Manufacturing in the Corporate Strategy*, Wiley, New York.
Skinner, W. (1984) Operations technology: blind spot in strategic management. *Interfaces*, **14**(1), 116–24.
Skinner, W. (1985) *Manufacturing the Formidable Competitive Weapon*. Wiley, New York.
Snyder, C.A., Cox, J.F. and Jesse. R.R., Jr (1982) A dependent demand approach to service organization planning and control. *Academy of Management Journal*, **7**, 455–566.
Starr, M.K. (1988) *Global Competitiveness: Getting the US Back on Track*, Norton, New York.

Steiner, C.A. (1969) *Top Management Planning*, Macmillan, New York.
Stobaugh, R. (1969) Where in the world should we put that plant? *Harvard Business Review*, **47**(1), 129–39.
Suresh, N.C. and Meredith, J.R. (1985) Justifying multimachine systems: an integrated strategic approach. *Journal of Manufacturing Systems*, **4**(2), 117–34.
Swamidass, P.M. (1986) Manufacturing strategy: its assessment and practice. *Journal of Operations Management*, **6**(3), 471–84.
Swamidass, P.M. (1988) *Manufacturing Flexibility*. Monograph of the Operations Management Association, No. 2.
Swamidass, P.M. and Newell, W.T. (1987) Manufacturing strategy, environmental strategy, and performance: a path analytic model. *Management Science*, **33**(4), 509–24.
Vernon, R. (1971) *Sovereignty at Bay*, Basic Books, New York.
Vernon, R. (1974) The location of economic activity, in *Economic Analysis and the Multinational Enterprise* (ed. J.H. Dunning), Allen & Unwin, London, pp. 89–114.
Vernon, R. and Wells, L.T., Jr (1976) *Manager in the International Economy*, 3rd edn, Prentice-Hall, Englewood Cliffs, NJ.
Voss, C.A. (1986) *Managing New Manufacturing Technologies*. Monograph of the Operations Management Association, No.1.
Weiss, A. (1984) Simple truths of Japanese manufacturing. *Harvard Business Review*, **62**(4), 119–25.
Wells, L.T., Jr (1972) *The Product Life Cycle and International Trade*, Division of Research, Harvard Business School, Boston.
Wheelright, S.C. (1978) Reflecting corporate strategy in manufacturing decisions. *Business Horizons*, February, 57–66.
Wheelwright, S.C. (1984) Manufacturing strategy: defining the missing link. *Strategic Management Journal*, **5**, 77–91.

Index